The Blue Economy

Somnath Hazra • Anindya Bhukta
Editors

The Blue Economy

An Asian Perspective

Springer

Editors
Somnath Hazra
School of Oceanographic Studies
Jadavpur University
Kolkata, West Bengal, India

Anindya Bhukta
Department of Economics
Netaji Mahavidyalaya
Arambagh, West Bengal, India

ISBN 978-3-030-96518-1 ISBN 978-3-030-96519-8 (eBook)
https://doi.org/10.1007/978-3-030-96519-8

This Springer imprint is published by the registered company Springer Nature Switzerland AG
The registered company address is: Gewerbestrasse 11, 6330 Cham, Switzerland

Professor Jaydeb Sarkhel

Who makes his student believe that economics as a subject is not as difficult as it is said?

Professor Jaydeb Sarkhel, an outstanding student of the University of Burdwan, served the Department of Commerce of the same university for over three decades as a very successful teacher. He retired from his service in the year 2014, but till now is very much active in writing textbooks. He has written nearly 40 textbooks in economics, and all of them are very popular for their lucid presentation. But to his students, he will be remembered for his fatherly affection and friendly guidance to their life and career.

An Introductory Note

In scientific jargon, the earth is known as the blue planet. The sky above and the ocean beneath are blue. But in economic jargon, the blue economy represents a sustainable ocean economy.

Oceans are the storehouse of marine living and non-living resources. From time immemorial, humankind has been using these resources for their economic benefit to such an extent that ocean economy occupies a distinct place in economic analysis. But later on, as environmental concerns for the unlimited use of natural resources started growing on, a new concept of blue economy emerged. The concept was initially developed by Professor Gunter Pauli in 2010. However, the idea occupied the centre stage of mainstream economics only after it received high acclaim in the Rio+20 conference in 2012. The policymakers, thereafter, started promoting the concept of blue growth, especially in the island economies and in the countries that have significant coastlines and maritime areas.

The term 'blue economy' is defined differently by different organisations and authors. What can be drawn from these definitions is that the blue economy encompasses all types of economic activities that directly or indirectly make sustainable use of coastal and marine resources. In fact, in introducing a new term, 'blue economy,' in place of an old one, 'ocean economy,' the objective was to draw the attention of the world to the fact that ocean resources are not used sustainably and the time has arrived when our utmost effort should be to ensure their sustainable use.

Use of ocean resources in a sustainable manner is essential not only for island and coastal economies, but for the rest of the world also. An estimate reveals that oceans can contribute nearly 3% of the global value-added each year and are responsible for about 80% of the economic activity. However, this is a rough estimate only, because what we have done so far is to classify contributions of ocean ecosystems to human life and environment into four categories, namely provisional services, supporting services, regulating services and cultural services. But till today, the entire mechanism of how oceans deliver different ecosystem services, what exactly the associated economic and ecological benefits are and how anthropogenic activities on marine resources and environment affect these services are yet to be fully explored. In consequence, the world economy is yet to take account of ocean-based and ocean-related activities in the national income accounting system. First of all, this requires a proper development of the evaluation methodology of marine ecosystem services. It also requires an effective administrative structure in this regard. Since Asian countries possess enormous ocean resources, they have a tremendous opportunity to include the blue economy in their national income accounting, and they may lead the global economy towards this direction.

The importance of incorporating the blue economy in the national accounting system, however, lies elsewhere. As soon as it can be incorporated, the economy will be gaining importance to the governments and national policymakers as well, because it will then open up a new avenue to alleviate the eternal problem of income and employment all over the world, especially in developing and underdeveloped world. Against this backdrop, we have planned to develop the present volume on various issues of blue economies keeping an eye on this objective. Global experts of different fields identify in their articles the misuses and overuses of ocean resources and focus on the way-outs to justify the renaming of 'ocean economy' into 'blue economy.' We hope that this volume will serve the purpose of bringing the blue economy under the purview of mainstream economics.

It is our immense pleasure that a publication house like Springer Nature has agreed to publish this priceless volume. We hope holding their hand tight we can reach global readers and draw their attention to this burning issue of today.

Kolkata, India
Arambagh, India
September 12, 2021

Somnath Hazra
Anindya Bhukta

Contents

Editors and Contributors

About the Contributors

Abhra Chanda is an Assistant Professor in the School of Oceanographic Studies, Jadavpur University. His research interest encompasses the greenhouse gas dynamics in the lentic and lotic ecosystems of the coastal and marine sectors. He researches issues of marine pollution and carbon dynamics in the blue carbon ecosystems of India and Japan.

Aneel Salman is currently associated with Comsats University as the Head of Faculty Development Academy (FDA), Principal Seat Campus. He has the unique distinction of initiating and setting up new departments at Pakistan's most prestigious and highest ranked universities. Internationally, he has been affiliated with some of the renowned universities. He possesses a diversified experience in teaching, research and management and has authored a large number of publications in several international journals and books. His areas of research expertise include climate change; behavioural, energy and marine economics; institutional governance; international trade; and public policy.

Anil Markandya is the former Director of the Basque Centre for Climate Change in the Basque Country, Spain. Professor Markandya has also acted as a consultant to a number of national and international organisations and has served as the Lead Economist at the World Bank. He has held academic positions at the Universities of Princeton and Harvard in the USA and at the University College London and Bath University in the UK. He was one of the lead authors for Chapters of the 3rd and 4th IPCC Assessment Reports on Climate Change, which were awarded a share of the Nobel Peace Prize in 2007. He has also been a Lead Author for the 5th Assessment Report published in 2014.

Anindya Bhukta has been teaching economics for over 30 years at the undergraduate level. He is a prolific writer and writes especially on issues in Indian economy and on environmental issues. However, he obtained his doctoral degree on Intellectual Property Rights law. He has authored a considerable number of books, most of which are published by the leading publishers at the regional, national and international levels.

Arnab Pramanik is a Scientific Staff of Jagadis Bose National Science Talent Search, Kolkata, West Bengal, India. His research interests include microbial community structure of coastal water and marine biotechnological study. He has published more than 40 peer-reviewed research articles.

Badri Narayanan is the lead adviser and head, trade and commerce, at NITI Aayog. He is a renowned economist and interdisciplinary scientist with extensive experience in producing and managing global economic data sets and models. He has a good number of international publications in the areas of trade, energy, environment, agriculture, labour and manufacturing. He is a peer reviewer and editorial board member for several international journals.

Clement A. Tisdell Emeritus Professor at the University of Queensland, was one of the forerunners in the field of environmental and ecological economics. During his professorship, he occupied various academic offices of Australia.

David Hoyos is the Director of ehuGune; Vice-Rectorate of Scientific and Social Development and Transfer; Assistant Professor at the Department of Quantitative Methods, Faculty of Economics and Business, University of the Basque Country, Universidad del País Vasco (UPV/EHU); Associate Researcher at EKOPOL (Research Group on Ecological Economics and Political Ecology), BC3 (Basque Centre on Climate Change) and HEGOA (Development and International Cooperation Studies Institute); and Member of the scientific advisory board of the UNESCO Chair on Sustainable Development and Environmental Education of the University of the Basque Country (UPV/EHU). His main area of research concerns the economic valuation of environmental and natural resources.

Elisa Sainz de Murieta is currently PI at BC3 of the H2020 project COACCH (coacch.eu) and the project Gobernadapt, in collaboration with IHCantabria, funded by the Basque Cooperation Agency. She has also been involved in several research projects related to the economics of climate change adaptation and the valuation of ecosystem services.

Estibaliz Treviño is a PhD scholar at the Faculty of Economics and Business, University of the Basque Country, Universidad del País Vasco (UPV/EHU).

Ghulam Samad is a senior research specialist at the CAREC Institute. He also served as an economist at the Planning Commission of Pakistan. He holds a PhD in Economics from Colorado State University, USA. He has written several international books and contributed to peer-reviewed national and international journals.

Heman das Lohano is the Professor and Program Director PhD/MS (Economics) and BS (EM) at the Institute of Business Administration, Karachi. He is also attached to the SANDEE as a resource person and associated with a lot of projects in SANDEE and other organisations. His main areas of research are environmental economics, climate change economics, non-market valuation, environment and health, applied microeconomics, impact evaluation, policy analysis for public services, etc. He is a Member of Drafting Team, Sindh Water Policy, Government of Sindh.

Ignacio Cazcarro (MSc, PhD in Economics) is ARAID researcher (Government of Aragon) and associate researcher at BC3 (Basque Centre for Climate Change). He participated in projects of the US-NSF, regional, national, EU and IDRC-DFID called DECCMA. His main interests are input–output analysis and modelling, water, resources, energy and GHG emissions.

Iñaki Arto graduate and PhD in Economics, and MSc in Environmental Engineering, is a Research Professor at the Basque Centre for Climate Change (BC3). Before, he occupied research positions at the University of the Basque Country and the JRC of the European Commission. His main research area is energy-environmental-economic modelling.

Jawaria Abbasi is a Research Scholar, Pakistan Institute of Development Economics (PIDE), Islamabad, Pakistan.

Jose A. Fernandes-Salvador is a scientist from computing and artificial intelligence with the skills and experience needed to translate data into valuable information for marine research. These skills are curiosity, multidisciplinary communicative, biological, economic, statistics and machine learning expertise. Current line of research focuses on long-term sustainability of fisheries.

Kapil Narula is a Senior Researcher at the Chair for Energy Efficiency within the Institute for Environmental Sciences (ISE), University of Geneva. He is an electrical engineer, development economist and an energy and sustainability professional. He co-teaches a course on 'Cities in Transition' and 'Energy Efficiency.' Kapil has earlier worked on board ships, as a faculty at academic institutes and in think tanks.

Muhammad Bilal Maqbool is a PhD scholar at IBA Karachi. He is working on extreme weather shock effect on poverty, unemployment and wage structure across different geographical locations in Pakistan. His research interests include investigating consumption/expenditure behaviour and environmental economics.

Rashmi Kundu is a Research Analyst with the Centre for Social and Economic Progress, New Delhi. Her area of research includes urban housing policy issues, trade policy, climate change, among others. She has written extensive data-driven columns, especially on Covid-19. She has been associated with How India Lives and Infinite Sum Modelling Inc, USA.

Rikhia Bhukta is presently working as a Research Fellow in the Department of Economics Science, Indian Institute of Technology, Kanpur, UP, India. She has had an outstanding career since her school life. She is now working on empirical development economics.

Sanjay Upadhyay (an India Visiting Fellow at the Boalt Hall School of Law, University of California; a legal intern to the Earthjustice Legal Defence Fund, San Francisco; and a Global Fellow in Marine Policy at the Duke University) is the founder of India's first environmental law firm. He has been in the drafting committees of several laws in India and abroad, including Wildlife Act, Forest Rights Act and Indian Forest Act. Sanjay practises in the Supreme Court of India.

Shekar Bose is an independent researcher. He has expertise in the technical analysis of socio-economic and regulatory aspects of marine resource management. One of his major works involves the economic analysis of a quota-managed fishery. Dr. Bose's research interests primarily entail areas such as natural resource management, sustainability, compliance and regulatory effectiveness.

Somnath Hazra currently working as a consulting economist and visiting faculty at the School of Oceanographic Studies, Jadavpur University, India, has more than 15 years of consulting and advisory experience with governments and bilateral and multilateral agencies. An ex-visiting researcher at the Basque Centre for Climate Change (BC3), Bilbao, Spain, Somnath has worked as an international CDM Consultant for the Ministry of Finance, Government of Kenya.

Somya Mathur is an Associate Fellow with National Council of Applied Economic Research, New Delhi. She is into Computational General Equilibrium Modelling, and her area of research include climate change and trade policy issues. She has served in various organisations including the Ministry of Panchayati Raj and Infinite Sum Modelling Inc, USA.

Sourav Das is a Postdoctoral Research Associate in the School of Oceanographic Studies at Jadavpur University, India. His research interests include biogeochemistry of marine and coastal water, carbon budgeting, optical properties of water (inherent & apparent optical properties) and blue economy.

Sourav Kumar Das teaches economics at Lalbaba College, Belur, Howrah. He works on rural development, tribal livelihoods, monetary policies and consumption patterns among refugees.

Sukanya Das an Associate Professor and Head in the Department of Policy Studies, TERI School of Advanced Studies, New Delhi, is currently involved with one international joint research Indo-EU project named SARASWATI 2.0 and completed an NSF project titled Investigating the Potential for Decentralized Institutions, Technologies, and Governance to Meet the Wastewater Challenge. She has the experience to work as PI in the socio-economic impact of waste water in two projects sponsored by DST and NSF. Apart from that, she has received grants from CPCB, MOEFCC, NTPC and IDRC. Besides several journal publications, she has several book chapters, reports, working papers and conference publications to her credit.

Sameer Guduru is currently a Consultant with the New and Emerging Strategic Technologies (NEST) Division, Ministry of External Affairs, Government of India. Prior to this, he was with the National Maritime Foundation, New Delhi, as a Marine Technology Associate. He pursued his Master's in International Studies from Symbiosis International University, India. He obtained his Doctorate in ultrafast laser-based microengineering from Politecnico di Milano, Italy, and was associated with CNRS in Marseille, France, as a postdoctoral research engineer.

Suvajit Banerjee is a PhD Scholar at the Department of Economics and Politics, Visva-Bharati University, Santiniketan, Bolpur, West Bengal, India. His areas of research expertise include energy economics and environmental economics, international trade and public policy.

Tuhin Ghosh currently works as a Professor at the School of Oceanographic Studies, Jadavpur University, and is a Coordinating Lead Author for the IPCC Special Report on Ocean and Cryosphere, published in 2019.

Valentina Lauriad (MSc, PhD) holds a position at the IRBIM-CNR, Institute for Marine Biological Resources and Biotechnologies, National Research Council (Italy). She has contributed to diverse European and international projects (CHARM 2, NEPHROPS, MERP-IMMERSE, DECCMA, RITMARE). Her main areas of research are spatial ecology, habitat modelling, fisheries and conservation.

About the Editors

Somnath Hazra is a trained ecological economist, with an expertise in climate change economics, forest economics, sustainable development, ecosystem services, environmental valuation and policy analysis. He has vast research experience in different Indian research centres as well as teaching experiences at UG and PG levels in different universities in West Bengal. He has experience in managing and executing development programmes in various capacities for governments, international development organisations and private sector. He has visited different countries all over the world for performing different project assignments and also for the presentation of research articles on climate change and environmental and development economics. He has more than 50 research articles and six edited volumes related to environmental economics and sustainable development to his credit.

Anindya Bhukta an Associate Professor of Economics in a college in rural West Bengal, is well known for his scholarly articles on economics and environment and also for a few outstanding textbooks in economics. He has so far written more than 300 articles on these two subjects. He has authored nearly 70 books in Bengali and English, among which the most remarkable are his two dictionaries in Bengali, one on economics and the other on environment. One of his Bengali books on the 'History of Indian Planning' has been highly acclaimed by the readers. He is also very popular to the readers in Bengal for his creative writings. The main essence of his writings is lucidity of language and easy presentation with an in-depth analysis of the subject matter. Presently, he is editing an online internationally peer-reviewed journal 'Economy Polity Environment.'

Part I
Blue Economy: An Overview

Chapter 1
Blue Economy: An Overview

Somnath Hazra and Anindya Bhukta

1.1 Introduction: Defining Blue Economy

Oceans are the storehouse of living and nonliving resources. From time immemorial, humankind has used these resources for their economic benefit to such an extent that the ocean economy occupied a distinct place in economic analysis. But later on, as environmental concerns for the unlimited use of marine resources started growing, a new concept, termed "blue economy," started emerging parallelly.

In 2012, the concept of blue economy was introduced first at the United Nations Conference on Sustainable Development held in Rio-de-Janeiro (UNCTAD 2014). Since then, various international bodies started promoting the concept to build up strategies for the protection and preservation of priceless ocean resources. However, initially this new term was often being confusingly used to mean "ocean economy" or "marine economy." In order to remove the ambiguity of these words, the UN in 2014 came up with a very distinct definition of the term blue economy. This definition of the United Nations states that the blue economy is an ocean economy that aims at "the improvement of human well-being and social equity, while significantly reducing environmental risks and ecological scarcities (U.N. 2014b, p. 2)."

In 2017, the World Bank came up with a new definition according to which the "blue economy" ensures "the sustainable use of ocean resources for economic growth, improved livelihoods, and jobs while preserving the health of ocean ecosystems" (World Bank 2017, p. 6).

S. Hazra (✉)
School of Oceanographic Studies, Jadavpur University, Kolkata, West Bengal, India

A. Bhukta
Department of Economics, Netaji Mahavidyalaya, Arambagh, West Bengal, India

S. Hazra, A. Bhukta (eds.), *The Blue Economy*,
https://doi.org/10.1007/978-3-030-96519-8_1

What blue economy is meant for is described, in this way, differently by different organizations and different authors. Let us first mention some of these definitions and then try to reach at our own.

> *The "blue economy" concept seeks to promote economic growth, social inclusion, and the preservation or improvement of livelihoods while at the same time ensuring environmental sustainability of the oceans and coastal areas. At its core it refers to the decoupling of socio-economic development through oceans-related sectors and activities from environmental and ecosystems degradation.* (World Bank 2017)

> *The blue Economy comprises a range of economic sectors and related policies that together determine whether the use of ocean resources is sustainable. An important challenge of the blue Economy is to understand and better manage the many aspects of oceanic sustainability, ranging from sustainable fisheries to ecosystem health to preventing pollution.* (World Bank and UN-DESA 2017)

> *Blue Economy encompasses all sectoral and cross-sectoral economic activities based on or related to the oceans, seas and coasts.* (E.U. 2020)

> *A sustainable ocean economy emerges where economic activity is in balance with the long-term capacity of ocean ecosystems to support this activity and remain resilient and healthy.* (Economist Intelligence Unit 2015)

The above definitions thus tell us, in a nutshell, that the blue economy encompasses all types of economic activities that directly or indirectly ensure sustainable use of coastal and marine resources. Among these definitions, the World Bank's one is relatively more acceptable in the sense that it tries to cover the multifaceted role of oceans and also focuses on sustainable use of their resources. A few more definitions available in the literature are mentioned in Table 1.1.

1.2 From Ocean Economy to Blue Economy: A Conceptual Evolution

Ocean economy describes the organizational structure of economic activities in the Ocean, receiving outputs from and providing inputs to the Ocean (Park and Kildow 2015). With continuous rise in global population, the entire world, especially coastal and island economies, started depending more and more on ocean resources mainly for food, medicines, and energy, and for other goods and services as well. Moreover, oceans are also being increasingly used as the ultimate dumping ground for waste disposal. Anthropogenic use of marine resources did not create any major problem initially, but eventually unrestricted the use of marine-based and marine-related goods and services led to deteriorating ocean and coastal health.

The concept of sustainable development developed only when it was realized that unrestricted use of natural resources for the sake of development ultimately stands on the way to the long-run development. In a similar fashion, the unrestricted use of marine-based and marine-related goods and services gave birth to a new term, "blue economy," which tries to describe a sustainable development framework of

Table 1.1 Various definitions of blue, ocean, and marine economy

Concept	Authors	Definitions and related concept
Blue economy	World Bank	The sustainable use of ocean resources for economic growth, the improvement of livelihoods and jobs, and the health of ocean ecosystems.
	Costa et al.	The concept of rethinking ongoing industrial processes and searching for a viable biological solution that reduces contamination.
	Phelan et al.	It has become synonymous with generating wealth from activities related to the oceans while protecting and supporting marine ecosystems.
	Graziano et al.	It arises from the growing worldwide interest in the growth of water-based activities.
	Schutter and Hicks	It seeks to curb biodiversity loss while stimulating economic development, thereby integrating environmental and economic interests.
	Kathijotes	It is the mainstream of national development and can integrate land and sea-based socioeconomic sustainable development.
	Kaczynski	It refers to the commercially sustainable development of the oceans.
	Hoegh-Guldberg et al.; Patil et al.; UNECA	It has emerged in the last two decades from various forums, but above all from within the policy and practice of environmental development.
Marine economy	Qi and Xiao	It is a dynamic and complex system that covers all industries and regions.
	Wenwen et al.	It is a new economic form that emphasizes a new development concept, a new operating mechanism, and a management model.
	Caban et al.	It is particularly exposed to dangers due to the environment of its operations. These risks are the result of deliberate and incidental actions (hydrometeorological, mechanical conditions, etc.).
	Bentlage et al.	A heterogeneous innovation system with enduring relevance to the spatial and functional development of European regions.
	Spammer	It simultaneously fosters social inclusion, environmental sustainability, strengthening maritime ecosystems, transparent governance, and economic growth and development.
Ocean economy	UNCTAD	A subset and complement of the evolving development paradigm emphasizing greener, more sustainable, and more inclusive economic pathways.
	Potgieter	It is considered a crucial factor for global economic growth and development, offering excellent opportunities, challenges, and risks.
	Colgan	They are marine construction, resource, shipping, and tourism and recreation industries whose establishments are located near ocean shorelines or large lakes.

Source: Martínez-Vázquez et al. (2021)

ocean economy. The term "blue economy" therefore is nothing but a new name for "ocean economy." The distinction lies in the fact that "blue economy" speaks of sustainable use of marine goods and services, whereas the erstwhile "ocean economy" did not.

The sustainable use of natural resources like marine resources, which are not confined to any geographical and political boundary, however, requires inter-country cooperation. In fact, different stakeholders use ocean resources differently. For example, different countries of Europe focus on aquaculture, marine renewable energy, tourism, recreation, and maritime transport. China has also identified the offshore production systems and focused on developing offshore aquaculture. New Zealand and Chile, on the other hand, put their emphasis on offshore aquaculture (FAO 2018; Potts et al. 2016). So, there is always a chance of conflict between these stakeholders (Voyer et al. 2018). The sustainability of development can be attained only by the successful elimination of these conflicts.

The decade 2021–2030 has been declared by the United Nations as the "Decade of Ocean Science for Sustainable Development." The objective is to promote the protection of ocean health by developing a common framework at the country level for sustainable development of the ocean. The World Bank has emphasized balanced development through the triple bottom lines of sustainable development (World Bank 2017, p. 4).

But in practice, balanced development cannot be attained unless our attitude toward nature and natural resources take a U-turn. For this, what is required first is the abandonment of the traditional "brown development" model where natural resources are practically considered as "free-good." This is much more true for the ocean economy because in general oceans are considered not only as a place for free resource extraction, but also as waste dumping ground. Therefore, the blue economy, unlike ocean economy, attempts to incorporate ocean values and services into economic modeling and decision-making so that the goods and services contributed by oceans can truly be recognized and measured.

1.3 Contributions of Blue Economy: Measurement and Valuation

Blue economy encompasses all the economic activities happening in and around the oceans, obeying sustainability rules. The oceans provide us with various types of ecosystem services. According to the Common International Classification of Ecosystem Services (CICES), these services can be grouped into three general categories, namely provisioning services, regulating and maintenance services, and cultural services (Potschin and Haines-Young 2011). Provisioning service benefits are obtained directly from the ecosystem (e.g., food, water, minerals, and energy). Regulating and maintenance service benefits are obtained from the regulation of ecosystem processes (e.g., climate regulation, carbon sequestration, and coastal protection). Lastly, non-material benefits like cultural service benefits (e.g., aesthetic, recreational, psychological, and spiritual benefits) are obtained directly from the ecosystem (Liquete et al. 2013).

Blue economy activities today are accounting for a significant share of GDPs of almost all the island and coastal economies. According to an estimate, oceans can contribute $1.5 trillion per annum to the global economy, which is near about 3% of the global value added. The oceans are responsible for about 80% of economic activity. Moreover, blue economy activities have enormous potential to occupy a significant share of international trade also. In brief, ocean economies are providing food and livelihood to a large section of the world's population. Apart from economic benefits, oceans provide us with enormous environmental benefits. Proper accounting for these activities is thus the need of the hour.

The SEEA framework provides a basis for ecosystem accounting (EA) and how ecosystem services (ES) can be accommodated in national income accounts. In the spirit of the System of National Accounting (SNA), the ocean economy is measured by the value-added method. However, ideally, the blue economy should be measured in terms of adjusted net value under the assumption of sustainability. The SEEA framework guides us to construct this adjusted net value added. Firstly, the measurement of the contributions of the blue economy on a sustainable basis requires the valuation of the stock of ecosystem assets and the valuation of the stock of ecosystem degradation/depletion. Secondly, it needs to estimate the value of ecosystem services (ES), namely non-SNA social benefits, which are not included in the GDP.

In 2012, the United Nations Statistical Commission adopted the System of Environmental Economic Accounting (SEEA) Central Framework. This was the first international statistical standard for environmental economic accounting, which considered the concepts, structures, rules, and principles of the System of National Accounts (SNA). In 2014, international bodies like the United Nations, European Commission, FAO, OECD, and World Bank developed SEEA Experimental Ecosystem Accounting based on the SEEA Central Framework (U.N. 2014a). According to SEEA, ecosystem services are the "*contributions of ecosystems to benefits used in economic and other human activity*." The SEEA Experimental Ecosystem Accounting describes a standard accounting approach for the ecosystem goods/assets (i.e., stocks) and the ecosystem services (i.e., flows). Ecosystem services and ecosystem assets are based on the spatial aspects by which one can separate the forest, wetland, and agriculture ecosystem goods and services.

The SEEA Central Framework emphasized the material benefits, which are received from the direct use of environmental goods, but the framework does not include the non-material benefits (e.g., water purification, carbon sequestration, and prevention of soil erosion). These non-material benefits are considered in the SEEA Experimental Ecosystem Accounting. The SEEA Experimental Ecosystem Accounting is used to estimate the contribution of two types of benefits that are meant for human well-being. These are (1) the SNA benefits, i.e., the benefits obtained from items produced by economic units and are considered in measuring GDP, and (2) the non-SNA benefits, i.e., the benefits obtained from items that are

Table 1.2 Correlation between blue economic activities and ecosystem services

Function	Economic activity	Associated ecosystem services
Food, nutrition, and health	Fishing	Provisioning services (wild fish)
	Aquaculture, blue biotechnology	Genetic resource provision of space, regulating services
Leisure and living	Tourism, living	Aesthetic attributes, opportunities for recreation
Energy and raw materials	Mining	Abiotic services (oil, gas, minerals, wind, etc.)
	Oil and gas, renewable energy Carbon capture and storage	Provision of space
Maritime shipping and shipbuilding	Transport, passenger services	Provision of space
Coastal protection	Protection against flooding and erosion Protection of habitats	Provision of space
Maritime monitoring and surveillance	Prevent and protect against illegal movement of people and goods Environmental monitoring	No direct link with ecosystem services

Source: European Commission (2012)

neither manufactured by the economic units nor traded in the market (e.g., clean air). Non-SNA benefits can also be estimated to compute the adjusted net value added of a blue economy. The relationship between the blue economy and ecosystem services is demonstrated by Table 1.2. The blue economy is divided into functions and activities, which contribute to SNA benefits and which are not.

In this regard, we also like to mention that different international organizations recommend different ecosystem services for the inclusion in N.I. accounting framework. We have explored all these recommendations and built up our own list for the development of framework of blue economy. A summary of these recommendations and our own list is presented in Table 1.3.

In the comparison Table 1.3, tick (✓) and cross (×) marks are used when the specific ecosystem service is considered or not by the particular institution/organization while estimating the blue economy. The ecosystem service "water storage and provision of desalination" is considered in UNESCAP Guidelines, but the World Bank and the European Union have included this in their estimation procedure. Based on the reports on freshwater availability in India and other Asian countries, we think that this ecosystem service should be included in the categories of ecosystem services (….).

The European Union adds two categories namely "life cycle management" and "biological regulation" under the section "regulatory services." The section "regulating and maintenance services" means the sources of benefits obtained from regulating the ecosystem processes. In terms of sectors and subsectors of the blue economy, these two categories fall under marine living and nonliving resources, coastal tourism, and global security. In the proposed need to estimate plan columns

Table 1.3 Comparison of ecosystem services considered for the estimation of blue economy

Ecosystem services listed in all documents	European Union	World Bank	Technical Guidance, UNESCAP	Need to estimate
Provision service				
Seafood	✓	✓	✓	✓
Energy resources (renewable)	✓	✓	✓	✓
Energy resources (oil and gas)	✓	✓	✓	✓
Water storage and provision of desalination	✓	✓	×	✓
Biotechnology and bio-prospecting	✓	✓	✓	✓
Trade of ocean-based industries	×	✓	×	✓
Transport	✓	✓	×	✓
Shipbuilding	×	✓	×	✓
Regulating service				
Air quality regulation	✓	×	✓	✓
Water purification	✓	×	✓	✓
Coastal protection	✓	✓	✓	✓
Climate regulation	✓	✓	✓	✓
Ocean nourishment	✓	×	✓	✓
Life cycle maintenance	✓	×	✓	×
Biological regulation	✓	✓	×	×
Waste disposal	×	✓	✓	✓
Cultural and supporting services				
Ocean-based and ocean-related tourism and recreation	✓	✓	✓	✓
Conservation and protection of biodiversity	×	✓	✓	✓
Shoreline protection by reefs and mangroves	×	×	✓	✓
Role of mangroves in serving as a breeding ground or nursery for offshore fisheries	×	×	✓	✓

Source: Prepared by the authors from the reports of different international bodies

of the above matrix, these two categories have already been included in the "regulating services" section under the biodiversity conservation. Hence, if we include these two ecosystem services separately in the estimation procedure again, we may commit double-counting error.

In the context of the above discussion, it should also be noted that for every economic activity in the marine ecosystem, there exists a lot of negative externalities. Hence, to calculate the adjusted net value added, one must consider the associated cost of depletion/degradation of ecosystem goods (e.g., stock depletion due to economic activities like fishing or mining in the marine ecosystem, generation of pollution, and biodiversity loss/degradation are some of the associated externalities associated with these economic activities). The Intergovernmental Panel on Biodiversity and Ecosystem Services (IPBES) emphasized the loss of natural resources and the occurrence of poverty as examples of these negative externalities. In a report of 2019, they showed that fishing had the highest negative impact on marine systems. Therefore, we can draw that in assessing the blue economy, assessing economic activity-related externalities is highly essential.

Unfortunately, however, the world economy is yet to consider the inclusion of marine living and nonliving resources in its GDP accounting. This is probably due to a lack of appropriate evaluation methodology of marine ecosystem services. Since Asian countries possess enormous marine resources, they can take initiative in this regard and may lead the global economy in this direction.

1.4 Opportunities and Challenges

The blue economy has opened up lots of opportunities before us. The ideas, principles, and different norms of the blue economy provide significant opportunities for poverty eradication, food and nutrition security, the reduction in the impact of climate change, and generating sustainable livelihood for the coastal communities. Therefore, a balanced approach toward utilizing the marine ecosystem services in a planned and sustainable manner is essential.

Most of the erstwhile ocean economies in Asia, which are now driving them to blue economies, need fundamental and systematic changes in their policy and governance framework. Since most of the services of the ocean ecosystem have colossal value and there exists a maritime value chain with potential forward and backward linkages, these framework changes are very essential. This is also highly significant because crucial economic activities take place in the core and the adjacent sectors. In addition, many financial functions like fishery, ocean trade and shipping, renewable energy, ocean tourism, and coastal protection are connected with the ocean ecosystem services, which are highly potential for human well-being. A complete account of all these functions across different sectors develops the value chains of ocean ecosystems.

Most of the marine resources are not harvested for the local markets. However, these items have colossal export potential as raw materials, intermediate goods, or in some cases even as final products. The demand for ocean ecosystem goods and services will increase as the population increases. Maritime trade and marine products will create a possibility of contributing to economic growth, export, and new investment opportunities in the coastal areas. Additionally, technological progress will improve the accessibility and feasibility of marine resources. Consequently, new economic opportunities of different ecological services will emerge, which will thereby generate new job opportunities through the inclusion of sustainable fishing and aquaculture, expansion of certain marine transport services and port management, the discovery of marine renewable energy, bio-prospecting of marine resources, extraction of sea-bed mineral resources, and marine tourism. These income-generating activities will improve social security services and the standard of living of the coastal communities. Some of the ecosystem services based on the human well-being and livelihood opportunities of the people have been identified by us (Table 1.3), which have enormous employment potential in East and South Asian countries with long coastlines and huge coastal and marine resources.

According to the geographical situation and climatic condition, the coastal area of Asia is known as one of the highly productive areas of the world. Asia has a rich biological diversity. One of the exceptional features of the Asian coastal areas is the influence of the mangrove forests, which support plenty of aquatic organisms, including fishes. Due to the geographical setting, the confluence is happening between several transboundary rivers in the Bay of Bengal in South Asia. Some of these unique features help form human habitation, socioeconomic structures, development priorities, and dependency on natural resources.

In this regard, we like to draw attention to another opportunity that can mitigate an emerging crisis the whole world is suffering from, namely the energy crisis. Oceans are beautiful breeding grounds of algae, which are now being used as biomass to produce energy. This novel source of energy is considered as the second and third generations of fuel. Sustainable aquaculture can provide a potential environment for the production of algae biomass. This will also generate employment opportunities through the creation of new value chains of renewable energy. Presently, the commercial production and trade of algae biofuels are not significant, but biofuel's commercial activity is expected to accelerate in the coming decade. This marine biofuel production can also be supplemented with bagasse or coconut wood, abundant non-edible by-products of sugar cane and copra, for electricity production. The production of marine biofuels can reduce dependency on hydrocarbon-based fuels for transportation and generation of electricity. To maximize the opportunities of blue economy, the coastal countries can explore the options of multilateral trade and sustainable development aspects with proper ocean governance and regulatory regimes.

Despite all these opportunities, future development of the blue economy is also quite challenging. From time immemorial, on the one hand, plentiful resources of ocean ecosystems have been extensively consumed and, on the other, these ecosystems are used as free waste repositories.

Over extraction of marine resources, like unsustainable harvesting of fish and other aqua fauna, is creating poor conservation and protection of resources. According to FAO, it has been observed that 57% of fish stocks globally are entirely exploited, and another 30% are over-exploited and degraded (FAO 2016). Furthermore, some illegal operations and under-reported fishing are accelerating and are responsible for roughly 11–26 million tons of annual fish catch, and in monetary terms, the value of which is approximately US$10–22 billion.

Frequent land-use land cover changes in marine and coastal landscapes primarily due to developmental activities, forest degradation, and mining are posing significant challenges. Due to sea-level rise, coastal erosion is destroying the coastal infrastructure and livelihoods. Unplanned development in the coastal boundary and frequent shoreline changes are leading to significant loss of infrastructure and loss of critical habitations. Unsustainable business activities and unfair trade are making the situation more fragile and ultimately vulnerable.

Untreated sewage, agricultural run-off, and plastic pollution are becoming crucial sources of marine pollution. The tendency of using oceans as the ultimate dustbin is creating enormous challenges before the sustainability of ocean economies and thereby stands on the way to turning the ocean economy to a blue economy. These threats to the blue economy are creating serious challenges before the livelihood and food security of coastal communities.

In this way, the overuse of marine and coastal resources, either as consumer goods and services or as waste bins, has been seen to create a negative impact on both life and livelihoods of coastal people and on climate change at the same time. Lack of governance, absence of legal regulations regarding conservation and protection of natural resources, and inappropriate operation of management tools have made the situation more and more deteriorating.

References

Economist Intelligence Unit (2015) The blue economy growth, opportunity and a sustainable ocean economy an economist intelligence Unit briefing paper for the World Ocean summit 2015. In: The economist intelligence Unit limited 2015, USA.

European Commission (2020): The E.U. Blue Economy Report. 2020. Publications Office of the European Union. Luxembourg. https://ec.europa.eu/maritimeaffairs/sites/maritimeaffairs/files/2020_06_blueeconomy-2020-ld_final.pdf

European Commission, (2012): Blue Growth, Scenarios and drivers for sustainable growth from the Oceans, Seas and Coasts. Third interim report. [pdf] Ecorys, Deltares, Oceanic Development for DG MARE. Call for Tenders No MARE/2010/01. Rotterdam/Brussels. Available at: http://ec.europa.eu/maritimeaffairs/documentation/studies/documents/blue_growth_third_interim_report_en.pdf

FAO (2016) The State of World Fisheries and Aquaculture 2016. Contributing to food security and nutrition for all. Rome. 200 pp.

FAO (2018) The State of World Fisheries and Aquaculture 2018 - Meeting the sustainable development goals, Rome, available at http://www.fao.org/3/i9540en/i9540en.pdf

Liquete, C. et al. (2013): "Current status and future prospects for the assessment of marine and coastal ecosystem services: a systematic review". PLoS ONE 8(7): e67737. DOI: https://doi.org/10.1371/journal.pone.0067737.

Martínez-Vázquez, R. M., Milán-García, J., & de Pablo Valenciano, J. (2021). Challenges of the Blue Economy: evidence and research trends. *Environmental Sciences Europe*, *33*(1), 1-17.

Park, K. S., and J. T. Kildow. (2015) "Rebuilding the Classification System of the Ocean Economy." Journal of Ocean and Coastal Economics 2014 (4).

Potschin, M., Haines-Young, R. (2011): Ecosystem services: Exploring a geographical perspective. Progress in Physical Geography 35(5): 571-574.

Potts, J., Wilkings, A., Lynch, M., McFatridge, S. (2016) State of Sustainability Initiatives Review: Standards and the Blue Economy. International Institute for Sustainable Development, Winnipeg available at https://www.iisd.org/sites/default/files/publications/ssi-blue-economy-2016.pdf

UNCTAD (2014): United Nations Conference on Trade and Development, The Ocean Economy: Opportunities and Challenges for Small Island Developing States, available at http://unctad.org/en/publicationslibrary/ditcted2014d5_en.pdf

United Nations (2014a): United Nations, European Commission, FAO, OECD, World Bank. System of Environmental-Economic Accounting 2012: Experimental Ecosystem Accounting.

United Nations (2014b) Blue Economy Concept Paper Available at https://sustainabledevelopment.un.org/concent/documents/2978BEconcept.pdf

Voyer, M., Quirk, G., McIlgorm, A., & Azmi, K. (2018). Shades of blue: what do competing interpretations of the Blue Economy mean for oceans governance?. *Journal of Environmental Policy & Planning*, *20*(5), 595-616.

World Bank (2017) The potential of the Blue Economy: Increasing long-term benefits of the sustainable use of marine resources for small island developing states and coastal least developed countries. World Bank, Washington DC.

World Bank and United Nations Department of Economic and Social Affairs. (2017): The Potential of the Blue Economy: Increasing Long-term Benefits of the Sustainable Use of Marine Resources for Small Island Developing States and Coastal Least Developed Countries. World Bank, Washington DC.

Chapter 2
Advancement of Science and Technology: Future Prospect of Blue Economy

Ghulam Samad and Jawaria Abbasi

2.1 Introduction

Marine ecosystems provide one of the biggest platforms for integration of innovative mobility and commercial sustainability. With the use of ocean from being the only source of cross-continental transportation to being at the heart of many global challenges, the dynamics of marine ecology have drastically changed. With such diverse pool of actions, the management of oceanic resources has become one of the most critical tasks. As we think about the management of marine resources, it is inevitable to integrate scientific evolution with sustainability prospects.

Science has a major role in developing socially stable and economically viable solutions to these issues. According to recent estimates published by Organization of Economic Cooperation and Development (OECD), a total of 3 trillion worth of value will be added to the oceanic economy by 2030. This presents us with the challenge of harboring and striking a balance between the prospects of economic growth and oceanic sustainability, and the major driver in the course of achieving this equilibrium can be achieved through science, technology, and innovation (SIT).

Sustainability works to achieve a growth pattern that makes consumption more efficient, production more subtle, and interventions more mechanized. The acceleration of the technological and digital integration in conventional methods has changed the scientific outlook of commercial activities across the globe. The process of organizational innovation has enabled the chances of integration of

G. Samad (✉)
CAREC Institute, Urumqi, China
e-mail: ghulam.samad@carecinstitute.org

J. Abbasi
Research Associate, Pakistan Institute of Development Economics, Islamabad, Pakistan
e-mail: Jawariaabbasi_18@pide.edu.pk

S. Hazra, A. Bhukta (eds.), *The Blue Economy*,
https://doi.org/10.1007/978-3-030-96519-8_2

technology in SME's cluster growth and industrial economies of scale. The viability of this integration rests in the networks that will be the result of academic discourse and scientific research that will be achieved through the network of academia, market, and public interventions.

Oceans are critical to human growth and therefore offer a very deep insight into how the patterns of sustainability may change over time. Oceans provide human well-being and economic wealth, but the cost is emerging in the multiple contexts of global warming, rising temperatures, and greenhouse gases. The cumulative impact of all these anthropogenic pressures is pushing the ocean to unprecedented conditions (OECD).

Ocean continues to support all existence on earth and remains to be one of the main sources of oxygen through phytoplankton. Moreover, other than just being the major source of oxygen, it is also a means for sustenance across the globe. About 40% of the world's population lives within 100 km of the coast and can be considered as coastal communities (CIESIN 2006). Ninety percent of the world trade is carried out through ocean (Westwood et al. 2001). The sea is also providing opportunities for renewable energy.

South Asia with its unprecedented known and unknown potentials of ocean economic prospects represents strong and valued resources at hand. The un-tapping of these resources is a loss of economic resource within itself. This chapter aims to remove the sea blindness and highlights the growing association of technology within the realm of oceanic resources. The evolutionary stature of technology in the oceanic resources has been one of the most rapidly growing and fast developing concern of the related authorities. Furthermore, this chapter adopts a sustainable implication of technology in order to adopt a more efficient allocation of these resources. This will allow the existing setup of the oceanic resources to be developed under public interventions whereby fully engaging the private sector for a holistic context.

South Asia, with a long coastline of 5.2 million km^2, has a huge potential of the ocean economy. This region accommodates approximately 24% of the total world population (Yuen and Kong 2009). Ocean economy is essential for the socioeconomic development and growth of the region (Bari 2017). The vast prospects of blue growth in the South Asian region can be unraveled through modern technology driven by innovation and cost-effectiveness.

The future of the ocean economy is linked with technological advancement and innovation in the field, thereby making it one of the most important issues that need to be tackled to ensure human existence. There are several advancements like aquaculture, biotechnology, marine mapping, and sea transportation that are already adopted by the ocean economies. James Bellingham, Director at the Woods Hole Oceanographic Institution, mentioned that future of ocean research and the economy is driven by technology, the majority of ocean exploration today is conducted by robots, and their role is incredible for future advancement in ocean economy (National Academies of Sciences, Engineering, and Medicine 2020).

Looking around, we can say that from the air we breathe to the food we eat, to the water we drink, to the recreation we want, and in short to the life we want, we owe it all to the presence of the ocean. According to the National Oceanic and Atmospheric Administration (NOAA), less than 5% of the seafloor has been explored in greater detail (How much of the ocean has been explored? 2020). The major part of the sea, i.e., 80%, is completely untapped and unexplored. The technological advancement contributed two ways in the ocean economy, one is an exploration of new resources and avenues and secondly management of these resources according to the principles of sustainability.

Sustainable development goals (SDGs): 14 is about "Life Below Water." This goal commits government and other relevant bodies to conserve natural resources and sustainably use ocean, sea, and marine resources for food and other development. The sub-point of SDG 14 is about reducing marine pollution, regulating fishing, conserving coastal resources, increasing scientific knowledge, developing research capacity, and transferring marine technology (United Nations 2018). The proposed actions by SDG 14 encompass initiatives in policy areas, regulatory areas, and development in science, technology, and innovation. These innovations have crucial roles to play for the protection of the ocean economy and its sustainable usage.

Ocean resources have been explored by human beings throughout history. However, the technological inventions in the early eighteenth and nineteenth century opened up new corridors of modern technology. The invention of Chronometer by John Harrison in the eighteenth century opened up the new avenues of prosperity and expansion for British. It was due to the invention of "chronometer" that they successfully calculated longitude of the ocean and navigated around safely. This not only became the savior of many lives but also helped the British gain hold of many countries. Thus, the technological intervention in maritime domains opened up myriads of new opportunities for the countries. The twenty-first century is known as the age of technological advancement.

Table 2.1 is a predicament of the current technological and scientific landscape of the major part of South Asia. South Asia has the lowest proportion of R&D investment globally. With India leading the block it still invests only 0.7% of GDP (WB 2019). Moreover, the access to basic facilities of research still remains the poorest among the world, thereby not being able to sustain a workable patent force to contribute in the economy. The USA is the leading country in patent registration with a nominal figure of more than 198,000 patents in the year 2019 only, whereas

Table 2.1 Science and technology landscape

	Pakistan	India	Bangladesh	Sri Lanka	Maldives
R&D (% of GDP)	0.2	0.7	NA	0.11	0.06
Patents	17	1218	3	412	0
Maximum value addition (USD nominal)	36 bn	389 bn	42 bn	14 bn	106 ml
Global Competitive Index Ranking	110	68	105	84	129
Infrastructure access (road density by population)	1.8	3.2	1.6	1.1	0.6

Source: In references (Nataraj 2007; Klaus 2019; UNIDO 2020)

the combined patents of the South Asian region accumulate to only 3213 patents (World Bank, 2019). The lack of basic facilities is backed by weak educational facilities, thereby not being able to develop a sustained framework of scientific development.

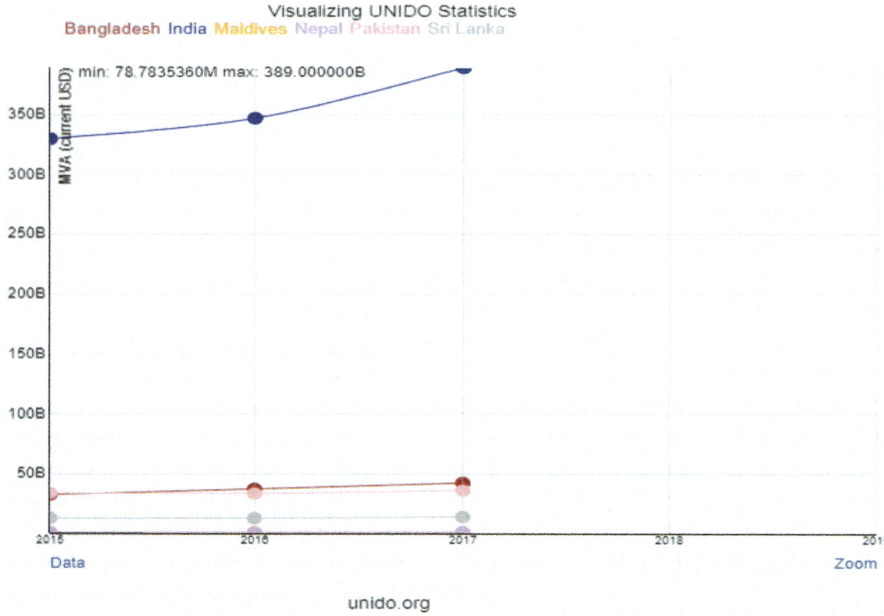

The ocean economic avenues are highly dependent upon the technological abilities of the country. The inability of South Asia to fully grasp their scientific potentials is one of the reasons that the sector still remains under-utilized. In order to fully develop the sector, it is very imperative to adapt a technological-led development paradigm that will enable the full potentials of the sector.

2.2 Potentials and Opportunities of the Blue Economy in South Asia

South Asia comprises seven countries, namely Pakistan, India, Afghanistan, Bangladesh, Bhutan, Maldives, Nepal, and Sri Lanka. Out of these seven countries, five have a long coastline and ocean resources, namely India, Pakistan, Maldives, Bangladesh, and Sri Lanka. These countries have a vast area for blue growth in their jurisdiction. As shown in Fig. 2.1, the exclusive economic zone or EEZ is greater in the area, especially for Sri Lanka and the Maldives as compared to the land area. Other countries like Pakistan, India, and Bangladesh have also a vast maritime area in their jurisdiction as per the laws of the United Nations Convention on the Law of

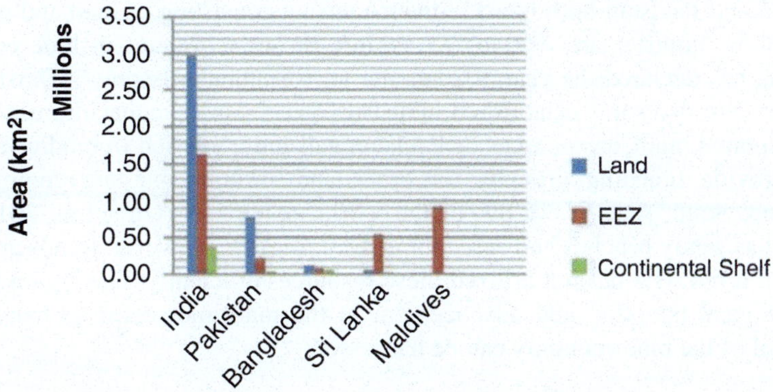

Fig. 2.1 EEZ of South Asia (Manikarachchim 2014)

the Seas (UNCLOS). The EEZs are exclusive economic zones and according to the UNCLOS Article Number 56, the single country has ownership rights to these areas extended up to 200 nautical miles from the baselines (UNCLOS, Article 56 1982). Article 77 of UNCLOS gives the sovereign right of living and non-living natural resources in EEZ to the owner country (UNCLOS, Article 77 1982).

This maritime zone and continental shelf of south Asia show the oceanic treasure for blue growth for the region and highlight its importance on the similar ground of land. These countries express a wide range of oceanic resources including fisheries, coastal communities or tourism, shipping, shipbreaking, maritime transportation, and blue energy initiatives. South Asia has the potential to improve its socioeconomic conditions by investing in the blue economy.

Bangladesh has a huge potential for the blue economy. The country has drafted 7 five-year plans and mentioned 12 actions to be undertaken for sustainable blue economy, i.e., fisheries, coastal tourism, climate actions, and renewable energy (Patil et al. 2019). According to the World Bank, blue economy has added gross value of 3.33% of the total economy in the year 2014–2015 (Patil et al. 2019). The blue economy in Bangladesh has living and non-living components. The major potential sectors from living components are fisheries, aquaculture, mineral, and non-living components like maritime transportation, marine surveillance, oil and gas ports, and related services like marine biotechnology, desalination (freshwater generation), shipbuilding and shipbreaking industry, waste management, renewable marine energy, coastal tourism, blue carbon, and seafood processing (Islam and Mostaque 2016).

Sri Lanka and the Maldives have more area of EEZ and less land or continental shelf area. They have the complete right to explore resources in their respective EEZ under the laws of UNCLOS. Almost 70% of the Maldives population is living nearby the ocean. Similarly, Sri Lanka has 7% larger sea territory than land. This strategic location of both these countries provides them with enormous opportunities for blue growth. Fisheries and coastal tourism are adding greater value to the national income of both countries. Additionally, both countries are also working on linear shipping.

India and Pakistan both boost a diverse terrain consisting of land and maritime area. Both countries are working to explore the new avenues in blue economy. Pakistan has declared the year 2020 as the year of "blue economy" (Dawn 2020). Pakistan has also some established industries like fisheries, aquaculture, shipping, and shipbreaking. Coastal areas of Pakistan and India are rich in biodiversity, and they provide breeding grounds for crabs and shrimps that are commercially imported. South Asia grows 8% of the world mangroves (Giri et al. 2015). The region has sandy beaches and areas for coastal tourism. However, trends are showing that South Asia has not utilized these resources efficiently. The region requires technological adoption and advancement in the maritime sector so that the full potential of the blue economy can be harnessed.

2.3 Existing and Established Sectors of the Ocean Economy in South Asia

The coastal states of South Asia have better opportunities to explore and more responsibilities. The South Asian countries like Maldives and Bangladesh have the highest GDP as compared to other regional countries. However, their earned revenue is less than their potential. The blue economy is not integrated into economic development, but it has other important elements like socioeconomic integration, attaining gender equality, and protecting oceanic resources from destruction. South Asian countries can get benefits from the blue economy by using technological advancement in all the mentioned domains.

According to the blue economy report from the European Commission, 2019 established blue economy sectors are fisheries, aquaculture, fish processing and distribution, coastal tourism, "maritime transport, port activities, shipbuilding, and marine extraction of oil, gas, and minerals" (Scholaert 2020). Some potential sectors of the blue economy are also identified by the Commonwealth. These sectors include marine energy, biotechnology, and blue carbon opportunities for climate change (Secretariat 2016). South Asian countries have some established industries out of the aforementioned and few industries that are emerging in blue growth. However, none of the regional countries has all the established industries despite having ocean resources (Table 2.2).

2.3.1 Fisheries and Aquaculture in South Asia

Fisheries is one of the major established industries in South Asia. This sector contributes largely to the food security and livelihood of South Asian countries. According to the World Fish globally, approximately 800 million people depend on fisheries and aquaculture for their livelihoods (World Fish 2019). The aquatic

Table 2.2 Blue economy industries in South Asia (Sources: Author's own)

Established industries in South Asia	New and emerging industries in South Asia
Fisheries	Aquaculture and multi-species aquaculture
Oil and gas extraction	Renewable marine energy
Shipping	Desalination
Shipbuilding	Bio-carbon
Port development	Biotechnology
Coastal tourism	Technology and R&D
Marine manufacturing and construction (port construction)	Protection of natural habitats
	Assimilation of nutrients
	Chemicals
	Deep seabed mining
	Maritime safety and surveillance
	High-tech marine products and services

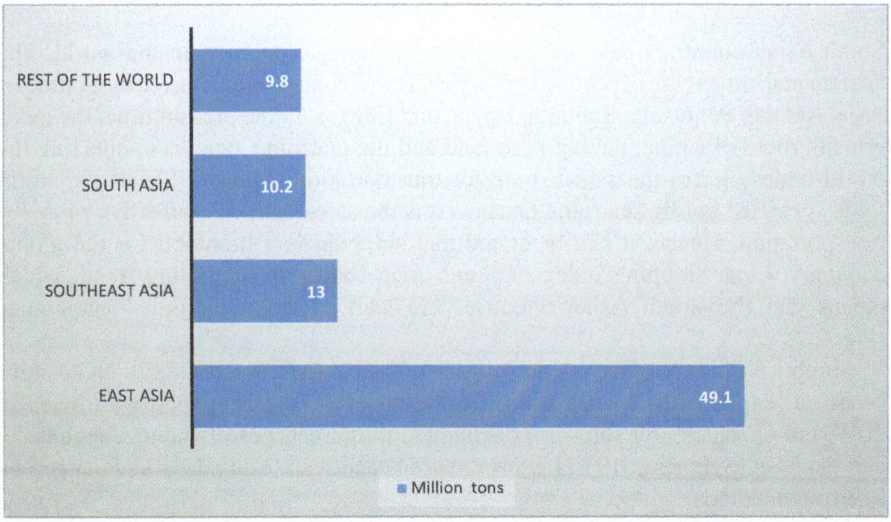

Graph 2.1 World aquaculture fish production by region. (Data sources: FAO 2019)

resources like fish are rich sources of healthy nutrients. Being a part of the food and marine industry, fisheries and aquaculture sector has a greater role to play as a sector of the blue economy. Fisheries trade provides opportunities for cash acquisition to the small traders and fishers. The global fish production has reached up to 179 million tons in 2018 and South Asia was a major contributor to this production and sale (Graph 2.1). The region contributed 34% of the total production. Aquaculture contributed 46% of the total production (FAO 2020).

In South Asia, Pakistan, Bangladesh, and India are a major producer of fish. Fisheries industry has three main methods of capture, inland, marine, and aquaculture. Majority of the catch is from the marine and then land in these three countries. According to the FAO, the fish production by sector among South Asian countries presented that India is the major producer of fish and then followed by Bangladesh and Pakistan. However, the sector-wise trend is the same in all three countries (FAO 2019) (Table 2.3).

Above-mentioned trends showed the potential of fisheries in South Asia. The small-scale fishers are a major contributor to the industry. However, the sector is facing many challenges such as the criminal networks to expand illegal, unreported, and unregistered (IUU) fishing (Bari 2017). Additionally, the lack of sustainable practices and bottom trawling are other problems faced by South Asia in the fisheries sector. The integration of South Asian fisheries and aquaculture sector with global SDGs is a solution to many problems of fisheries in the region.

2.3.2 Shipping and Shipbreaking in South Asia

South Asian countries have an important geostrategic position in the world. The ancient maritime route of second century BCE was connecting South Asia, Southeast Asia, Arabian Peninsula, Somalia, Egypt, and Europe. In the present time, the modern silk route of China, having both land and the maritime zone, is connecting the world trades. Sea is the safest route for transportation. Almost 90% of the world trade is carried out by sea route because it is the safest and cost-effective route for transportation. Hence, it can be stated that shipping is a lifeblood for the global economy. Liner shipping index of South Asia shows an increasing trend, which means that the South Asian countries are well linked with global economies (Fig. 2.2).

Shipbreaking is another major industry of South Asia. According to NGO shipbreaking, a total of 674 ships and offshore vessels were dismantled worldwide in 2019, out of which 469 ships are dismantled in the beaches of India, Bangladesh, and Pakistan (Schuler 2019). However, these beaches are using dirty and dangerous beaching methods.

The global shipping industry is moving toward technological advancement. Digitalization and software-based services are providing accessibility for the new

Table 2.3 Sector-wise regional comparison of fish production (Sources: FAO 2019)

Fisheries sector	Pakistan	Bangladesh	India
Inland	276,501	2,821,266	6,181,000
Marine	346,841	588,988	3,414,821
Aquaculture	148,266	1,859,808	4,881,000
Total	771,608	5,270,062	14,476,821

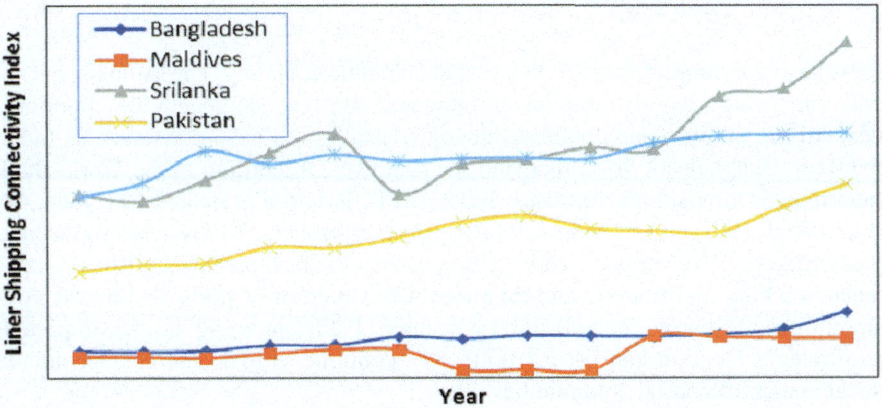

Fig. 2.2 Sources (World Bank 2017)

registration of ships. Shipbreaking is a hazardous industry. According to ILO, shipbreaking workers have very limited access to health, housing, welfare, and sanitary services (ILO 2020). Modern technological advancement can help increase efficiency and decrease the hazardous effect of industries.

2.3.3 Coastal Tourism

Tourism has a major share in the economy of the world. According to the World Economic Forum (WEF), coastal and maritime tourism will grow at a rate of 3.5% annually by 2030. In the Maldives, tourism accounts for 17% of total GDP and 70% of total revenue are generated in foreign currency (Firaag 2019). Similarly, Sri Lanka earned $4.8 bn revenue from tourism in 2019 as per the data of Sri Lankan Tourism Development Authority. India has been promoting tourism under its Incredible India 2.0 campaign.

In the year 2019, Indian tourism contributed 106.9 billion to the national GDP (Frost and Sullivan 2019). Pakistan is also a tourist country with historic heritage, coastal lines, and serene places. According to World Trade and Tourism Council, the tourism revenue of Pakistan was $19.4 billion in 2017–2018. Like other countries, Bangladesh has a long unbreakable coastline and country can also earn revenue from this industry. However, sustainable practices are necessary to adopt so that the environment and ocean biodiversity remain safe.

2.3.4 *Port Development*

Ports have very important roles in a sustainable blue economy. The sailing ships not only carry cargo but also provide employment and revenue generation opportunities. All the South Asian countries having coastal zones have ports such as India, Pakistan, Bangladesh, Sri Lanka, and the Maldives. According to the World Bank International Logistic Performance Index (LPI), Pakistan is ranked 122, India 44, Bangladesh 100, and Sri Lanka at 94 for ports performance. LPI is measured around the indicators like customs, infrastructure, international shipments, logistics competence, tracking and tracing, and timelines. Construction of ports also boosts construction and marine manufacturing sectors. Pakistan being at the important geostrategic location can use its ports for economic growth and socioeconomic development of coastal communities.

2.3.5 *Marine-Based Energy*

Energy and economic development have a direct link. Research has indicated that GDP growth and energy consumption are positively related (Smeets et al. 2007). The South Asian region has the lowest per capita energy consumption at the household level, while commercial energy demand in the region is increasing by the annual rate of 4.2% (Gupta and Jaswal 2006). The South Asian countries like India and Pakistan export a large part of their oil consumption. According to the sources, the world crude oil and natural gas demand will increase in the coming year and highest peak demand is forecasted in 2022. Furthermore, demand will be high in Asian countries along with China (DNV-GL 2017).

South Asian region has proven oil and gas reserves. Though they are less in global percentage except for India, they are enough for domestic usage. India is the fourth-largest oil consumer country and third largest importer with 9.7% of the world total (Workman 2019). Similarly, other South Asian countries like Pakistan, Bangladesh, and Sri Lanka are also an importer of oil. The geographical location of Pakistan has favorable oil and gas reserves, but they are unexplored and untapped. South Asian countries with their long coasts can further explore their oil and gas resources.

2.4 Existing Scientific and Technological Innovation in the Ocean Economy

Ocean has become the center of interest more than ever, and countries are looking for sustainable growth options. Like many other countries, South Asian states are also willing to harness the vast ocean resources. However, preserving the environment while harnessing the resources and fully utilize the potential of the parallel economy for growth is a big challenge. The advancement of science, technology, and innovation is a key to attain sustainable growth and aforementioned objectives. South Asian countries are developing countries, and they are not much advanced in science and technology. Therefore, these states need to look into the existing innovation and technological advancement in the ocean economy of developed countries. Application of scientific innovation and practices helps in many ways, exploring the new potential of ocean, improving efficiency, productivity and cost structure, biotechnology, exploring sea minerals, marine energy, and maritime spatial planning.

Organization for Economic Cooperation and Development (OECD), in his report: "the Ocean Economy in 2030," has mentioned that science, technology, and innovations are the major drivers of the ocean economy. Apart from the existing ocean technology, there would be a string of technologies to stimulate the sectors of the ocean economy. These technologies can include satellite technologies, big data, subsea engineering, and physical sensors; all of the existing and promising technologies have one common goal of economic and environmental sustainability.

2.4.1 *Marine Aquaculture*

Fish is a common food and source of livelihood in South Asian countries. It can be produced naturally and in farms, i.e., aquaculture. According to World Fish, approximately 800 million people depend on fish globally. Aquaculture deals with the farming of fish. In South Asia, aquaculture has expanded in recent years and contributed 40% of total fish production. India and Bangladesh are major aquaculture producer countries (Hossain and Shrestha 2019), while other countries like Pakistan, Nepal, Sri Lanka, and Maldives are far behind. Globally, aquaculture production stands out at 59 million tons. However, some risks like the environmental footprint of aquaculture, coastal water rising, and scarcity of material to feed fish are involved in aquaculture. The technological innovations are needed to resolve these issues and increase the production of fish.

Earth observation is in practice for aquaculture or fish farming. It has the potential to support the management of fish farming through, site selection, environmental monitoring, mapping of farm locations, and water tests for algae, etc. (Kim et al. 2017). Apart from earth observation, GIS-based modeling and mapping are other tools for the selection of location for fish farming. GIS-based mapping also provides information on oceanography, animal production, and growth and environmental effects of the intervention (OECD 2019). Algae blooming is another issue that affects the breeding and harmful for aquaculture. The South Asian countries can combine technologies like earth observation, remote sensing, GIS-based modeling and mapping, and satellite data for analyzing the aquaculture suitability in any country.

Aquaculture production loss due to diseases is another issue. Viral infectious diseases, sea lice, and muscles infection and inflammation are creating a negative impact on growth and health of marine and farm environment. In South Asian region, transboundary diseases are also spreading. Therefore, South Asian states require new technologies, health strategies, and biosecurity innovation to cater to these issues. South Asian region has implemented some strategies such as the implementation of international codes like code of conduct for the responsible fisheries, the convention of biodiversity, and OIE aquatic animal health code. At the regional level, there is a comprehensive document The Asia Regional Technical Guidelines on Health Management for the Responsible Movement of Live Aquatic Animals and the Beijing Consensus and Implementation Strategy by FAO. This document describes technical health problems to aquaculture and guideline for the management of those problems. FAO has suggested laboratory observation techniques like the use of parasitology, bacteriology, mycology, and histopathology for the disease diagnostic (Bondad-Reantaso et al. 2005).

South Asian countries have enormous potential of aquaculture and import of fish in the international market. The adoption of modern scientific and innovative technologies in all regional countries can ensure sustainable blue growth in the region.

2.4.2 Technologies for Exploration and Decommission of Offshore Oil and Gas Platforms

Energy is a basic driver of the economy, and in developing countries like South Asian states, need of oil and gas is increasing. India is among the top five importers of oil. Similarly, other countries like Pakistan, Bangladesh, Maldives, Nepal, and Sri Lanka also import oil and gas. The South Asian countries have proven reserves of oil, gas, and coal. These countries are working toward the exploration of oil and gas and renewable energy resources. The exploration of oil and gas has some global and local environmental issues.

These issues can be loss of biodiversity, carbon emission, marine and freshwater discharge, oil spill, and water containment. After the exploration and extraction of oil and gas resources, the process of decommission starts. The empty well and abandoned installations are removed. The decommission of those filed is very essential so that the environmental hazards to the flora and fauna of the ocean can be minimized. Different activities are used for offshore oil and gas exploration and decommission, such as exploration, location choice, engineering design, drilling, production, and decommissioning. All of these activities require different products, services, and technological innovation including wind, wave, current, bathymetric information, biotechnology, robotic information, and deep-sea technologies.

In the case of offshore oil and gas exploration, new reservoir monitoring and management techniques are used such as seismic imaging and interpretation, 3-D and 4-D seismic reservoir monitoring, 4-D drilling analysis, and the fitting of well sensors for real-time permanent monitoring (OECD 2019). Certain technologies can be used to minimize the environmental impact of oil and gas extraction and maximize the utilization of resources. The drilling fluid has high toxicity and low biodegradability (Cordes et al. 2016). There should be a restriction on the discharge of toxic fluid, and permits should be given for the drilling and cutting. Similarly, the management techniques should be used in decommissioning.

In South Asian countries, the exploration process is usually done on a contractual basis and the decommission is ignored in the whole process. For instance, in India, there are many areas where attempts are made to extract oil and gas from the well, and later on, those wells are abandoned and infrastructure is still there. The European countries have OSPRA decision, which has prohibited leaving disused infrastructure or dumb in place and asked for the removal of the installation immediately after the project completion (Kirk et al. 1999). The South Asian countries can also use such legislations to ensure timely decommission. Furthermore, the seismic operation can be carried out to identify the potential reserves of oil and gas. However, these seismic operations should replace with observation at the areas with active marine mammals.

Another technological innovation that offers great benefits and support in the ocean and offshore wind energy is a support structure. Support structure allows access to deeper water. It enables development in sites with greater average wind speed (Leimeister et al. 2018). The South Asian countries can not only explore their resources and meet the energy demand but also handle the environmental hazards by opting technological and innovative options.

2.4.3 Ballast Wastewater Management Technologies

The world is more connected than ever. Today, more than 90% of the cargo is shipped through ocean routes across the globe. Shipping is one of the safest and economically friendly options of transportation. The shipping industry is equally important for South Asian countries. India has its largest port at Mumbai; Pakistan has Gwadar Port, and Maldives Port, Sri Lanka Port, Banglabanda Port of Bangladesh, and Tianjin Port of Nepal are few ports South Asian ports along with other ports also. These ports have active shipping activity. On the one hand, these shipping activities are beneficial for the socioeconomic growth of the country. On the other hand, the ballast water from the ship comes up with environmental hazards. The ballast water is water, which is carried by ship to maintain its balance and stability and ship discharge ballast water when it loads cargo. Approximately, ships discharge 3 to 5 million tons of ballast water internationally globally (Endresen et al. 2004). Ballast water carries alien species and microorganisms, and they cause infectious diseases and pose threats to the human beings and ecosystem.

The International Marine Organization (IMO) has adopted a ballast water convention. According to this convention, all ships must be equipped with a ballast water management system by 2024. The ballast water management (BWM) requires the ships to carry ballast water record and international water management certificate. Apart from BWMS, the ships use different technologies for wastewater management. The major technologies that are in practice for ballast water treatment are electro-chlorination, UV radiation, oxidation, and deoxygenating and filtration (OECD 2019). These technologies are in practice and continuously getting better through the modern technological innovation that mitigates their negative impact on the ecosystem. IMO is testing some other technologies like advanced lab-on-chip detector that can help in identifying the type and size of microorganisms easily.

South Asian countries are a signatory of IMO convention. However, the South Asian countries arranged a workshop "South Asia Regional Ballast Water Management Strategy Development Meeting" in 2012 (Hemachandra 2012). The basic objective of this meeting was to ensure regional cooperation and implement the IMO BWM system in the South Asian region. A national strategic plan to assess the implementation of IMO ballast water management was designed. Regional countries like Bangladesh, Sri Lanka, India, and Maldives agreed to follow a regional strategic plan for the ballast water treatment. The inclusion of modern and approved technologies and innovation by IMO for the treatment of ballast water in the South Asian region can help mitigate the impact on the environment.

2.4.4 Oil Spill Management Technologies

The concept of the blue economy is to promote economic growth, social inclusion, and improvement of livelihood while preserving the environment. South Asian countries like India, Pakistan, Bangladesh, Sri Lanka, and the Maldives are exporting much of their oil consumptions. Apart from this, the highest oil transportation region globally is also South Asia, because of its location at the crossroad of oil-rich and oil-consuming countries.

Additionally, the South Asian region is growing economically and so is the demand for goods and oil, which means higher transportation of oil and a higher risk of oil spill are due to the increase in vessel movement. The South Asian countries have faced oil spill incidents many times in the past years. Pakistan has its worst oil spill with extreme environmental hazards that happened in 2003, named as Tasman Spirit oil spills. Around 33,000 tons of oil spill in this incident and polluted the marine water and coastal lines (ITOPF 2010).

Another oil spill incident happened in India named an Ennore oil spill incident of 2017. This incident affected 34,000 sq. foot area. Bangladesh has also a long history of the oil spill in 1991, 1992, and 1989, and then in Sundarbans, oil spills of 2014 damaged the heritage site of UNESCO. Sri Lanka has a recent oil spill of 2006, which affected the area of 13 km long coastline. Most of the oil spill in Sri Lanka is reported by the cargo ships. The Maldives has no recent history of an oil spill, but the country has the biggest coral reef that needed to protect. The recent incident of fire in Panama-registered vessels near Sri Lanka has alarmed the Maldives about the environmental and ecological danger caused by oil spills.

Table 2.4 Oil spill risks profile resources (Gunasekara 2011)

	India	Pakistan	Bangladesh	Maldives	Sri Lanka
Likelihood	High	Medium	Low	Low	Medium
Risk	High	Medium	Medium	Medium	Medium
Consequences	High	High	Very high	Very high	High

The South Asian countries have a medium to high risk of oil spills. Consequences of oil spills are very high for all South Asian countries. However, the risk and likelihood of oil spill vary from medium to high (Table 2.4).

South Asian countries have legislation, preparedness, and supporting resources. The use of modern technology is a way forward to deal with the oil spill risks. Preparedness to meet the disaster and risks assessment is necessary for the region. The role of software tools for planning is increasing among nations. Artificial neural networks (ANNs), case-based response (CBD), genetic algorithm (GA), and mathematical models are used for the environmental emergency preparedness and risk assessment (OECD 2019). Similarly, for monitoring oil spills, satellite data are used. Advancement in sensory and imagery technology has removed the doubt of inaccuracy during aerial views. AVIRIS approach that was first used in the Deepwater Horizon incident is used to measure oil thickness, and MODIS camera is used for the ocean images. These technologies help in differentiating between algae blooming and oil thickness in the ocean. Additionally, the sensors like visible, infrared, microwave, radar, position indication, and UAV are used to monitor the oil spill (Ali Cemal et al. 2016).

There is also a technological response for oil spill remedy that is used to clean the ocean and recover from the disaster. These technologies involved chemical treatments like the use of iron-rich salty soap and dispersants. The recent technology of salty soap pulls the oil through a magnetic force to the surface of the ocean. These dispersants do not decrease or remove oil, but they limit the oil slick in the ocean (Shata 2010). Boom and skimmers are other cleaning technology. Skimmers suck the oil and temporarily store it. Bio-cleaner concept is the relatively modern and more efficient skimmer. It has an inner space that stores oil and earns the marine species about the hazards (Seahow 2015).

Pertaining to the high risk of oil spills, the South Asian counties need early assessment and risk analysis for the preparedness. The modern technologies along with the policy framework and funding are necessary to mitigate the risks and to deal with the problem of oil spill efficiently.

2.4.5 Digital Navigation and Sea Traffic Management

Advancement and innovation in technology have also changed the course of work for ocean traffic, shipping, and navigation. Sensor technology has increased the efficiency of the ships and vessels. Similarly, other technologies like robotic automation, Internet of the things, big data, blockchain technology, artificial intelligence, and navigation system are steps toward the digital navigation and smart sea traffic management. These technologies have changed the plan, design, interaction, communication, and operation of the transportation system.

Digitalization of navigation and introduction of technologies like automatic radars, electronic navigation charts, identification and tracking, and global maritime safety system are used for decision-making. Global satellite navigation system (GNSS) is operational on all ocean traffic like ships and vessels. International Maritime Organization (IMO) developed e-navigation for the better organization of ship data. According to IMO, e-navigation is a collection, integration, exchange, and analysis of maritime information and data for the safety and protection of the marine environment." E-navigation improves the reliability of the information, ensures standardized reporting, and improves communication services (OECD 2019).

Robotic automation is another technique that is used to aid delivery, packaging, checking, and inspection in the shipping industry. The new technology of robot consists of sensor and electronic navigation that can easily identify data. IoT or Internet of the thing is also connected with the robots through a wireless network. IoT helps the shipping industry to increase the efficiency of delivery and improve customer services. Similarly, the use of modern technology at the ports ensures environmental sustainability and efficiency. Big data, e-navigation, and IoT reduce the time of ships on the port, and this leads to the reduction of carbon emission at the ports. Artificial intelligence is also used in maritime transportation for logistics and communication. The innovations are using it for increasing the cybersecurity and safety of the vessels.

South Asian counties have immense potential in shipping. There is an improvement in ports and infrastructure building. But the ships still cost more and took more time. Therefore, the continuous effort is needed for improvement of efficiency. Marine transportation is also a source of marine pollution. South Asian region is high at risk due to the increase in the number of vessel movement. The use of modern technology that can improve efficiency and save time and cost will also preserve the environment. Therefore, for the sustainable blue economy, South Asian countries need digitalization of maritime transportation.

2.4.6 Other Incremental Ocean Technologies

The OECD has highlighted some other technologies that are growing rapidly and are beneficial for the ocean economy. These technologies are used for the exploration, construction, capacity building, ecosystem analysis, and improvement in efficiency and productivity. Some of these technologies are biotechnology, nanotechnology, animal telemetry, and autonomous technology.

The ocean has lots of mysteries and species. According to the National Ocean Services of the USA, more than 80% of the ocean is still unexplored. The biotechnology is used to explore the new biodiversity of the ocean. Aquaculture is also using biotechnology. The advancement of the pharmaceutical enzymes also depends on the exploration through biotechnology. Similarly, other marine industries like bio-fuel are also dependent on biotechnology (OECD 2019). According to the European Union, biotechnology helps in exploring the oceanic biodiversity while ensuring environmental safety and protection.

Seafood is the major source of protein. Nanotechnology has the potential to the revolutionized seafood industry. The nanotechnology has implications on seafood processing, detecting bacteria, and monitoring air-based system. The optical fiber detection is used to assess the quality of seafood products. Similarly, the nanotechnologies are used for packaging the seafood (Alishahi 2015). The use of animal telemetry, a nano-sensor-based tag to monitor and study the marine species is another technology that has increased its footprints in ocean observing. This technology collects data from the unreachable parts of the ocean (Block et al. 2016). This is an example of an autonomous device to collect the data of the ocean. The higher level of autonomous methods is used through the algorithm, 3D techniques, and autonomous or semi-autonomous devices for the ocean observation.

Abovementioned technologies and emerging innovations are powerful tools toward the sustainable ocean economy. The concept of the blue economy is not only concerned with economic development, but it is also about social inclusion and environmental sustainability. South Asian nations are blessed with marine resources. However, they are far behind in ocean exploration. Some industries are still emerging, and some are premature in south Asia. The region has approximately 24% of the global population. To feed these people, socioeconomic development and environmental preservation are ensured, and South Asia needs modern and advanced technologies. These technologies will provide the region with a sustainable way forward of ocean development.

2.5 Conclusion: Progression of Future Technologies—Future Prospect of Ocean Economy in South Asia

Over time, technology has exploited the ocean resources and results in the depletion of ocean resources, For instance, nearly 90% of the fish are exploited or depleted globally (Kituyi 2018). According to FAO, out of total global employment in fisheries, 84% is from South Asia. The ocean health is also affecting the exploration of oil and gas reserves. There is a need for innovation that not only provides economic benefits but also ensures environmental sustainability. The strategic concept of blue growth attracts innovation and technology that has a low impact on the environment. Modern technologies of aquaculture and practices are not only beneficial for food safety but also promote environmental protection and collaboration (Soma et al. 2018).

The advanced technology used in shipping and other ocean economy sector has increased the income value of industries. According to the EU, maritime technology has created an additional value of $21 billion net income in shipping sectors (European Commission 2020). Technology development helps to reduce the environmental cost and improve safety and reliability. According to European Union report of 2020 on blue economy, innovation and technology impact ocean economy and facilitate by providing surveillance of pollution and illegal activities, intervention for preservation and protection and communication through satellites and sensors, etc.

The South Asian region has a huge potential of the blue economy, and there is much talk about the ocean economy. The regional countries have some legislative and policy documents for blue growth, but the progress is continuing in the policy domain. However, the South Asian region is lagging in technology and innovation (Mostaque 2020). According to the global innovation index (GIT), only India is ranked among the top 100 countries. India, Pakistan, and Bangladesh are majorly focusing on fishing and port building. India is using biotechnology for exploration of the ocean species. India is also installing technology on its ports. Bangladesh under its 7 five-year plan has mentioned the blue economy as a major field for growth.

The country has also identified a lack of technological development and innovation as a key issue in the ocean economy. Pakistan has also founded its Maritime Science and Technology Park intending to start technological advancement in the nation's blue economy. Similarly, Sri Lanka on its report on "Sustainable Sri Lanka Vision 2030" has identified innovation and technology as an area to work for the ocean economy. South Asian countries are on their way toward sustainable growth through technological innovation in the blue economy.

The South Asian region needs to invest more in ocean technology and exploration. They should invest in aquaculture techniques for securing food safety. They also need to invest on satellite, sensor, and imagery techniques. These techniques are used in exploring the sea, and South Asian region is still at the exploration phase. Additionally, the region has active ports, but there is a need for technology like big data, IoT, and digitalization to improve service delivery and efficiency. The South Asian countries have vast scope in the field of biotechnology. They should invest more in biotechnology and find new pharmaceutical and beauty resources. For the successful implementation of the blue economy plan, the South Asian region needs collaboration, innovation, technology, capacity building, and integrated policies.

References

Alishahi, A. (2015). Application of nanotechnology in marine-based products: a review. Journal of Aquatic Food Product Technology, 24(5), 533-543.

Ali Cemal, T. O. Z., Koseoglu, B., & Sakar, C (2016). Marine Environment Protection: New Technologies On Oil Spill Response Industry.

Bari, A. (2017). Our oceans and the blue economy: Opportunities and challenges. Procedia Engineering, 194, 5-11.

Block, B. A., Holbrook, C. M., Simmons, S. E., Holland, K. N., Ault, J. S., Costa, D. P., ... & Mahmoudi, B. (2016). Toward a national animal telemetry network for aquatic observations in the United States. Animal Biotelemetry, 4(1), 6.

Bondad-Reantaso, M. G., Subasinghe, R. P., Arthur, J. R., Ogawa, K., Chinabut, S., Adlard, R., ... & Shariff, M. (2005). Disease and health management in Asian aquaculture. Veterinary parasitology, 132(3-4), 249-272.

Cordes, E. E., Jones, D. O., Schlacher, T. A., Amon, D. J., Bernardino, A. F., Brooke, S., ... & Gates, A. R. (2016). Environmental impacts of the deep-water oil and gas industry: a review to guide management strategies. Frontiers in Environmental Science, 4, 58.

Center for International Earth Science Information Network (CIESIN) of Columbia University. 2006. CSD Coastal Population Indicator: Data and Methodology Page. http://sedac.ciesin.columbia.edu/es/csdcoasta

DNV-GL. (2017). OIL AND GAS FORECAST TO 2050 (p. 76). DNV-GL. Retrieved from https://www.ourenergypolicy.org/wp-content/uploads/2017/09/DNV-GL_Energy-Transistion-Outlook-2017_oil-gas_lowres-single_3108_3.pdf

Endresen, Ø., Behrens, H. L., Brynestad, S., Andersen, A. B., & Skjong, R. (2004). Challenges in global ballast water management. Marine pollution bulletin, 48(7-8), 615-623.

European Union (2020). 2020 Blue Economy Report: Blue sectors contribute to the recovery and pave way for EU Green Deal. European Commission. Retrieved 5 November 2020, from https://ec.europa.eu/jrc/en/news/2020-blue-economy-report-blue-sectors-contribute-recovery-and-pave-way-eu-green-deal.

FAO. (2019). Food and Agriculture Organization of the United Nations. Retrieved from http://www.fao.org/faostat/en/#data

Frost & Sullivan. (2019). The outlook of the Tourism Industry in India, 2019 (p. 55). India: Research and Markets. Retrieved from https://www.researchandmarkets.com/reports/4768763/outlook-of-the-tourism-industry-in-india-2019

Firaag, -. (2019). Tourism and the Environment: Current Issues for Management. Fao.org. Retrieved 17 October 2020, from http://www.fao.org/3/x5623e0p.htm.

FAO. (2020). THE STATE OF THE WORLD FISHERIES AND AQUACULTURE. FAO. Retrieved from http://www.fao.org/3/ca9229en/CA9229EN.pdf

Gupta, M. D., & Jaswal, P. (2006). Asia's Energy Security Challenges: Need for Collaboration. A Newsletter, (17), 12-22.

Giri, C., Long, J., Abbas, S., Murali, R. M., Qamer, F. M., Pengra, B., & Thau, D. (2015). Distribution and dynamics of mangrove forests of South Asia. Journal of environmental management, 148, 101-111.

Gunasekara, A. J. M. (2011). Oil spill contingency management, its financial arrangement and implications in the South Asian region.

Hemachandra, L. (2012). South Asia Regional Ballast Water Management Strategy Development. In south Asia Regional Ballast Water Management Strategy Development (p. 40). Mumbai; IMO. Retrieved 29 October 2020, from http://www.sacep.org/pdf/BWM/2012.05.29-30-SA_Regional_Ballast_Water_Management_Strategy_Development_Meeting.pdf.

Hossain, M. A., & Shrestha, R. B. (2019). Fisheries in South Asia: Trends, Challenges and Policy Implications. Agricultural Policy and Program Framework: Priority Areas for Research & Development in South Asia, 332.

How much of the ocean has been explored?. (2020). Retrieved 5 October 2020, from https://oceanexplorer.noaa.gov/facts/explored.html

Islam, M., & Mostaque, L. Y. (2016). Blue Economy and Bangladesh: Lessons and Policy Implications. Economics, 2(2), 1-21.

ITOPF (International Tanker Owners Pollution Federation Limited), (2010). Pakistan Country Profile. Retrieved November 4, 2030, from the: http://www.itopf.com/_assets/country/pakist.pdf

Kim, H. C., Son, S., Kim, Y. H., Khim, J. S., Nam, J., Chang, W. K., ... & Ryu, J. (2017). Remote sensing and water quality indicators in the Korean West coast: Spatio-temporal structures of

MODIS-derived chlorophyll-a and total suspended solids. Marine Pollution Bulletin, 121(1-2), 425-434.

Kirk, E. A., Warbrick, C., & McGoldrick, D. (1999). Ospar Decision 98/3 and the Dumping of Offshore Installations. The International and Comparative Law Quarterly, 48(2), 458-464.

Klaus (2019). The Global Competitiveness Report. Insight Report, World Economic Forum.

Kituyi, M. (2018). 90% of fish stocks are used up. Unctad.org. Retrieved 5 November 2020, from https://unctad.org/news/90-fish-stocks-are-used-fisheries-subsidies-must-stop.

Leimeister, M., Kolios, A., & Collu, M. (2018, November). A critical review of floating support structures for offshore wind farm deployment. In Journal of Physics: Conference Series (Vol. 1104, No. 1).

Mostaque, L. Y. (2020) Science, Technology and Innovation for Sustainable Ocean Economy.

National Academies of Sciences, Engineering, and Medicine. (2020). The Role of Research and Technology in the Changing Ocean Economy: Proceedings of a Workshop—in Brief.

Manikarachchim, I. (2014). Stepping up from the green revolution to the blue economy: a new paradigm for poverty eradication and sustainable development in South Asia.

Nataraj, G. (2007). *Infrastructure challenges in South Asia: the role of public-private partnerships* (No. 80). ADB Institute Discussion Papers.

New 'Blue economy policy' to help save foreign exchange, hopes PM. (2020). Dawn.

OECD. (2019). Rethinking innovation for a sustainable ocean economy. OECD Publications Centre.

Patil, P., Virdin, J., Colgan, C., Hussain, M., Failler, P., & Veigh, T. (2019). Initial Measures of the Economic Activity Linked to Bangladesh's Ocean Space, and Implications for the Country's Blue Economy Policy Objectives. Journal Of Ocean And Coastal Economics, 6(2). Doi: https://doi.org/10.15351/2373-8456.1119

Patent applications, residents | Data. (2019). World Bank Data. https://data.worldbank.org/indicator/IP.PAT.RESD

Seahow launches new generation skimmer systems for coastal oil spill response. Neftegaz.ru. (2015). Retrieved 4 November 2020, from https://neftegaz.ru/en/news/companies/410282-seahow-launches-new-generation-skimmer-systems-for-coastal-oil-spill-response/.

Shata, A. A. M. (2010). Recovery of oil spills by dispersants in marine arctic regions (Master's thesis, University of Stavanger, Norway).

Smeets, E. M., Faaij, A. P., Lewandowski, I. M., & Turkenburg, W. C. (2007). A bottom-up assessment and review of global bio-energy potentials to 2050. Progress in Energy and combustion science, 33(1), 56-106

Soma, K., van den Burg, S. W., Hoefnagel, E. W., Stuiver, M., & van der Heide, C. M. (2018). Social innovation–A future pathway for Blue growth?. Marine Policy, 87, 363-370.

Secretariat, C. (2016). The Blue Economy and Small States: Commonwealth Blue Economy Series no. 1. London, United Kingdom: pdf.

Scholaert, F. (2020). The blue economy Overview and EU policy framework. European Parliamentary Research Service. Retrieved from https://www.europarl.europa.eu/RegData/etudes/IDAN/2020/646152/EPRS_IDA(2020)646152_EN.pdf

Schuler, M. (2019). Shipping Industry Continues to Favor South Asia Beaches for Shipbreaking. gCaptain. Retrieved 17 October 2020, from https://gcaptain.com/shipping-industry-continues-to-favor-south-asia-beaches-for-shipbreaking/.

Ship-breaking: hazardous work. ILO. (2020). Retrieved 17 October 2020, from https://www.ilo.org/safework/areasofwork/hazardous-work/WCMS_110335/lang%2D%2Den/index.htm.

UNIDO | United Nations Industrial Development Organization. (2020). United Nations Industrial Development Organization MVA Database. https://www.unido.org/

United Nations. (2018). Report of the Inter-Agency and Expert Group on Sustainable Development Goal Indicators. New York: UN Economic and Social Council. Retrieved from https://unstats.un.org/unsd/statcom/49th-session/documents/2018-2-SDG-IAEG-E.pdf

UNCLOS-Article 77. (1982). Retrieved October 7, 2020, from Oceans and Law of the Sea: http://www.un.org/depts/los/convention_agreements/texts/unclos/part6.htm

UNCLOS-Article 56. (1982). Retrieved October 7, 2020, from Oceans and Law of the Sea: http://www.un.org/depts/los/convention_agreements/texts/unclos/part6.htm
WB (2019). Research and development expenditure (% of GDP) | Data. Retrieved 11 March 2022, from https://data.worldbank.org/indicator/GB.XPD.RSDV.GD.Z
Westwood, J., Parsons, B., & Rowley, W. (2001). Global ocean markets. Oceanography, 14(3), 83.
Workman, D. (2019). Crude Oil Imports by Country. World's Top Exports. Retrieved 21 October 2020, from http://www.worldstopexports.com/crude-oil-imports-by-country/
World Fish. (2019). Annual report 2018. World Fish, Penang, Malaysia.
Yuen, B., & Kong, L. (2009). Climate change and urban planning in Southeast Asia. SAPI EN. S. Surveys and Perspectives Integrating Environment and Society, (2.3).

Chapter 3
Marine Ecosystems and the Blue Economy: Policies for Their Sustainable Exploitation

Anil Markandya

3.1 Background to Oceans and Coasts

The ocean is a critical part of Earth's life-support system and vital for the well-being of humanity. Nearly three billion people rely on fish as a major source of protein and fisheries and aquaculture assure the livelihoods of 10–12% of the world's population (WWF 2015).

This chapter reviews the information on the value of marine ecosystems, how these ecosystems have been changing in recent years and their likely changes in the future. It goes on to look at the role of different policies and measures to prevent their degradation. There is great interest in using the marine resources to promote the "blue economy and blue growth". The combination of high economic value and a declining and at risk asset base makes it critically important to have a sound understanding of where ecosystem service values are at risk, how they can be exploited sustainably, and what can be done to arrest the losses and overuse of the ocean that we currently observe.

3.2 The Use of the Ecosystem Approach

Much of the estimation of the value of the marine environment is based on work initiated by the Millennium Ecosystem Assessment (MEA 2005), which classifies values based on *ecosystem services* (*ESS*). These are derived from the complex

A. Markandya (✉)
Basque Centre for Climate Change in the Basque Country, Leioa, Spain

University of Bath, Bath, UK
e-mail: anil.markandya@bc3research.org

S. Hazra, A. Bhukta (eds.), *The Blue Economy*,
https://doi.org/10.1007/978-3-030-96519-8_3

biophysical systems and are classified under four headings: provisioning, regulating, cultural, and supporting, with a number of sub-categories under each heading. These services are provided by a range of different ecosystems within which different habitats can be found. An ecosystem where several habitats are present is referred to as a biome. The literature contains ten broad categories of which the ones relating to the marine environment are listed in Table 3.1.[1]

Before proceeding to look at the values of services provided by the ecosystems or biomes, two factors are worth noting.

First, as noted, one finds the planet has experienced major losses in the services derived from these ecosystems. During the last century, for example, the planet has lost 50% of its wetlands and around 60% of global ecosystem services have been degraded in just 50 years (Ten Brink 2011). Today 60% of the world's major marine ecosystems that underpin livelihoods have been degraded or are being used unsustainably. By 2100, without significant changes, more than half of the world's marine species may stand on the brink of extinction.[2]

Second, the benefits provided by ESS can be concentrated to a few people at one extreme or they can be global at the other. Fish food is an example of the former, while limiting GHGs is an example of the latter. The estimates of the benefits given later need to take account of who benefits, as that is important in determining the appropriate policies to protect the resources and particularly where the sources of finance to implement the policies should be mobilized. This issue also arises when the physical areas of the ESS are under national jurisdiction or beyond national jurisdiction. The methods of valuation for both types of areas are the same but it is important to have information on this question when determining the right policies.

Table 3.1 Marine biomes used in the ecosystem valuation literature

Marine (open oceans)
Coral reefs
Coastal systems
Coastal wetlands
Inland wetlands

Note: Coastal systems include estuaries, continental shelf areas, and sea grasses but not wetlands such as tidal marshes, mangroves, and salt water wetlands
Source: De Groot et al. (2012)

[1] Marine biomes include those in the polar regions although our understanding of these is even more limited than that of other regions because of their remoteness, hostile weather, and the multi-year (i.e., perennial) or seasonal ice cover. Coastal systems include and indeed pay special attention to those of small islands, where their role is of heightened importance.

[2] http://www.unesco.org/new/en/natural-sciences/ioc-oceans/priority-areas/rio-20-ocean/blueprint-for-the-future-we-want/marine-biodiversity/facts-and-figures-on-marine-biodiversity/

3.3 Global Estimates of the Value of Marine and Coastal Ecosystems

Going from the unit values of marine and coastal ESS presented above, there are two studies that estimate the global value of these services. The WWF has estimated these to be USD 2.5 trillion, from an asset base of at least USD 24 trillion (Hoegh-Guldberg et al. 2015). The authors note that only market based ESS are included, made up of direct output from the ocean (marine fisheries, mangroves, coral reefs, and seagrasses) of USD 6.9 trillion in asset value; shipping lanes, valued at USD 5.2 trillion; productive coastlines valued at USD 7.8 trillion; and carbon absorption of the ocean, thereby reducing global warming, valued at USD 4.3 trillion. Non-market values not covered include ecosystem services such as water filtration by mangroves, seagrass, and wetlands, and the value generated by ecosystems in terms of human culture and lifestyle. Furthermore, market based services may be undervalued. The economic importance of fisheries, for example, is often underestimated, because many of these fisheries are small-scale in nature, spatially dispersed and therefore poorly documented and/or under-reported.

The other major estimate of global values is from the work of Costanza et al. (1997) and Costanza (2014). The 1997 estimates were made for 17 ESS for 16 biomes, including open oceans and coastal areas. The result was a value of USD 11.6 trillion for open oceans and 17.3 trillion for coastal areas, making a total of USD 28.9 trillion. For 2011 the numbers made use of much more information on values for similar sites and included more ESS. The results showed a value of ESS provided of USD 49.7 trillion, of which USD 21.9 trillion was for ESS from the open ocean and USD 27.7 trillion from the coastal areas. The authors note that although the total value of ESS has gone up, a like for like comparison of values taking account of loss of area shows a decline in ESS values between 1997 and 2011 of 18%.

There is thus a big difference between these two estimates, with Costanza et al., being 20 times greater. It is more comprehensive (including more ESS), but it is also subject to more potential error, given the difficulty in valuing ESS across very different locations.

Numbers in trillions of dollars are difficult to comprehend, so it may help to put them in perspective. Global GDP in 2014 was about 78 trillion, which means the WWF figure would imply marine ESS to be worth 3% of GDP, while the Costanza et al., figure would indicate that these services were worth 64%, a figure some might find hard to believe. Indeed, a number of criticisms of the Costanza estimates have been made (see Pendelton et al. 2016).

Global estimates such as these serve to raise public awareness and interest but are of little use in designing policies for conservation or protection of the marine assets. They are also subject to considerable errors and can be controversial, as the above discussion indicates. In order to support such policies what is needed is a detailed location-specific and thematic-specific valuation of ESS over time, so it can be compared with possible costs of policies that would encourage conservation.

3.4 Trends in Services Provided by Coastal and Marine Ecosystems and Reasons for the Trends

There is a lot of evidence that a range of the services provided by marine and coastal biomes has been declining in physical and value terms. The reasons are habitat degradation on account of encroachment/reclamation and general air and water pollution and climate change. The underlying factor behind these primate causes, however, is the fact that decisions on the use of these biomes do not place a value on the ESS they provide. The underlying policy indicators, such as GDP are flawed to the extent that they do not take these values into account. This section looks at the overall trends in services in physical and value terms and see the extent to which they have been valued and where the values can be linked to these phenomena.

3.4.1 *Trends in Global Values*

The work of Costanza and colleagues estimates a decline in the value of ESS over the period 1997–2011 (Costanza 2014). The global value from all marine and coastal ESS in 1997, based on 2011 unit values was estimated at USD 60.5 trillion; by 2011 this had fallen to USD 49.7 trillion, a decline of 18%. The main cause for the fall is the area of coral reefs, which are estimated to be less than half their 1997 levels. The other declines are in the area of estuaries, which are estimated to be 28% lower by 2011. As noted above, however, there is some dispute about these figures, especially the trends in coral.

3.4.2 *Trends in Biomes*

There is information on trends in wetlands and mangroves (Worm et al. 2006), marine populations (FAO 2011; Dulvy et al. 2003), coastal ecosystems (Jackson et al. 2001; Lotze et al. 2006), and sea grasses (Waycott et al. 2009). Worm et al. (2006) compiled long-term trends in regional biodiversity and services from a detailed database of 12 coastal and estuarine ecosystems. They examined trends in 30–80 (average, 48) economically and ecologically important species per ecosystem. Records over the past millennium revealed a rapid decline of native species diversity since the onset of industrialization.

The decline in coral reefs has been well documented. UN, 2016, notes that these reefs have been in a state of continual decline around the world over the past 100 years, and especially over the past 50 years. A recent study by the World Resources Institute calculated that more than 60% of the world's coral reefs are under immediate threat. Indeed, the latest Intergovernmental Panel on Climate Change (IPCC 2014) report suggests that "coral reefs are one of the most vulnerable ecosystem on Earth" and will be functionally extinct by 2050, without adaptation (worst case scenario), or by 2100 with biological adaptation of the whole ecosystem. Presently the level of threats varies considerably in different geographical regions; reefs of the Pacific Ocean are least threatened, but those throughout Asia and the wider Caribbean and Atlantic regions are under greater threats.

This serious and deteriorating status of coral reefs around the world is due in part to climatic factors, especially temperature rise, ocean acidification, and sea level rise; and in part to damaging stresses driven initially by new technologies commencing in the 1970s. The major threats include extractive activities, pollution, sedimentation, physical destruction, and the effects of anthropogenic climate change (UN 2016).

Kelp forests and seagrass meadows are important marine ecosystems that are suffering losses. Both provide food and habitat to many economically exploited species, have high productivity. Brown seaweeds, which are composed primarily of kelps, contribute about half of the total world seaweed production from aquaculture of about 6.8 million tons a year. Seagrasses are not presently harvested commercially but they are critical food sources for large herbivores, birds, and for many other species.

Seagrass beds are reported to be among the most threatened ecosystems on earth with an estimated disappearance rate of 110 km^2 per year since 1980; the rates of decline accelerating from a median of 0.9% per year before 1940 to 7% year—1 since 1990 (Waycott et al. 2009). According to their assessment, 29% of the known areal extent has disappeared since seagrass areas were initially recorded in 1879.

Apart from overexploitation, a number of factors are identified as causes. (UN 2016). Kelp forest distribution worldwide is reported to be affected by overfishing of high value predators that causes explosions in herbivore populations, such as sea urchins, which feed on kelps, resulting in massive reduction of kelp cover and consequently affecting other trophic levels. In addition, changes in the distribution of species have been reported due to increased seawater temperatures.

The environmental degradation of seagrasses has been valued in economic terms by Waycott et al. (2009), to be an estimated USD 1.9 trillion per year in the form of nutrient cycling; an order of magnitude enhancement of coral reef fish productivity; a habitat for thousands of fish, bird, and invertebrate species; and a major food source for endangered species such as the green turtle.

Fortuna and Wilkie (2007) complied a review of trends in mangroves from 1980 to 2005 under the auspices of FAO. They estimated global mangrove areas in 2005 to be 15.2 million hectares, with the largest areas found in Asia and Africa, followed by North and Central America. An alarming 20%, or 3.6 million hectares, of mangroves had been lost between 1980 and 2005. Human pressure on coastal ecosystems and the competition for land for aquaculture, agriculture, infrastructure, and tourism are major causes of the decrease in area reported. The relatively large negative change rates that occurred in Asia, the Caribbean, and Latin America during the 1980s were caused primarily by large-scale conversion of mangrove areas to aquaculture and tourism infrastructure. UNEP (2014) notes that this degradation and loss are predicted to continue into the future if a business-as-usual scenario prevails.

In addition, mangroves are now threatened by climate change which could result in loss of a further 10–15% of mangroves by 2100.

A review of the state of the world's marine fishery resources (FAO 2011) notes the large increase in total fish production, which was only 19.3 million tonnes in 1950, but it increased dramatically to 163 million tonnes by 2009. Of the fish stocks assessed, 57.4% were estimated to be fully exploited in 2009. These stocks produced catches that were already at or very close to their maximum sustainable production. They have no room for further expansion in catch, and even some risk of decline if not properly managed. Among the remaining stocks, 29.9% were overexploited, and 12.7% non-fully exploited in 2009. The World Summit on Sustainable Development (WSSD) goal demands that all these overfished stocks be restored to the level that can produce MSY by 2015. This review suggests that this goal is very unlikely to be achieved, notwithstanding the good progress made in some countries and regions.

Looking at coastal ecosystems Lotze et al. (2006) observe that transformation of such areas is as old as civilization but has accelerated dramatically over the past 150–300 years. Looking back at 12 once diverse and productive estuaries and coastal seas worldwide, they find similar patterns of loss. Human impacts have depleted more than 90% of formerly important species, destroyed more than 65% of seagrass and wetland habitat, degraded water quality, and accelerated species invasions. The value of such losses is not assessed in detail, nor are the costs of conservation compared with the potential losses they arrest or recovery they achieve.

3.4.3 Potential for Use of Marine Ecosystems to Promote Sustainable Blue Growth

An important part of the discussion about marine ESS relates to the role they can play in extending the value of the services to develop new ones, thereby contributing to the sustainable development through the promotion of "blue growth".[3]

Before going into the options it is important to address the question of what is meant by blue growth and how it can be made operational. A useful definition is that provided by FAO, in its development of the *Blue Growth Initiative* (BGI), where it defines Blue Growth as "Sustainable growth and development emanating from economic activities in the oceans, wetlands and coastal zones, that minimize environmental degradation, biodiversity, loss and unsustainable use of living aquatic resources, and maximize economic and social benefits" (FAO 2014b). This requires growth to be measured in the right way, with actions that cause a loss of ESS to be debited with the value of that loss and with wealth accounts to reflect the changes in

[3] FAO has developed the *Blue Growth Initiative* (BGI) which is based on the principles enshrined in its Code of Conduct for Responsible Fisheries (CCRF), a key programme embedded in FAO's Strategic Objectives.

all forms of capital, including natural capital, which is the basis for the ESS that Blue Growth seeks to promote.

Critical to this interpretation of Blue Growth is an understanding of both the potential for using marine ecosystems to generate new services as well as possible damages to the natural capital from these services are derived. It is important to have information on the costs of different methods of exploiting the marine environment, so that it can be done sustainably. Areas where new or increased use of the marine environment is taking place that could be of interest to South Asia include:

- Multi-use offshore platforms
- Algae for biofuels
- Oil and gas extraction
- Aquaculture

3.4.4 *Multi-use Offshore Platforms (MUOPs)*

The world's oceans are being subject to massive development of marine infrastructure in the near future. This will include energy facilities, e.g. offshore wind farms, exploitation of wave energy, and also development and implementation of marine aquaculture. A key component of this infrastructure is the multi-use offshore platform. Such platforms require effective marine technology and governance solutions. There are around 16,600 oil and gas platforms.[4] Aquaculture, a fast growing sector has been increasing at an average annual rate of 6.2% in the period 2000–2012 (9.5% in 1990–2000), with corresponding growth in offshore platforms where it is located.

It is expected that the multiple functions of MUOPs have several environmental effects on Marine Ecosystem Services, directly or indirectly. Potential negative impacts range from loss of area and disturbance of biota, potential risk to affect the seabed, risks to jeopardize native habitats and species (biodiversity), including fish, mammals and birds, visual and noise impacts, use of marine space (otherwise used by marine communities), water or fish pollution because of toxic materials, coast modifications, etc. On the other hand, there are also some possible positive impacts created by the MUOPs which should be taken into account, such as the reef effects of the MUOPs' structures that can attract species and enhance biodiversity. In addition, MUOPs can help to mitigate for global warming, since they incorporate energy extraction technologies that do not emit greenhouse gases and substitute non-environmental friendly technologies. Accordingly, by going offshore coastal space is available for other uses (i.e. added value of open space), while offshore aquaculture does not affect the coastal water quality by creating eutrophication. The excess of continental nutrients in coastal waters causes eutrophication. Moving to the open

[4] http://www.infield.com/oil-gas-database/fixed-floating-platform-facilities

sea has naturally less nutrient values from coastal areas where the topography is more shallow and complex, restricting easy water exchange.

These factors have been reviewed by Koundouri et al. (2016). They set out the steps in an Environmental Impact Analysis for evaluating a potential platform site, which considers anticipated changes in the conditions, biology, and morphology before the platform and after. Factors also to be evaluated are biological diversity, existence of non-indigenous species, food web, eutrophication levels, seabed integrity, contaminants, marine litter, commercial fishing, and noise pollution.

3.5 Algae for Industry and Biofuels

Microalgae are currently cultivated commercially for human nutritional products around the world in several dozen small- to medium-scale production systems, producing a few tens to a several hundreds of tons of biomass annually. Total world production of dry algal biomass for these algae is estimated at about 10,000 tons per year. About half of this produced takes place in mainland China, with most of the rest in Japan, Taiwan, USA, Australia, and India, and a few small producers in some other countries (Benemann 2008).

Algae biofuels may provide a viable alternative to fossil fuels; however, this technology must overcome a number of hurdles before it can compete in the fuel market and be broadly deployed (Hannon et al. 2010).

3.5.1 Oil and Gas Extraction

Oil and gas extraction is a growing activity, with a number of impacts on marine ESS, including those relating to oil discharges from routine operations, the use and discharge of chemicals, accidental spills, drill cuttings, low level naturally occurring radioactive material, noise, and to some extent the placement of installations and pipelines on the sea bed (OSPAR 2009).

Protection of the marine assets can be partly but not wholly provided by the use of appropriate technology. For example, for exploration one can switch from oil based and synthetic based drilling fluids to water based drilling, which is less harmful when discharged. However, the discharge of water based fluids and associated drill cuttings are still a concern in areas with sensitive benthic fauna, for example, cold water corals. The desirable level of protection can be determined by comparing the costs in terms of damages to the costs of more protective methods. In making such comparisons, it is important to recognize the uncertainty in the state of knowledge and to apply the precautionary principle in setting the regulations.

3.6 Aquaculture

As noted under the multi-use platform discussion aquaculture has grown continuously in the past decades, increasing its global share of total fish production to close to half total production in 2012 (42%). It has already overtaken wild caught species in Asia. World aquaculture production can be categorized into inland aquaculture and mariculture. Inland aquaculture generally uses freshwater, but some production operations use saline water in inland areas (such as in Egypt) and inland saline-alkali water (such as in China). Mariculture includes production operations in the sea and intertidal zones as well as those operated with land-based (onshore) production facilities and structures (FAO 2014).

The environmental effects of aquaculture have been assessed in some detail in physical but rarely in economic terms. They include some positive and some negative factors:

- *Mangrove clearance*: In the past this was a major issue with respect to shrimp farming but the practice has practically been stopped. In fact, it has been estimated that less than 5% of mangrove areas have been lost due to shrimp farming, most losses occurring due to population pressures and clearing for agriculture, urban development, logging, and fuel (Da Silva and Soto 2009).
- *Effects on wild fish and habitats*: Aquaculture can diminish wild fisheries indirectly by habitat modification, collection of wild seed stock, food web interactions, introduction of exotic species and pathogens that harm wild fish populations, and nutrient pollution (Naylor et al. 2000).
- *Wild fish overexploitation*: Expanding aquaculture production can alleviate pressure on wild fisheries stocks; for example, increasing the production of farmed fish that compete directly with wild fish (such as shrimp, salmon, and molluscs) could reduce prices and create conditions that can lower investments in fishing fleets and fishing effort over time. Other farmed fishes, such as tilapia, milkfish, and channel catfish, provide alternatives to ocean fish such as cod, hake, haddock, and pollock (Naylor et al. 2000).
- It has also been argued that high fixed costs of fishing fleets, inelastic supplies of labour in the fishing industry, and continued subsidies to the fisheries sector (that approach 20–25% of gross revenue globally) may mean that increased aquaculture production will not result in lower catches of wild fish in the short term. Examples by Naylor et al. (2000) showed little obvious effect of aquaculture production on capture rates of wild fish.[5] In summary, aquaculture is a possible

[5] In the case of salmon, increased farm production has not resulted in reduced capture levels despite 30–50% declines in international prices for four of the five main species of wild salmon (chinook, coho, pink, and chum) during the 1990s. Salmon catches worldwide actually rose by 27% between 1988 and 1997. Similarly, despite rapid growth in alternative farmed fish like tilapia, wild capture of hake and haddock remained relatively stable during the past decade (Naylor et al. 2000).

solution, but also a contributing factor, to the collapse of fisheries stocks worldwide (Naylor et al. 2000).

A positive aspect of aquaculture involving mussels and oysters is that they are filter feeders and grow only on the basis of available nutrient and available carbon in the environment. A large amount of research and monitoring has demonstrated that mussel culture has a positive effect on the environment by removing the excess of nutrients from the water column by biofiltration (Massa et al. 2016). An example of a success story in carbon trading took place in the North Adriatic Sea, in the district of Venice and Emilia Romagna, where in the last few years about 50 mussel farmers have collectively introduced a system of ISO 14064 certification for carbon credits. In 2011, they produced about 32,000 Mt of mussels, worth about 20 million Euro, and were able to generate 4269 Mt of CO_2 credit in 2011 and 5883 Mt in 2012 (*data from MAA—Mediterranean Aquaculture Association*).

Thus while the jury is out on the overall environmental effects of aquaculture on fisheries and the marine environment, it is clear that site selection is a key process in any aquaculture development and suitable locations to undertake farming activities require sites with appropriate environmental characteristics, good water quality and enabling legal and economic conditions and where social acceptability and social responsibility are two essential components of aquaculture development (Massa et al. 2016). Through the establishment of specific zones dedicated to marine aquaculture, for example, allocated zones for aquaculture (AZA), the aquaculture site selection process would be improved while reducing negative aquaculture externalities, thus protecting aquaculture itself from adverse environmental conditions. The adoption and implementation of AZA would also improve the integration of aquaculture with other coastal activities thus preventing conflicts among stakeholders on the use of the marine resources.

Unfortunately, hardly any economic assessments are available of the trade-offs between the economic benefits of different sites and potential economic costs. Such an analysis would need to include possible effects of climate change on different locations, as well as considerations raised above of impacts on wild stocks. A wider economic analysis looking at the interactions between farmed and capture fisheries, including the use of fishmeal and fish oil in aquaculture (Massa et al. 2016).[6]

[6]Compound aqua feeds for farmed high-trophic level finfish and crustaceans are still strongly dependent on fishmeal and fish oil. In 2012 FAO estimated that, although on a declining trend, 14% of world fisheries production was destined to non-food uses, of which 75% (16.3 Mt) was reduced to fishmeal and fish oil (FAO 2014a). The use of fish-derived products in feed formulas raises the issues of whether this fish could directly be utilized as human food, or that a rising demand for fish as animal feed would eventually lead to an even higher overexploitation of marine resources. Sustainability efforts by industry and research are being made to identify more cost-effective dietary fishmeal and fish oil substitutes.

3.7 Policies that Impact on Marine Ecosystems

This section looks at the kinds of policies and investments that can be undertaken to manage marine ecosystems, so as to get the most sustainable use out of them in the long term. It also considers current inhibitors of development and drivers of change that need to be addressed, such as subsidies to damaging actions, subsidies to fisheries and others. In some cases, these instruments can have a bigger effect on marine ecosystems than conservation policies. The discussion here is not intended to be comprehensive in covering the policies, but rather it focuses on how the valuation of marine ecosystem services can help to better design such policies.

The review is divided into two sections. The first looks at policies specifically targeting conservation and management of the marine environment. These include:

- Regional governance
- Private governance
- Design of specific measures: MPAs; Closing high seas to fisheries; Co-management of fishing regimes

The second subsection looks at the role of policies that have as a primary goal an economic target, such as employment protection or growth, but that also impact on marine ESS. There is a role for valuing these impacts so that the policies can be designed to be more effective for both the environmental and economic goals of the government. Cases examined include:

- Fiscal reforms (Mohammed et al. 2016)
- Subsidies for fisheries (World Bank and FAO 2009)

In each case the discussion focuses on what role, if any, economic valuation has had or could have in policy design and decision-making and how greater use of economic valuation could strengthen the decision-making process.

3.8 Policies Targeting Conservation

3.8.1 Regional Governance

A major effort has been made by national and regional governments to address marine and coastal conservation. Given that most marine ecosystem cross national boundaries and include international waters, the focus is often at the regional level.

Billé et al. (2016) note that the regional seas programmes have had limited success, attributed to a lack of systematic implementation of agreements, problems of coordination between the three mechanisms, lack of finance for the programmes, and a lack of capacity. Notwithstanding these problems a few programmes have used economic valuation at least in part to obtain information on where priorities for action should lie and designed action plans based on that (implementation may, however, still be a problem). The following are some examples.

South China Sea and Gulf of Thailand A review of measures to reverse environmental degradation trends has been carried out in the nine countries in this region.[7] UNEP (2009) undertook a major exercise to value the degradation by setting up an economic valuation Task Force, which collected economic data assembled by national focal points. It concluded that the information gathered was superficial and inadequate from the perspective of conducting a cost-benefit analysis of the costs of action versus no action in implementing the regional Strategic Action Programme. The group then took a pragmatic approach to the problem and developed an initial listing of all the goods and services provided by specific coastal habitats. On the basis of an extensive dataset of national economic values for coastal goods and services it developed a method for determining regional economic values that could be used in a cost-benefit analysis of regional programmes or activities and it used that framework to support its decisions. The latter represents a significant intellectual input to economic analysis of ecosystem goods and services at the regional level since the values are derived through application of a formula that takes account of both local and intra-regional variations in market price and relates prices to the total stock.

The Guinea Current Ecosystem This ecosystem extends from the Bijagos Archipelago (Guinea Bissau) in the north to Cape Lopez (Gabon) in the south. The program had as a primary focus the priority problems and issues identified by the 16 affected countries that have led to unsustainable fisheries and use of other marine resources, as well as the degradation of marine and coastal ecosystems by human activities. As part of this it undertook a valuation of marine and coastal ESS under present conditions and under possible degradation. Using a combination of local valuation studies and benefit transfer from studies in other regions, an estimation was made of the use and non-use values of marine ecosystems (sustainable fishery, biodiversity, and non-use values); as well as coastal ecosystems (valuing timber and non-timber forest products, tourism, carbon sequestration, coastal protection, coastal protection, sewage treatment, drinking water, fish nurseries, and biodiversity). Values per hectare per year were obtained and based on those estimates made of the national value of ESS from these areas. The data, however, have not been used in designing specific actions, policies, and measures. As the report notes, decision makers need to make use of this information, taking account of its strengths and weaknesses.

3.8.2 Private Governance

For oceans, a range of sustainability governance arrangements have emerged in the last decade that see new kinds of interaction between public and private actors (Groeneveld et al. 2016). These interactions have arisen, in part, from the realization that ocean governance involves more than just management. It also requires diverse institutions that support sustainable practices. Prominent examples of such

[7] Cambodia, China, Indonesia, Malaysia, Philippines, Thailand and Vietnam.

initiatives are fisheries certification and seafood recommendation lists, where consumers are informed on sustainability aspects of fisheries and aquaculture products, and traceability schemes, where consumers can obtain detailed information on how and where their fish was caught. Rather than being strictly commercial or idealist, these sustainability initiatives are often the result of cooperation between private companies and civil society. Certification and traceability are now major activities involving private data providers and public agencies that promote sustainable fisheries.

There remains a risk that such practices limit the trade of some products for producers who cannot afford to obtain the necessary certification and some developing countries see the procedures as a form of protection on the part of the richer consumer markets. Developing countries are also creating national standards, but unlike Iceland and the US these efforts are motivated by concerns that segments of their fishing and aquaculture industries are unable to comply with the international standards, and as such be excluded from the market. These issues affect countries in South Asia and South East Asia as well. For example, in Thailand the government has developed the government-run and certified Thai Shrimp Label, while other governments in Southeast Asia have invested in Better Management Practice (BMP) and/or Good Aquaculture Practice (GAP) standards. Both these schemes are more inclusive of small-holders than the international schemes, but they are also constrained by a lack of recognition in export markets. Moreover, as noted by Vandergeest and Unno (2012), the standards imposed on fisheries and environmental policy by international certification bodies (often based in developed countries) have resonated with notions of an extension of protectionism.

The private sector is becoming increasingly involved in management and operation of marine protected areas, where its financial resources are a welcome contribution. The factors governing private sector choices are a combination of profit and a desire to contribute to sustainability through corporate social governance. It is not always the case, however, that these factors ensure the greatest social good. Hence the public sector has a responsibility to ensure that sites where private sector investments are made comply with practices that meet overall national and global sustainability goals. In doing this the valuation of social costs and benefits in economic terms has an important role to play.

3.9 Fiscal Reforms

The links between the state of marine ESS and macroeconomic and sectoral policies are very important; arguably such policies can have a bigger effect on these ESS than measures focussing on conservation.

Though fiscal reforms are not necessarily designed to meet the three dimensions of sustainability (economic, social, and environmental), if well designed and used in combination with other policy instruments, they can play an important role in sustainable management of fisheries (Bostock et al. 2004; Slunge and Sterner 2012, p. 107).

Three types of fiscal reforms are of particular importance for ESS—taxation, subsidies, and ecological fiscal instruments (Mohammed et al. 2016).

3.9.1 Taxation

Taxation is often used as a control instrument in fisheries: to regulate input (fishing effort) and output (fish landing). In practice, the experience of taxation in the marine area is mixed. On the one hand, it can provide valuable revenue to be re-invested for better marine management, for example, with fisheries, for regulating fishing input and output. On the other hand, taxation often prioritizes short-term budgetary needs over sustainable resources management. In Morocco, for example, the tax regime has led to under-declaring of catch levels and increased sales in informal markets—making the instrument less efficient in terms of resource management. This points to the fact that taxation is often not popular among fishers and therefore politically costly for many national governments to pursue.

A more successful example is that of Pacific license fees for tuna. Fishery taxes for the rich tuna fishery of the Pacific are governed by the 1982 Nauru Agreement among eight Pacific Island countries. The Nauru Agreement members moved to a minimum fee for fishing per vessel day which was set at a minimum amount of USS$6000 effective in January 2014. Initial data suggest that overall fishing license fee revenue almost quadrupled, from about US$60 million in 2010 to US$230 million in 2012 have increased fourfold to $230 million in 2012. This contributes to better management of the fishery in the area.

3.9.2 Subsidies

Subsidies are direct or indirect financial contributions made by governments to promote a specific activity or policy. Global fisheries subsidies are estimated at US$30–34 billion annually, with fishing equipment and fuel subsidies accounting for US$20–24 billion of that sum (Mohammed et al. 2016). Fish stock depletion globally has been driven in part by high levels of fishing subsidies (Sumaila et al. 2014; FAO and World Bank 2009). In many regions, subsidies are provided when costs exceed revenue, making too many fishing activities artificially viable financially, leading to overfishing. Such capacity enhancing subsidies are called *harmful* subsidies—from a sustainable management of fisheries point of view. A retreat from this approach would make a difference to the conservation and sustainable use of fish stocks.

Harmful subsidies provided by high income countries and their impacts on low income countries' fisheries are significant. For example, the EU provides up to €3 billion (USD 3.3 billion) of harmful subsidies annually to the fisheries sector. Such subsidies have enabled many fishing fleets to exploit fisheries beyond the territorial jurisdiction of the EU.

However, not all subsidies are harmful. There are some positive subsidies that can be used to promote sustainable management of marine fisheries. Such 'good' subsidies may include financing for monitoring and assessment of fisheries resources, effective policing and enforcement, and research and development for sustainable fishing gears and post-harvest loss reduction. According to Sumaila et al. (2014), the proportion of good subsidies to total subsidies varies from region to region. In their assessment of global fisheries for the year 2003, bad subsidies made up USD16.2 billion, while just USD 8 billion constituted good subsidies. The balance of subsidies were classified as ambiguous. This shows that despite the encouraging trend in the increase of 'good' subsidies, bad subsidies still dominate. The same paper notes that the developed world provides most of the world's subsidies. Since most of the world's small-scale fishers are in the developing world, it follows that small-scale fishers generally receive relatively less subsidies compared to large-scale fishers.

It should be noted, however, that some "good" fishery subsidies provided in the developing world have made positive contributions to the overall well-being of fisher communities and poverty alleviation. Bangladesh provides a good learning example to demonstrate how some EFTs can be successfully implemented. In this case the government needed to restrict the catch of Hilsa (the largest single species of fish in Bangladesh) to allow stocks to recover. Since Hilsa is very important in the livelihoods of the poor, the government also compensated them for lost earnings, by providing "affected" fisher communities (more than 210,000 households) with 40 kg of rice per household and alternative income generating activities. Even though no impact evaluation of the scheme has been done, increased fish catch levels suggest that the compensation scheme has had positive impacts both on hilsa population and the livelihoods of thousands of fishers in the lower Meghna Basin.

Overall, there has been limited progress in moving away from harmful subsidies, towards beneficial subsidies such as incentives for monitoring and enforcement. Therefore, international frameworks that tackle the wide use of subsidies particularly by high income countries need to be strengthened. At national level, governments should have clear targets to curb harmful subsidies and eventually eliminate them. This is timelier now than it has ever been, as the Open Working Group proposal for Sustainable Development Goals calls up on countries "by 2020, (to) prohibit certain forms of fisheries subsidies which contribute to overcapacity and over-fishing, eliminate subsidies that contribute to illegal, unreported and unregulated fishing and refrain from introducing new such subsidies".

3.9.3 Ecological Financial Transfers

While taxes and subsidies mainly target the behaviour of individuals or private agents, neither of these instruments necessarily support local administrators in providing incentives for the promotion of sustainable marine management. This is particularly important in countries where natural resources management falls under the jurisdiction of sub-national administration levels such as provinces and districts and costs of marine management are borne by local administrations.

Ecological fiscal transfers (EFTs) have been proposed and introduced in a number of countries (e.g. Germany, Brazil, Switzerland, and India) to compensate decentralized jurisdictions for the costs of providing ecological goods and services which generate spill over benefits beyond their boundaries. This is done by incorporating an environmental performance indicator to fiscal transfers from central governments to local or sub-national levels of the administrative hierarchy.

In marine and costal ecosystems management context, EFTs can be introduced in multiple ways. For example, imposition of no-take-zones or marine protected areas (MPAs) may impose loss in revenue to adjacent local governments which could be compensated by EFTs. One of the main reasons given for limited use of EFTs in developing countries is the limited financial capacities of national governments. However, there are existing social safety net programmes such as conditional social transfers in many developing countries (e.g. Bangladesh, Brazil, and India). Adding an ecological performance indicator to such existing programmes could be a cost-effective way of delivering both social and ecological objectives.

Lastly, we have the case of Payment of Environmental Services as a fiscal mechanism that can help sustainable management of environmental resources (PES). In the marine context one recent example is a study by Barr (2012) who examines its use in the context of artisanal fisheries. She notes that small-scale artisanal fisheries are identified as amongst the world's most vulnerable and display a high occurrence of poverty; many still live on the margins of human dignity and 20% are thought to earn less than $1 a day. At the same time small-scale fisheries are one of the major factors affecting coastal and coral reef health. Persistent overfishing and a rising use of destructive fishing gear—in an effort to catch whatever fish remain—result in the untiring and increasing degradation of these areas.

In areas of prevalent poverty, justifying interventions which serve to reduce fishers' effort, catch and ultimately income proves to be difficult. Indeed, in the past, many 18 marine conservation efforts met with high resistance and low compliance for failing to deal with the socioeconomic aspects of many of these fishing communities.

The use of PES to capture 'blue carbon' in coastal wetlands and mangroves has also been studied recently. Such beginning to feature prominently on the international agenda, under programs such as the International Blue Carbon Initiative, coordinated by the *International Union for Conservation of Nature*, UNESCO, and Conservation International. Local schemes have also emerged, such as Mikoko Pamoja, a 107-hectare mangrove conservation project in Kenya. PES schemes and proposals are also emerging to protect a range of other coastal and marine ecosystems for services associated with fisheries, marine biodiversity, and coastal protection. Payments are made in such cases for avoided destruction and thereby the continued storage of carbon. Friess et al. (2015) note, however, that a number of difficulties can arise from a number of external stressors that can result in damage to the PES site and a failure of the scheme. These include changes in sediment input, agricultural pollution, and pest infestations. Such risks effect the market price of the payments that buyers of the ESS are willing to pay and measures to address them would make the schemes more effective.

3.10 Conclusions

This chapter has reviewed the trends in marine ecosystems from an economic perspective. The value of services derived from these ecosystems, while subject to considerable uncertainty, is significant. Unfortunately, the value has been declining in recent years, in spite of higher levels of exploitation, because of the degradation to the ecosystems in many locations. The causes of the degradation are over exploitation, misuse of the marine biomes as a waste sink, conversion of coastal systems, and loss of habitat and climate change.

This degradation is not inevitable and can be reversed. A key role in doing so is to value the costs and benefits of marine-related activities and the incorporation of those costs in setting regulations and designing policies to manage the marine environment. If this is done, increased value can be obtained from the marine environment to promote blue growth.

The chapter has reviewed a range of instruments, to see how they can help to move the use of these ecosystems towards a more sustainable use. At present, while there are some developments that indicate a shift in a more encouraging direction, we also have government interventions, particularly some subsidies, that are harmful to the marine environment. There is also a greater need to increase cooperation in setting regulations that apply across extra-territorial jurisdictions. Finally, management in this area has to take account not only of the overall costs and benefits of different interventions, but also pay special attention to the way in which they impact on the many vulnerable groups that depend on marine resources for their livelihoods. Sadly, their conditions are worsening in many countries.

References

Barr, R.F. (2012). Marine Payments for Environmental Services in an artisanal fisheries context. Doctoral Dissertation to the London School of Economics.

Benemann, J.R. (2008). "Opportunities and Challenges in Algae Biofuels Production", Position Paper, FAO: Rome.

Billé, R. et al., (2016). "Regional Oceans Governance: Making Regional Seas Programmes, Regional Fishery Bodies and Large Marine Ecosystem Mechanisms Work Better Together" in P. Nunes et al. (eds.), Handbook on the Economics and Management for Sustainable Oceans, Edward Elgar: Cheltenham.

Bostock T, Cunningham S and Bennett E (2004) Political economy of fishery fiscal reforms, Fishery fiscal reforms and poverty reduction, http://www.odi.org/sites/odi.org.uk/files/odi-assets/publicationsopinion-files/3115.pdf.

Costanza, R., et al., (1997). The value of the world's ecosystem services and natural capital. Nature 387, 253–260.

Costanza, R. et. al. (2014). "Changes in the global value of ecosystem services", *Global Environmental Change*, Volume 26, May 2014, Pages 152–158.

Da Silva, S.S. and Soto, D. (2009). "Climate change and aquaculture: potential impacts, adaptation and mitigation". In K. Cochrane, C. De Young, D. Soto and T. Bahri (eds). Climate change implications for fisheries and aquaculture: overview of current scientific knowledge. FAO Fisheries and Aquaculture Technical Paper. No. 530. Rome, FAO. pp. 151-212.

Dulvy, N.K. et al., (2003). Extinction vulnerability in marine populations, *Fish and Fisheries*, 4, 1, 25-64.

de Groot, R.S. et al. (2012). "Global estimates of the value of ecosystems and their services in monetary units", *Ecosystem Services* 1 (2012) 50–61

FAO (2011). "Review of the state of the world marine fishery resources", FAO Paper 569, FAO: Rome.

FAO (2014a). "The State of World Fisheries and Aquaculture: Opportunities and challenges." Food and Agriculture Organization of the United Nations, Rome, 2014.

FAO (2014b) Global Blue growth initiative and Small Island Developing States - Fisheries and Aquaculture Resources Use and Conservation Division (FIR) 7 pp. http://www.fao.org/3/a--i3958e.pdf.

Friess, D.A., Phelps, J., Garmendia, E., Gómez-Baggethun, E. (2015). Payments for Ecosystem Services (PES) in the face of external biophysical stressors, Global Environmental Change, 30, 31-42.

Groeneveld, R.A. et al., (2016). "Private Governance of Ocean Resources" in P. Nunes et al. (eds.), Handbook on the Economics and Management for Sustainable Oceans, Edward Elgar: Cheltenham.

Hannon et al., (2010). "Biofuels from algae: challenges and potential", *Biofuels*, 1, 763-784.

MEA (Millennium Ecosystem Assessment), (2005). *Millennium Ecosystem Assessment: Ecosystems and Human Wellbeing: Current State and Trends*. Washington DC, Island Press.

Fortuna, S. and M.L. Wilkie, (2007). "The Worlds Mangroves: 1980-2005", FAO Forestry Paper 153, FAO, Rome.

Hoegh-Guldberg, O. et al. 2015. Reviving the Ocean Economy: the case for action - 2015. WWF International, Gland, Switzerland., Geneva, 60 pp.

IPCC, (2014). *Climate Change 2014: Impacts, Adaptation, and Vulnerability. Contribution of Working Group II to the Fifth Assessment Report of the Intergovernmental Panel on Climate Change* [Field, C.B., V.R. Barros, D.J. Dokken, K.J. Mach, M.D. Mastrandrea, T.E. Bilir, M. Chatterjee, K.L. Ebi, Y.O. Estrada, R.C. Genova, B. Girma, E.S. Kissel, A.N. Levy, S. MacCracken P.R. Mastrandrea, and L.L. White (eds.)]. Cambridge University Press, Cambridge, United Kingdom and New York, NY, USA.

Massa, L. et al. (2016). "Aquaculture in the Mediterranean and the Black Sea: a Blue Growth perspective" in P. Nunes et al. (eds.), Handbook on the Economics and Management for Sustainable Oceans, Edward Elgar: Cheltenham.

Mohammed E.Y. et al., (2016). "Fiscal reforms for sustainable marine fisheries governance: Institutional frameworks and design essentials" in P. Nunes et al. (eds.), Handbook on the Economics and Management for Sustainable Oceans, Edward Elgar: Cheltenham.

Jackson, J.C.B. et al. (2001). "Historical Overfishing and the Recent Collapse of Coastal Ecosystems", *Science*, Vol. 293, Issue 5530, pp. 629-637.

Koundouri, P. et al., (2016). "A Socio-Economic Framework for Integrating Multi-Use Offshore Platforms in Sustainable Blue Growth Management: Theory and Applications" in P. Nunes et al. (eds.), Handbook on the Economics and Management for Sustainable Oceans, Edward Elgar: Cheltenham.

Lotze, H.K. et al. (2006). "Depletion, Degradation and Recovery Potential of Estuaries and Coastal Seas", *Science*, 312, 1806–1809.

Naylor, R. L. et al. (2000). "Effect of aquaculture on world fish supplies". Nature, 405: 1017-1024

OSPAR Commission (2009). "Assessment of impacts of offshore oil and gas activities in the North-East Atlantic": OSPAR, London

Pendelton, L.H. et al. (2016). "Has the value of global marine and coastal ecosystem services changed?" *Marine Policy*, 64, 156-158.

Slunge, D. and Sterner, L. (2012). Environmental fiscal reform in East and Southern Africa and its effects on income distribution. In Castelluci, L. and Markandya, A. (eds.) Environmental taxes and Fiscal reform. Central Issues in Contemporary Economic Theory and Policy. Palgrave Macmillan. London, UK.

Sumaila UR, Dyck A, Baske A (2014) Subsidies to tuna fisheries in the Western Central Pacific Ocean. Mar Policy 43:288–294.

Ten Brink, P. (ed.) (2011). *The Economics of Ecosystems and Biodiversity in National and International Policy Making*, Earthscan, London and Washington.

United Nations (2016). First Global Integrated Marine Assessment (First World Ocean Assessment). Available at http://www.un.org/Depts/los/global_reporting/WOA_RegProcess.htm.

UNEP (2009). Reversing environmental degradation trends in the South China Sea and Gulf of Thailand. UNEP.

UNEP (2014). The Importance of Mangroves to People: A Call to Action. van Bochove, J., Sullivan, E., Nakamura, T. (Eds). United Nations Environment Programme World Conservation Monitoring Centre, Cambridge. 128 pp.

Vandergeest, P. and A. Unno (2012). A new extraterritoriality? Aquaculture certification, sovereignty, and empire. *Political Geography* 31:358-367.

Waycott, M. et al. (2009). "Accelerating loss of seagrasses across the globe threatens coastal ecosystems", Proceedings of the National Academy of Sciences, 106, 12377-81.

World Bank and FAO. (2009). "The Sunken Billions: The Economic Justification for Fisheries Reform".

Worm, B. et al. (2006). "Impacts of Biodiversity Loss on Ocean Ecosystem Services", *Science*, Vol. 314, 787-790.

WWF (2015). Living Blue Planet Report: Species, Habitats and Human Well-Being. WWF, Gland, Switzerland.

Part II
Provisioning Services

Chapter 4
Analysis of Trade Liberalization of the Blue Economy in Indian Ocean Rim Association

Rashmi Kundu, Somya Mathur, and Badri Narayanan

4.1 Introduction

As the world evolves, the rules for industrialization written 300 years ago have become outdated. Sustainability is the need of hour. Now the priorities are not just limited to economic growth or human development but also toward environmental sustainability. Over the years, due to mankind's personal quest to quench their greed, the environment has suffered. It has been subjected to degradation and damage that have become irreversible. As industries move beyond the land, they aim to take advantage of the vast scale of the oceans and its resources.

Oceans comprise more than 70% of earth's surface and have not been left unscathed. Natural events such as global climate change and earthquakes, in addition to pollution, seabed mining, and even industrial fishing in large quantities have quite a large negative impact on the ocean. As toxic and biodegradable waste makes its way into our oceans, fish, corals, and other types of wildlife, it eventually makes its way into the human bodies as well. The Great Barrier Reef, off the coast of Australia that stretches about 2100 km, is one prime example of degradation of ocean resources. The loss of the marine biodiversity threatens millions of livelihoods. It has a long-term impact on the region's economic and food security.

In order to address this issue, the concept of "blue economy" came into being, which is promotion of ocean ecology in coherence with economic development. In

R. Kundu
Research Analyst, Centre for Social and Economic Progress, New Delhi, India

S. Mathur
Economist, NCAER, New Delhi, India

B. Narayanan (✉)
Lead Adviser and Head, Trade and Commerce, NITI Aayog, Government of India, New Delhi, India

S. Hazra, A. Bhukta (eds.), *The Blue Economy*,
https://doi.org/10.1007/978-3-030-96519-8_4

the most general of definitions, a blue economy seeks to promote economic growth while prioritizing the preservation and sustainability of the ocean. Developing economies with a coastline, especially the island economies, derive most of their income security from ocean resources. Tourism creates many opportunities for economic growth. The sustainability of this growth is, however, is compromised by various ways, for example, overfishing and rising ocean pollution. Three years back, in 2018, Maya Beach, one of the most popular beaches in Thailand, was shut down indefinitely. Increasing tourism has caused irreversible damage to the marine biodiversity in the area. This has jeopardized job and food security and economic growth. Thus, the implementation of blue economy is of utmost importance as it aims at improving quality of life and reducing ecological risks, while steering the economy onto the path of sustainability.

Groups and associations all over the world have held meetings and conferences discussing the implementation of a blue economy. Many have created so-called pillars of a blue economy, exemplifying its importance for both the ocean itself and for the people. These pillars include things like fisheries, trade and shipping, tourism, biotechnology, and new technologies such as marine renewable energy and aquaculture.

Indian Ocean comprises 14.4% of the earth's surface, and the region has 28 major emerging and developed economies, 21 among them are members of the Indian Ocean Rim Association (IORA). This region is home to a staggering 35% of the world's population. Geopolitically speaking, it is a key strategic basin, with abundance of natural resources. The growth within this region has mostly been trade-led. It has a share of nearly 12% of the World trade and "commands control of major sea-lanes carrying half of the world's container ships, one third of the world's bulk cargo traffic and two thirds of the world's oil shipments, the Indian Ocean remains an important lifeline to international trade and transport."

The Indian Ocean Rim Association (IORA) has promoted the concept of "blue economy" in the Indian Ocean region, by promoting trade and cooperation among its member countries. In 2014, the IORA found that a blue economy could generate employment and food security and help in poverty alleviation and ensuring sustainability in business and economic models in the region. The IORA adopted the Blue Economy Declaration at the first Ministerial Blue Economy Conference in 2015 (IORA). The goal of this declaration was to generate jobs, economic growth, and promoting environmental sustainability, by preserving the ocean and its resources wisely. At the second Ministerial Blue Economy Conference in Jakarta, Indonesia, the need for developing innovative financial tools to enhance the blue economy in the IORA states was highlighted. The Blue Economy Working Group (WGBE) is part of the 2017–2021 plan of implementing a blue economy. The IORA priority areas concerning a blue economy are fisheries and aquaculture, renewable ocean energy, seaports and shipping, offshore hydrocarbons and seabed minerals, marine biotechnology, research and development, and tourism. "The 'blue economy' concept seeks to promote economic growth, social inclusion, and preservation or improvement of livelihoods while at the same time ensuring environmental sustainability. At its core it refers to the decoupling of socioeconomic development through oceans-related sectors and activities from environmental and ecosystems degradation" (World Bank 2017). "A blue economy is low-carbon, efficient, and clean".

"It is also an economy that is based on sharing, circularity, collaboration, solidarity, resilience, opportunity, and interdependence". "Its growth is driven by investments that reduce carbon emissions and pollution, enhance energy efficiency, harness the power of natural capital—such as the oceans—and halt the loss of biodiversity and the benefits that ecosystems provide".

In this chapter, we attempt to capture the impact of the global trade, its input–output linkages, and the subsequent economic behavior in the blue economy. This would help in understanding whether the objectives of increased economic growth and employment opportunities can be attained through trade facilitation. As there has been very limited comprehensive study on this aspect so far, this chapter attempts to fill this gap.

4.2 Literature Review

Much of the literature highlights the benefits of a blue economy, especially in terms of employment growth, slowing climate change, and sustainability for fisheries. It emphasizes what may take to achieve a blue economy and the challenges and investments that may be necessary to implement the policy.

Against a terrible backdrop of ever-rising population and ever-increasing need for new sources of food, employment, and energy, ocean-related economic activities have come to the forefront. For example, "by 2030 two thirds of the fish for food consumption is expected to be farmed, much of it at sea". "Offshore wind capacity is forecast to rise to become the leading power generation technology by 2030", "and seaborne trade is expected to quadruple by 2050". Such economic activities will experience a surge in investment as focus shifts to the development of coastal infrastructure, industry, and tourism as migration to cities and coasts deepens. Likewise, they will also experience risks from rising sea levels, as a result of climate change. This will drive the need for a wave of defensive and sustainable development.

According to research, few countries and association have so far successfully implemented steps and policies toward the establishment of blue economy. However, existing research is not very accurate as to how far along they are in the process and what kinds of steps that remain to be taken in order for a successful implementation. "A blue economy is supported by a trusted and diversified knowledge base and complemented with resources, which helps to inspire and support innovation".

> The European Union proposed a 'Blue Growth' strategy in 2021, thus implementing different initiatives and policies related to the sustainability of their oceans (UNEP 2012).

They have over the years seen vast amounts of growth and sustainability in their economies, especially in employment. This included the Blue Economy Innovation plan, proposed in 2014, with a goal of developing sustainable jobs and growth, providing knowledge and security, and cooperation between countries. According to the 2019 Blue Economy Report by the European Union, the gross value added by the blue economy sectors is around 179 billion euros, leading contributions made by coastal tourism (32.6 billion euros) and port activities (34.4 billion euros), respectively. A total of about 42 million people are directly employed by various sectors of the blue economy in the European Union.

One component of a blue economy, such as trade, can vastly grow seaborne trade and strengthen international relations. Seafood is one of the most highly traded foods in the world with an extremely high export value compared with other highly traded goods (World Bank Group 2016). This is just one example. Another important aspect of a blue economy would be to reverse unfair trade. The island states rely heavily on exclusive economic zones, or zones that only certain states can utilize and use marine resources from CIEL. Often, this is overlooked, and a key aspect is to promote fair trade for these developing countries.

Over 80% of trade all over the world is done by the sea, and the majority of it is done by developing countries (UNEP 2012). Food and other goods are most often transported through oceans, fish being the largest single trade item for developing countries. As trade is projected to grow at record numbers in the near future, it is highly important to consider the sustainability of trade and its effect on pollution, highlighting the importance of a blue economy for trade, as trade is important for ocean growth (The Ocean Foundation). Reducing greenhouse gas emissions, pollution, creating more possibility of trade, and therefore creating opportunity for employment are all benefits of implementing a blue economy. As volume of trading grows, it is important that coastal countries are able to match up to it in the most efficient ways possible and take advantage of it. "Key opportunities for growth in trade in the ocean is directly related to consumption of seafood products, including an increase in the demand for food and cosmetic and pharmaceutical products" (World Bank and United Nations Department of Economic and Social Affairs 2017).

Other key drivers in the growth of a blue economy in relation to trade are various maritime transport industries, international regulations, ability to harness ocean energy, among others. The ensuing paragraphs highlight the various quantitative studies that have been recorded related to a blue economy. These highlight the importance of prioritizing sustainability for our oceans, as it may stimulate major economic growth and have long-term benefits relating to the well-being of the oceans and of the people. Most quantitative studies do not specifically involve trade and commerce between countries or around the world.

BenDor et al. (2011) evaluated how variations in the rules governing the aquatic ecosystem market shift risk between regulators and entrepreneurs to promote ecological restoration as market-based regulations become increasingly common. Exploratory data analysis and simple linear regression are used to understand broad relationships between market geography, phased credit sale policies, and banking prevalence at the district level to determine regulatory variability. It is concluded that with increased use of market mechanisms for environment management, future of conservation may be impacted more by distorted incentives, asymmetric information, and local regulatory discretion than species interaction.

Analysis of federally funded coastal restoration projects was conducted by Conathan et al. (2014) to study its potential economic benefits, where an extensive interdisciplinary research and cost–benefit analysis were undertaken, which also included appropriate survey methods like the "contingent valuation." The combined economic output and the long-term ecosystem service benefits outweigh the cost of investment.

Developing a data-driven understanding of the water economy resulted from policy developments that touch on economic and ecological impacts of water use is pertinent. A mathematical framework was built by Mayer et al. (2016) using economic production and trade datasets with water consumption data and models of surface water depletion in the region under study. The economy uses more surface water than groundwater, and water productivity should be considered as a factor in regional or local planning for future, as there are stark differences in the water use profiles of different economic sectors.

Discussing the experiences of African countries with regard to developing blue economy strategies and studying scope for Asia-Africa co-operation to maximize welfare, Mohanty et al. (2020) undertook trend analysis of various macroeconomic data points that are conducted, especially those related to resource endowments in African Oceans. It was said that as Africa is emerging as one of the fast-growing continents of the world, adopting blue economy strategy will help accelerate the growth process.

In "Stakeholder Perspectives on Opportunities and Challenges in Achieving Sustainable Growth of the Blue Economy in a Changing Climate," Hemer et al. (2018) identified challenges pertaining to the development of the emerging ocean renewable energy (ORE) industry and explore possible pathways to grow ocean energy in Australia. Potential strategies and research priorities were discussed in a symposium, where needs to develop the industry were categorized under technical; policy and regulation; education and awareness; investment; or other. A coordinated and motivated community is required to raise awareness and attract investment to face the current challenge of realizing the potential of the emerging ocean renewable energy industry.

In 2016, Gillett started a time-sensitive comparative study using macroeconomic data of 22 pacific Island countries and territories, to provide a regional perspective to contribution of fisheries to GDP, fisheries-related employment, exports of fishery production, among others. More than half of employment generated in the region is directly related to the fishing industry, and although the total amount of fishery exports in the region fell by 42% from 2007 to 2014, access fees for foreign fishing increased by 279%.

To study the monetary and non-monetary benefits of coral reef food fisheries in order to develop strategies for long-term sustainability, Grafeld et al. (2017) did a value and supply chain assessment with a qualitative mixed-methods approach that includes value chain tracing, value-added estimations, food provisioning estimates, cultural valuation, and field interviews. It was concluded that improved management that results in the sustainable use of fishery resources is required for ensuring food security in the region.

Toward defining the blue economy, Keen et al. (2018) conducted an exploratory study to assess sustainable oceans development initiatives, with a particular focus on fisheries as an example of an important sector within a blue economy. A case study approach was adopted for an in-depth understanding, duly undertaken through a combination of observation, interviews, and document analysis. It was observed that the pre-existing literature and studies tended to forget sociopolitical elements needed for sustainable ocean governance such as power, agency, and gender.

In Current Trends in the Philippines' Shrimp Aquaculture industry: A Booming Blue Economy in the Pacific (2017), Vergel correlated Shrimp Aquaculture Industry and the Blue Economy in the Philippines. In order to assess the current situation of the Philippines' Shrimp Aquaculture Industry, related data metrics like volume of production and import–export numbers are considered for a comparative study. Co-operation among research institutions, policymakers, government, and local players is necessary for the progress of the aquaculture and fisheries sector, a key industry under blue economy, which contributes toward achieving poverty alleviation and sustainable marine resource management.

To introduce a model framework, suitable for the impact assessment of blue economy innovations, Varga et al. (2013) computed a multisector computable general equilibrium (CGE) model, which provides the empirical frame for studying the economic impacts of new technologies developed under blue economy, in the frame of the geographic macro and regional (GMR) model. These technologies usually use those products as inputs that are otherwise considered waste, reducing production waste and need for raw material. Such innovations reveal transformed relationships among various economic sectors.

Exploring the potentials of blue economy for enhancing economic sustainability in Bangladesh, Sarker et al. (2018) aimed to identify the economic value, potential, and challenges to develop a framework for a blue economy. A qualitative approach is taken by reviewing policy documents and consulting stakeholders related to the growth of the blue economy. They reported that the biggest challenge that the blue economy faces is that of climate change, and that even though there is a huge potential, Bangladesh lacks strategic planning to make blue growth a reality.

To study the impact of climate change on marine fisheries that constitute an important part of the blue economy and to determine the high-risk countries, Blasiak et al. (2017) built a vulnerability index, involving 147 countries, with focus on exposure, sensitivity, and adaptive capacity. It is seen that there is a negative correlation between vulnerability and per capita carbon emissions. Due to low adaptive capacity, the developing countries are most vulnerable to climate change.

Identifying, measuring, and assessing impact of the marine resource congestion on sustainable development of the marine economy, Cao et al. (2020) take up China as a case study to examine the impact of excess use of marine resource inputs that restrict sustainable development and output efficiency. An index system is constructed to evaluate the input level of marine resources, by analyzing its spatiotemporal evolution and others factors that primarily influence the input congestion. By using targeted strategies, long-term marine resource congestion can be alleviated, which will lead to reduction in inefficiencies.

As rehabilitating the ocean ecosystem in a bid to create a sustainable blue economy takes center stage, it is upon the countries and its citizens to adhere to the conservation initiatives, while keeping in mind the gains in the economic front that comes with them.

4.3 Methodology and Data

This paper uses GTAP database version 10.A and the standard GTAP model to analyze welfare and macroeconomic and trade impacts. The computable general equilibrium modeling framework of the Global Trade Analysis Project (GTAP) has been used to explore the impacts of trade liberalization on IORA countries for the expansion of blue economy. A GTAP model is a multi-region, multisector, computable general equilibrium model, with perfect competition. GTAP helps to run simulations interactively in a Windows environment using the GTAP model. Results and complementary information for the analysis were provided in a Windows environment and were accessed.

4.4 Model Structure

The standard GTAP model has been used to analyze trade liberalization impacts. Version 10A of the GTAP database covers 65 disaggregated sectors, 141 regions/countries, and 5 factors of production, which are aggregated into appropriate version for the simulations (Table 4.1).

In the model, the economies are aggregated into 16 sectors and 18 countries/regions. The regions and their codes are given in Table 4.1.

Table 4.2 gives details of all the sectors included in the model. The sectors included in this model are the ones whose interests align with the objectives of the blue economy to promote economic growth and generate employment opportunities while conserving the ocean resources. Some other sectors have also been included in the model to know the impact on them due to trade liberalization in some of the key sectors of the blue economy.

Fish is the vital resource of this region. As per FAO Fisheries and Aquaculture Statistics, Indian Ocean region's share is 12.74% in capture fish production of the world in 2018 as shown in Fig. 4.1. Being a key source of food security and livelihood, it is crucial that under the blue economy, the ocean resources are conserved from depletion, due to unregulated and illegal fishing, pollution, and climate change. The fisheries and sea food product industries could be developed, so that it can even further contribute toward income generation, food security, job creation, and, most importantly, poverty alleviation. In the chapter, we have simulated the impact of improvement of trade facilitation in the blue economies' fisheries and sea food product industries. Fisheries is coded as "fsh," and seafood product industry is coded as "Ofd."

"The Indian Ocean as a major transit area for international trade is evident in the fact that half of the world's container ships, one third of the world's bulk cargo traffic and two thirds of the world's oil shipments cross its waters annually." Thus, it is revealed how important it is to develop sea-borne transport sector for further

Table 4.1 Countries included in the model and their codes

Sl. No.	Code	Region description
1	Aus	Australia
2	Bgd	Bangladesh
3	Ind	India
4	idn	Indonesia
5	irn	Iran
6	ken	Kenya
7	mdg	Madagascar
8	mys	Malaysia
9	mus	Mauritius
10	moz	Mozambique
11	omn	Oman
12	sgp	Singapore
13	zaf	South Africa
14	lka	Sri Lanka
15	tza	Tanzania
16	tha	Thailand
17	are	United Arab Emirates
18	xtw	Rest of World

Table 4.2 Sectors included in the model and their codes

Sl. No.	Code	Sector description
1	Afs	Accommodation, and food and service
2	Fsh	Fishery
3	Ofd	Necessary food products
4	GrainsCrops	Grains and crops
5	HeavyMnfc	Heavy manufacturing
6	LightMnfc	Light manufacturing
7	MeatLstk	Livestock and meat products
8	Extraction	Mining and extraction
9	OthServices	Other services
10	ProcFood	Processed food
11	Osg	Public administration and defense
12	Ros	Recreational and other services
13	TextWapp	Textiles and clothing
14	TransComm	Transport and communication
15	Util_Cons	Utilities and construction
16	Wtp	Water transport

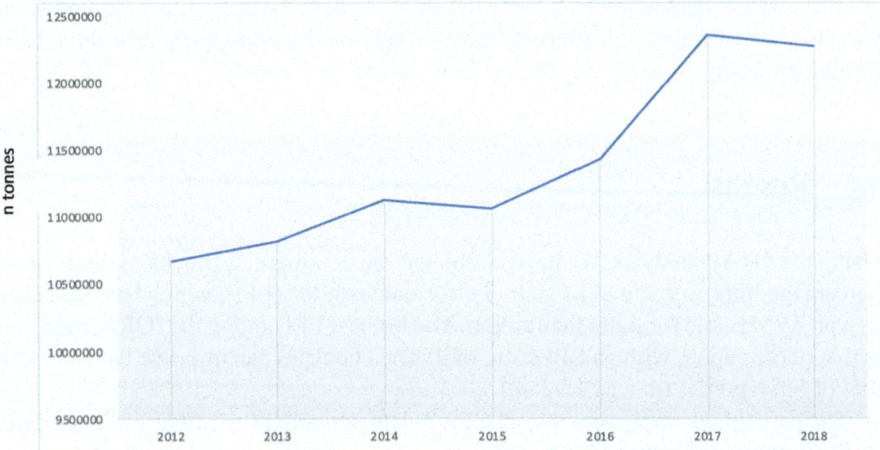

Fig. 4.1 Capture fish production in India Ocean. (Source: *FAO*, Fisheries and Aquaculture Department)

development of the region. Hence, water transport has been included in the model structure and has been coded "Wtp."

Another significant sector under the blue economy is promoting maritime safety and security among member states from pirates, terrorism, weapon trafficking, IUU fishing, unlawful exploitation of marine resources, and climate change. The public administration and defense sector, which comes under GTAP model, has been coded "Osg." Tourism is an area of priority as this region comprises beautiful island nations and countries with vast coastlines. Accommodation and food and service have been coded as "Afs," and recreational and other services have been coded as "Ros."

Tariffs are custom duties, which are levied on the imports of a country. They are thus essentially a cost burden. They make the imports expensive and give a price advantage to the locally produced goods over similar goods that are imported. Trade facilitation implements practices that are beneficial to both the exporting and importing countries. They reduce cost burdens, with an aim to maximize efficiency, while safeguarding requisite regulatory objectives.

The commercial shocks have been simulated in the form of improvement in Logistic Performance Index. The World Bank created this benchmarking tool, to help countries identify advantages, improvements, and opportunities, and they might face in a bilateral trade logistics. In case of blue economies, it helps to promote fisheries and seafood product industries and boosts tourism and regional cooperation in scientific research, ensuring maritime safety and security and development of water transport.

The rest of the sectors as given in Table 4.2 have been included to know the transmission of impact of improvement of trade facilitation in key priority sectors of blue economy.

4.5 Results

Using the GTAP analysis, we have simulated the commercial shocks in the form of percentage improvement in Logistic Performance Index (LPI), which has been used as a proxy variable for trade facilitation. The highest LPI among the IORA countries is that of Singapore with an LPI score of 4. The countries can improve their score to that of Singapore = (4 – LPI Score).

The percentage improvement in LPI is calculated as [((4 – LPI Score of country A)/(LPI Score of Country A)) * 100].

Since its bilateral trade between country A and B:

$$\%\text{Improvement in LPI} = \%\text{Improvement in LPI of Country A} + \%\text{Improvement in LPI of Country B}$$

The impact of this commercial shock on the macro variables has been simulated in this model.

An improvement of trade facilitation in the blue economies of some of the IORA countries such as Maldives, Seychelles, and Comoros has not been included as they were not disintegrated from their respective composite regions and can in hindsight lower the prices of products being traded—one of the heavily traded sectors being that of fisheries and fish food products. Domestic consumers may immediately substitute away using competing imports, thereby increasing the aggregate demand for imports. The composite price of imports of other sectors may also fall. Thereby, the aggregate demand for imports in IORA may increase. The simulated result is reflected in Fig. 4.2 where the imports are seen to rise for most of the regions, most significantly in developing and small island economies like Sri Lanka and Mauritius, which are majorly dependent on their ocean resources.

When imports become cheaper, it may lead to a decrease in the composite prices of intermediate products. As a result, excess profits are incurred at the current price. This leads to an increase in the volume of output, generating an expansion effect. It is aptly reflected on Tables 4.3 and 4.4, which are obtained from the model simulation—where it is noted that overall both country-wise and industry-wise outputs have risen. As in Fig. 4.3, it is clear that every country has had a positive impact with trade facilitation, and imports in absolute volume are now higher than it was before the facilitation of the blue economy. While trading, countries generally apply various para-tariff, port restrictions, pre-shipment inspections, SPS and TBT restrictions, and quality standards. With trade facilitation, various non-tariff barriers imposed through import restriction, procedural delay, non-recognition of SPS certificates, and differentiated port procedures have been eased.

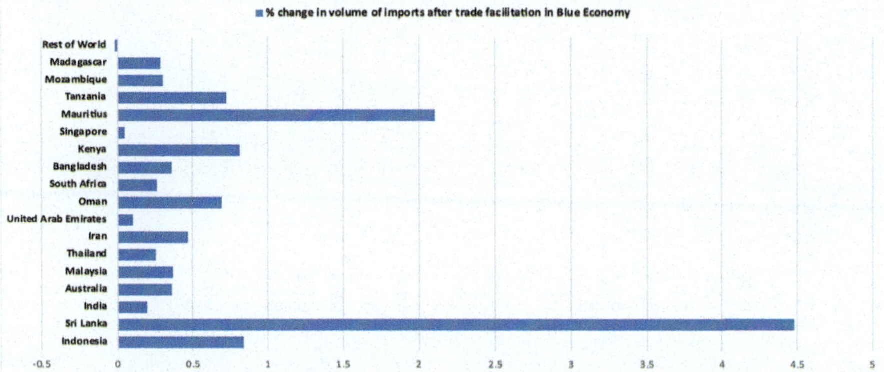

Fig. 4.2 Change in volume of imports after trade facilitation in blue economy

Table 4.3 Change in imports due to trade facilitation

Country	% Change in volume of imports after trade facilitation in blue economy	Absolute volume of imports in (million $) before trade facilitation enforced due to promotion of blue economy	Absolute volume of imports in (million $) after trade facilitation enforced due to promotion of blue economy	Absolute change in volume of imports (in million $) after trade facilitation enforced due to promotion of blue economy
Indonesia	0.84	194,031.88	195,660.3	1628.42
Sri Lanka	4.48	27,056.69	28,269.55	1212.86
India	0.2	548,031.25	549,135.44	1104.19
Australia	0.36	267,659.63	268,630.56	970.94
Malaysia	0.37	208,582.22	209,346.11	763.89
Thailand	0.25	258,147.89	258,798.27	650.38
Iran	0.47	105,165.61	105,660.13	494.52
United Arab Emirates	0.11	301,983.31	302,329.5	346.19
Oman	0.69	43,152.82	43,449.13	296.3
South Africa	0.26	104,635.7	104,912.88	277.17
Bangladesh	0.36	52,774.88	52,962.82	187.95
Kenya	0.81	21,240.01	21,411.24	171.23
Singapore	0.05	323,400.84	323,567.47	166.63
Mauritius	2.11	7194.75	7346.55	151.79
Tanzania	0.72	12,395.74	12,484.78	89.04
Mozambique	0.3	9271.33	9298.85	27.51
Madagascar	0.29	4270.05	4282.27	12.22
Rest of World	−0.02	18,684,524	18,679,914	−4610

An even more interesting metric in this case is the percentage change. Sri Lanka has gained the most with an almost 5% increase in the volume of imports after trade facilitation in blue economy, as shown in Table 4.3. It is followed by Mauritius and Indonesia with increments of almost 2% and 1% respectively. Sri Lanka's imports

Table 4.4 Change in the industrial output of sectors in the countries due to trade facilitation within blue economies

Country	Accommodation, and food and service	Fishery	Necessary food products	Grains and crops	Heavy manufacturing	Light manufacturing	Livestock and meat products	Mining and extraction
Australia	0.18	−0.26	−1.2	−0.09	−0.2	−0.12	−0.25	−0.04
Indonesia	0.19	0.27	0.43	−0.08	−0.52	−0.18	0.34	−0.24
Malaysia	1.06	0.7	8.33	0.01	−0.7	−0.17	1.83	−0.18
Singapore	1.26	0.45	7.61	−0.18	−0.42	−0.59	1.81	−0.19
Thailand	0.81	1.79	6.2	0.09	−0.72	−0.48	0.08	−0.29
Bangladesh	−0.16	−0.03	−4.33	0.07	0.14	0.13	0.13	0.05
India	0.15	−0.45	0.92	0.08	−0.13	−0.12	0	−0.1
Sri Lanka	1.92	0.67	2.27	−0.97	−4.61	−4.58	0.72	−3.89
Iran	0.76	0.05	−1.61	−0.02	−0.01	0.04	0.11	−0.07
Oman	2.53	1.75	31.51	3.98	−0.46	−0.81	5.37	−0.21
United Arab Emirates	0.99	0.78	−0.29	0.83	−0.19	−0.23	2.75	−0.03
Kenya	−0.05	0.17	−0.17	−0.11	−0.61	−0.45	0.06	−0.63
Madagascar	0.48	0.39	−6.36	−0.18	−0.26	−0.13	0.6	−0.22
Mauritius	4.82	0.13	2.8	−0.65	−2.14	−0.75	0.09	−0.93
Mozambique	0.47	1.06	−14.4	0.22	−0.9	0	0.92	−0.06
South Africa	−0.01	−0.16	−0.16	−0.03	−0.18	−0.15	0.07	−0.03
Tanzania	1.35	0.39	−1.19	−0.1	−0.42	−0.35	0.24	−0.13
Rest of World	−0.02	−0.11	−0.29	−0.01	0.04	0.03	−0.01	0.02

Country	Other services	Processed food	Public administration and defense	Recreational and other services	Textiles and clothing	Transport and communication	Utilities and construction	Water transport
Australia	0.02	−0.01	0.06	0.15	−0.2	−0.01	0.13	0.41
Indonesia	0.11	−0.38	0.14	0.25	−0.76	0.05	0.48	−0.15
Malaysia	0.1	0.07	0.34	1.83	−0.48	0.02	0.1	0.74
Singapore	−0.11	1.07	0.16	0.52	−0.53	−0.02	0.26	0.54
Thailand	−0.03	0.13	0.16	1.11	−0.78	−0.1	0.37	3.41
Bangladesh	0.17	0.15	0.07	0.07	0.09	0.04	0.09	−0.15
India	−0.04	0.09	0.06	1.35	−0.23	0	0.07	1.27
Sri Lanka	−0.08	−1.08	0.58	1.98	−8.88	0.67	2.76	23.5
Iran	0.05	0.18	0.17	1.73	−0.12	0.05	0.11	2.32
Oman	−0.14	2.97	0	0.87	−1.38	0.32	0.86	−0.19
United Arab Emirates	0.02	1.67	0.11	0.2	0.07	0.08	0.14	0.32
Kenya	0.2	0.03	0.56	−0.01	−0.34	−0.09	0.24	7.8
Madagascar	0.29	0.6	0.18	−1.9	−0.59	0.36	0.38	2.51
Mauritius	0.08	−1.46	1.16	2.16	−4.21	0.07	2.21	26.46
Mozambique	0.85	0.26	0.29	−1.48	0.76	−0.15	−0.24	5.37
South Africa	0.03	0.01	0.09	0.61	−0.04	−0.06	0.11	−1.42
Tanzania	0.14	0.11	0.11	3.05	−0.39	0	0.06	−2.48
Rest of World	0	0	−0.01	−0.04	0.1	0	−0.02	−0.4

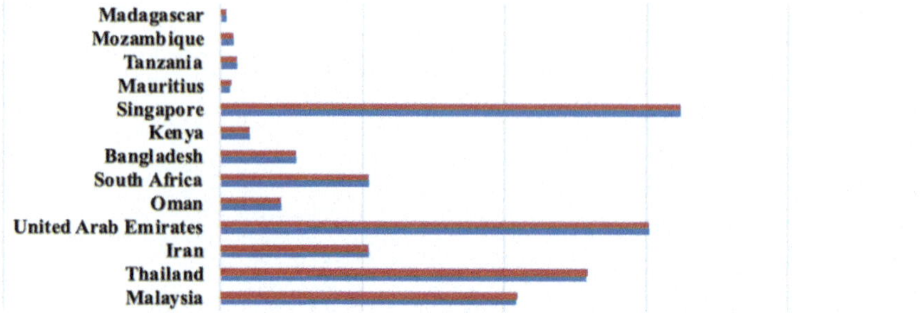

Fig. 4.3 Absolute volume of imports before and after trade facilitation in blue economy

in absolute volume increased 1.04 times from 27,056.69 million dollars before the promotion of blue economy to 28,269.55 million dollars after it. In terms of the gains through imports, pre- and post-trade facilitation, Indonesia with an absolute gain of $1628.42 million dollars occupied the top spot, followed by Sri Lanka ($1212.86 million) and India ($1104.19 million).

Table 4.4 generated out of the model simulation reflects the expansion effect as seen in the various industrial sectors. Out of the various sectors in consideration, industries like fishery, necessary food products, water transport, recreational and other services, utilities, and constructions, saw the highest gains. There are some exceptions like mining and extraction, heavy manufacturing, and textiles and clothing, where there has been a decrease in the absolute volume of imports, consistently across all countries, after the trade facilitation.

The expansion effect may induce excess demand for factors of production. As full employment is assumed in this general equilibrium model, and the excess demand may bid up the prices of the mobile endowments. This phenomenon might percolate to other sectors as well. Thereby, factors of production witness increased efficiencies, for they now may have to compete with the more competitive foreign imports. Inter-sectoral transfers of factors of production may also take place. In addition, economic benefits may expand dynamically through capital formation mechanisms and productivity improvements. There may be an expansion of production, higher consumption, and more capital investments. This leads to an overall increase in welfare as reflected in Fig. 4.4. A country like India which is the leading importing country in the region has had an absolute gain of about $1500 million dollars in welfare measures after trade facilitation was enforced due to promotion of blue economy. However, the expansion phase in these economies due to elimination of tariffs may be offset by adverse terms of trade effects and trade diversion effects. Due to this, we notice only a marginal increase in the absolute volume of imports, as seen in Fig. 4.3.

The trade liberalization increases profit margins and diversion of factors of production from higher capital-intensive sectors like mining and extraction, textiles and clothing to comparatively lower capital-intensive sectors like tourism and recreation, fishery and food products, and water transport. The economic activity in the

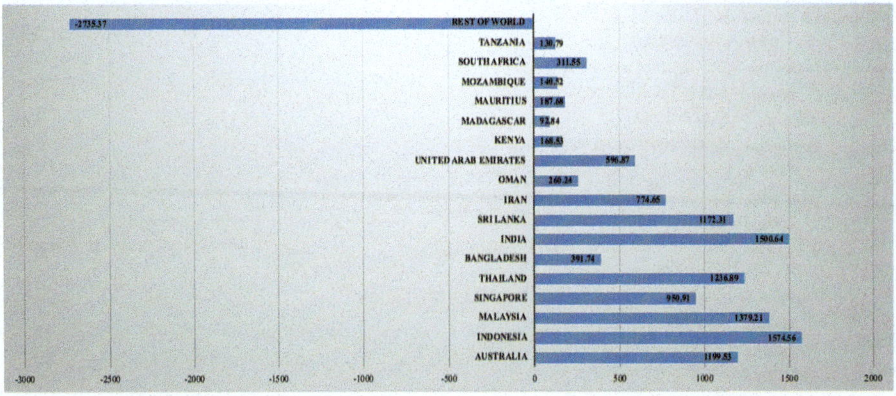

Fig. 4.4 Absolute change in welfare measure (in million $) after trade facilitation due to promotion of blue economy

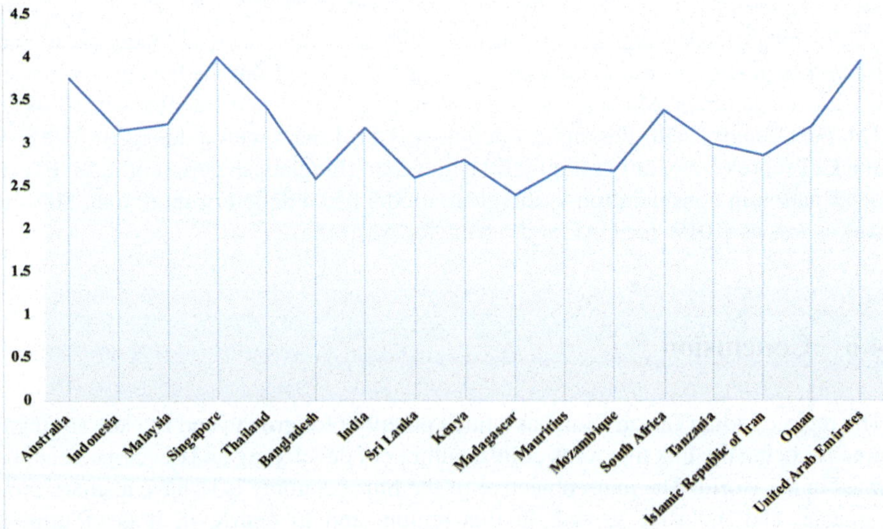

Fig. 4.5 Logistic Performance Index (LPI) score (proxy for trade facilitation)

economy is thus stimulated. Figure 4.5 shows the Logistic Performance Index score of the various countries. The percentage change in LPI has been considered a proxy for trade facilitation. Using data from World Bank, it is seen that Singapore has the highest LPI among all the countries considered, and thus, its score has been considered the benchmark for comparison in this model, as is shown in Fig. 4.6. The LPI score ranged from 2.4 to 4. Singapore was closely followed by the United Arab Emirates with a score of 3.96, Australia (3.75), Thailand (3.41), and South Africa

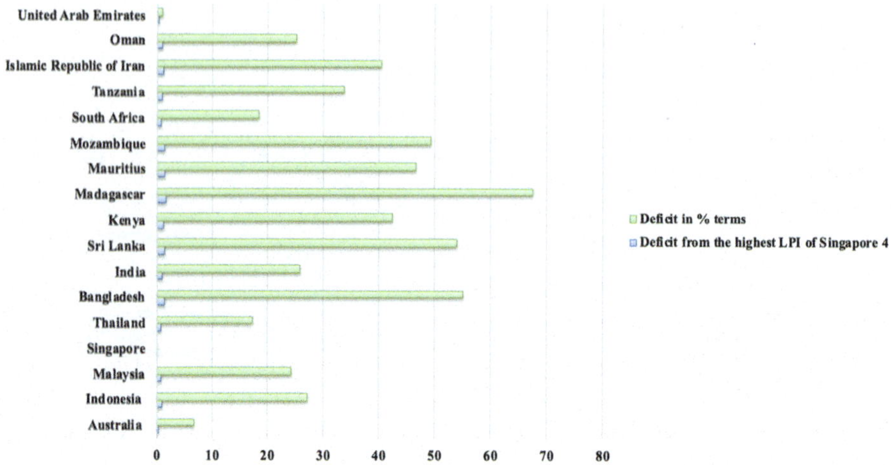

Fig. 4.6 Country-wise deficit in Logistic Performance Index (LPI) score (proxy for trade facilitation) from the highest score (Singapore: 4)

(3.38), respectively. The two countries with the lowest scores, Madagascar and Bangladesh, witnessed deficits as high as 1.61 and 1.42 from the benchmark score. In percentage terms, Madagascar had a deficit of 67.34% and Bangladesh, 55.04%. The next country with the highest deficit was Sri Lanka with a deficit of 53.85%, at a LPI score of 2.6, only marginally better than Bangladesh deficit of 2.58. Every other nation in consideration in the given model had a deficit of more than 50%. It was as low as 1.01% for UAE and 6.67% for Australia.

4.6 Conclusion

The region, Indian Ocean Rim, has a huge potential for growth and is a key strategic area for being the link that joins quite a number of developing and developed economies in the world. The main objective of the blue economy is to have a smart, sustainable, and inclusive growth in this region, and to realize it, it is of utmost importance that trade liberalization is brought about in the region, among different trading partners. Using the computational general equilibrium model, we have simulated the effects of elimination of tariffs and non-tariff barriers in the blue economy sectors on key macroeconomic variables. The results show a rise in the volume of imports and overall welfare and much greater increase in trade. Some countries witness increase in trade to the tune of 1% or higher, while others have smaller changes. There have been gains in industrial output of industries like water transport, defense, fisheries, food products, recreation and tourism, utilities like gas manufacture and its distribution, and construction. Thus, on the policy front the paper brings out that, for the sustainable growth of the blue economy, there should be more trade facilitation among various trading partners. This will help them to identify the kind of challenges and opportunities, one may face more efficiently, and as a result, not only will there be a rise in volume of trade, but also will there be a rise in overall welfare.

There are, however, some caveats and limitations of our study. We assume full employment in our model, which reduces the magnitude of impact, because the employment gains in the blue economy sectors are offset by losses in employment in other sectors. Therefore, our estimates are conservative in nature, and the real impact may be higher than what we predict. Furthermore, some countries like Maldives, Seychelles, and Comoros could not be included in the study, as they have separated from their respective composite regions. Even some sectors like mining and textiles appear to lose in the trade liberalization scenario, because of almost zero dependence on ocean resources. This loss might also be mitigated in real world by an overall expansion in employment. Furthermore, our chapter does not capture dynamics and capital accumulation, which can have even greater impact. In other words, our chapter just takes the first step in this direction and provides sufficient evidence that the blue economy trade policies can have a profound positive economic impact even in such a conservative setting.

References

Todd K. BenDor, J. Adam Riggsbee, and Martin Doyle (2011) Environmental Science & Technology 2011 45 (24), 10322-10330 https://doi.org/10.1021/es203201n. https://pubs-acs-org.offcampus.lib.washington.edu/doi/pdf/10.1021/es203201n

Michael Conathan, Jeffrey Buchanan, and Shiva Polefka (2014) The economic case for restoring coastal ecosystems. https://www.americanprogress.org/wp-content/uploads/2014/04/CoastalRestoration_report.pdf

Mayer, Alex, Mubako, Stanley, Ruddell, Benjamin L. (2016) Developing the greatest Blue Economy: Water productivity, fresh water depletion, and virtual water trade in the Great Lakes basin. Earth's future. 2016;4(6):282-297. doi:https://doi.org/10.1002/2016EF000371. https://agupubs.onlinelibrary.wiley.com/doi/full/10.1002/2016EF000371

Mohanty et al (2020) Economic Opportunities in Blue Economy in Africa. Springer Singapore Pte. Limited;

Hemer, Mark A, Manasseh, Richard, McInnes, Kathleen L, Penesis, Irene, Pitman, Tracey. Perspectives on a way forward for ocean renewable energy in Australia. Renewable energy. 2018;127:733-745. https://doi.org/10.1016/j.renene.2018.05.036. https://www-sciencedirect-com.offcampus.lib.washington.edu/science/article/pii/S0960148118305573

Gillett R. Fisheries in the economies of the Pacific island countries and territories [Internet]. 2nd ed. Asian Development Bank; 2016. https://www.spc.int/DigitalLibrary/Doc/FAME/Manuals/Gillett_16_Benefish.pdf

Grafeld, Shanna, Oleson, Kirsten L L, Teneva, Lida, Kittinger, John N. Follow that fish: Uncovering the hidden blue economy in coral reef fisheries. PloS one. 2017;12(8):e0182104-e0182104. https://doi.org/10.1371/journal.pone.0182104

Keen, Meg R, Schwarz, Anne-Maree, Wini-Simeon, Lysa. Towards defining the Blue Economy: Practical lessons from pacific ocean governance. Marine policy. 2018;88:333-341. https://doi.org/10.1016/j.marpol.2017.03.002. https://www-sciencedirect-com.offcampus.lib.washington.edu/science/article/pii/S0308597X16308235

Carlo V Vergel J. Current Trends in the Philippines' Shrimp Aquaculture Industry: A Booming Blue Economy in the Pacific. Oceanography & fisheries open access journal. 2017;5(4). https://doi.org/10.19080/OFOAJ.2017.05.555668. https://juniperpublishers.com/ofoaj/pdf/OFOAJ.MS.ID.555668.pdf

Varga, Attila, Hau-Horváth, Orsolya, Szabó, Norbert, Járosi, Péter. Blue Economy Innovation Impact Assessment with the GMR-Europe Model. Ji shu yu tou zi. 2013;4(4):213-223. https://doi.org/10.4236/ti.2013.44025. https://core.ac.uk/download/pdf/42932303.pdf

Sarker, Subrata, Bhuyan, Md. Aminul Haque, Rahman, Muhammad Mizanur, et al. From science to action: Exploring the potentials of Blue Economy for enhancing economic sustainability in Bangladesh. Ocean & coastal management. 2018;157:180-192. https://doi.org/10.1016/j.ocecoaman.2018.03.001. https://www-sciencedirect-com.offcampus.lib.washington.edu/science/article/pii/S0964569117307809

Blasiak, Robert, Spijkers, Jessica, Tokunaga, Kanae, Pittman, Jeremy, Yagi, Nobuyuki, Österblom, Henrik. Climate change and marine fisheries: Least developed countries top global index of vulnerability. PloS one. 2017;12(6):e0179632-e0179632. https://doi.org/10.1371/journal.pone.0179632. https://go-gale-com.offcampus.lib.washington.edu/ps/i.do?p=AONE&u=wash_main&id=GALE%7CA496023740&v=2.1&it=r

Cao, Qiang, Sun, Caizhi, Zhao, Liangshi, Cao, Weiwei, Yan, Xiaolu. Marine resource congestion in China: Identifying, measuring, and assessing its impact on sustainable development of the marine economy. PloS one. 2020;15(1):e0227211-e0227211. https://doi.org/10.1371/journal.pone.0227211. https://go-gale-com.offcampus.lib.washington.edu/ps/i.do?p=AONE&u=wash_main&id=GALE|A612163971&v=2.1&it=r

UNEP, FAO, IMO, UNDP, IUCN, GRID-Arendal (2012). Green Economy in a Blue World. ISBN: 978-82-7701-097-7.

World Bank and United Nations Department of Economic and Social Affairs (2017). The Potential of the Blue Economy: Increasing Long Term Benefits of the Sustainable Use of Marine Resources for Small Island Developing States and Coastal Least Developed Countries. *World Bank.*

World Bank Group (2016). Blue Economy Development Framework: Growing the Blue Economy to Combat Poverty and Accelerate Prosperity.

World Bank, (2017). https://sustainabledevelopment.un.org/content/documents/15434Blue_EconomyJun1.pdf.

Chapter 5
Role and Prospect of Marine Biotechnology in Blue Economy

Arnab Pramanik, Sourav Das, and Tuhin Ghosh

5.1 Introduction

The blue economy has developed as a progressive model, which has effectively blended economic progress with sustainable development. EIU's (2015) report showed that the global argument empowered the world to recognize the usefulness of the blue economy since 1990. The opinion has been endorsed by both developed and developing countries. This model delineated that oceans and ocean-related actions are significant for the commercial and social growth of the coastal nations. These practices form the basics of the blue economy model. Marine trade, shipping, marine fishing, etc., are not the only sectors of the blue economy, but also the following segments from the complete model of the blue economy, e.g., agriculture, energy, minerals, manufacturing, construction, and services (EIU 2015). Furthermore, the blue economy are combining both goods and services activities of each sector. It is developing as the most dynamic segment in the world economy. Planning to take policy action, preservation, protection, and sustainable usage of ocean capitals are the central issues of the blue economy. To report inefficiency in marine resource usage and increase the frontiers of the oceanic economy, the blue economy model provides substantial explanations as it adopts sustainability in a universal and jointly reinforcing pathway (Hurst et al. 2016).

In the present day, marine biotechnology is a developing segment of the blue economy. Its exercise also splurges into economy of the society. Nowadays, it is proved that living organisms and biological species in the seas are precious sources of biomolecule or drugs, or medicine. New derived compounds can be developed

A. Pramanik
Jagadis Bose National Science Talent Search, Kolkata, West Bengal, India

S. Das (✉) · T. Ghosh
School of Oceanographic Studies, Jadavpur University, Kolkata, West Bengal, India

S. Hazra, A. Bhukta (eds.), *The Blue Economy*,
https://doi.org/10.1007/978-3-030-96519-8_5

for medicine in suitable cultural conditions. The effective formulations derived from the ocean biological resources inspire investment in biosynthesis and bioprospecting business (FAO 2017). Along with the finding of new medicines, marine biotech has much broader efforts in aquaculture, tissue culture, and engineering of medical gadgets. Derived enzymes and protein from marine beings are of great usage to the pharmaceutical industries and food and beverage industries. However, biopolymers are a valuable product of use as recyclable plastics to food additives, drug, medical polymers, dental biomaterials, tissue regeneration, etc. There may be several other applications of marine-derived tools and biological processes, which will finally grow the blue economy of the respective countries. In this regard, the present book chapter will focus on the role of marine biotechnology in the blue economy.

5.2 The Strategic Importance of the Blue Economy toward Achieving Sustainable Development Goals (SDGs)

In the present day, the blue economy is considered as a marine economy that intends at the "improvement of human well-being and social equity, while significantly reducing environmental risks and ecological scarcities." The perception of the blue economy is indicating many interpretations because of the reportage of different activities, subdivisions, and geographical settings. From the existing literature, a suggestive list of segments and the actions that fall in those areas are shown in Table 5.1.

The present chapter conceptualized the relationship between different subdivisions of the blue economy and SDGs toward achieving objectives by 2030 (Table 5.2). The present effort has synthesized that blue biotechnology can be a help to achieving the following SDGs, i.e., ending poverty in all its ways everywhere (SDG-1), healthy livelihood, and encouraging welfare for all (SDG-3), developing sustainable industrialization, and promoting innovation (SDG-9). A brief description of the above has been illustrated in Sect. 5.4. The sincere clarification of energy security is not a matter of coincidence, but a more realistic outlook is needed. The modern sustainable development goals (SDGs) have grown the need for reasonable, consistent, environmental, and modern energy requirements for all by the year 2030. Besides, SDG-14 goals at the sustainable usage of oceans, and marine properties. Both these areas can be proficient together while replying the untapped possibility of offshore renewable energy. Maritime-based renewable energy has the potential to improve the effectiveness of harvesting non-conventional capitals, decrease carbon emissions, and diminish the usage of land for power production.

Table 5.1 Categories of marine biotechnology (Source: Kim 2019)

	Category	Content
Marine biotechnology	Marine organism source technology	• Foundational technology for marine organism resource management and use • Technology for bio-uses of marine organisms • Technology for identifying vital phenomena and physiological functions of marine life • Technology for discovering and identifying marine genes • Technology for omic analysis of marine organisms
	Marine food resource development technology	• Technology for breeding and development of new species of marine organisms • Technology for controlling and monitoring diseases • Technology for advanced farming and mass production • Technology for bio-safety assessment
	Marine new materials development technology	• Technology for development of new industry materials • Technology for developing new pharmaceutical materials • Technology for developing new functional food materials • Development of renewable bio-energy
	Marine ecosystem and environment preservation technology	• Technology to ensure biodiversity • Technology to monitor and predict environmental changes • Technology to control and remove marine pollution

5.3 Role of Marine Biotechnology for Building the Blue Economy

Marine biotechnology can significantly contribute to economic recovery and may add an important part toward solving new challenges of the current century. Providing new information may facilitate access to products and services by executing innovative technologies (Hurst et al. 2016). Marine biotechnology is a fresh technology that supports and develops the expansion of the bio-economy. It is capital-intensive and multidisciplinary machinery that is most significant throughout the value chain and encompasses different subdivisions as shown (Fig. 5.1). Blue biotechnology can create new jobs in society (Rampelotto and Trincone 2018). But the major encounter of the coming period is to build the basic knowledge of the marine environment and its uses. Marine biotechnological developments have already evolved in the arena of human health, fisheries, and environmental restoration. In-depth understanding of the convolution of the ocean ecosystem, humans are proficient in protecting the marine ecosystem (Trincone 2014). The blue economy

Table 5.2 Taxonomy of blue economy sectors and activity linkage with SDGs (*sources*: Accumulated from Morrissey et al. (2010), EIU (2015), Govt. of Ireland (2012), and Marine Institute (2005))

Sector	Activity	Linkage with SDGs
Fishing	Seafood processing, aquaculture, capture fishery, etc.	SDG 1, 9, 14
Marine biotechnology	Marine-derived bio-products, seaweed harvesting, seaweed products, pharmaceutical and drug production, etc.	SDG 1, 3, 9
Marine manufacturing	Boat and ship manufacturing, boat and ship repair, net manufacturing, sail making, marine instrumentation, marine industrial engineering, aquaculture technology, etc.	SDG 1, 9
Marine renewable energy	Wind energy production (offshore), tidal energy production, wave energy production, etc.	SDG 7
Minerals	Deep-sea mining (exploration of rare-earth metals, oil and gas, hydrocarbon	SDG 14
Marine tourism & leisure	Sea angling from the shore, sea angling from boats, boating at sea, sailing at sea, sea kayaking, water skiing, bird watching in coastal areas, jet skiing, scuba diving, trips to the beach, seaside and islands, whale/dolphin watching, visiting coastal natural reserves, etc.	SDG 1
Marine commerce	Marine insurance, marine financial services, ship finance and related services, marine legal services, charterers, media and publishing, etc.	SDG 1, 8
Shipping, port, and maritime logistics	Shipbuilding and repairing, shipping agents and brokers, ship owners and operators, liner and port agents, container shipping services, ship management, port companies, ship suppliers, stevedores, roll-on roll-off operators, custom clearance, safety and training, etc.	SDG 1, 8, 9
Marine construction	Marine engineering and construction, etc.	SDG 9
Marine ICT	Marine engineering consultancy, environmental consultancy, meteorological consultancy, project management consultancy, hydro-survey consultancy, geo-informatic services, ICT solutions, submarine telecom, etc.	SDG 8
Education and research	Research and development, education and training, etc.	SDG 4

also produces baseline ocean database systems (part-wise) to help the future researcher (Vieira et al. 2020) by using different mathematical modeling and software. The finding of new drugs by the use of marine biotechnology has many widespread applications in tissue culture techniques and aquaculture. Similarly, marine microorganisms derived new protein and unusual enzymes, which are of great usage in the food and pharmacological industries. Very recently biopolymers are developed for the uses of recyclable plastics to food extracts, pharmaceutical, and therapeutic polymers, bio-glues, dental biomaterials, tissue revival, and 3D skin culture. Public support for the development of aquatic science edification and implementing sea research may be useful to stimulate better applications of inventions in the field of ocean biotechnology (Mohanty et al. 2015). Therefore, the general focus of blue

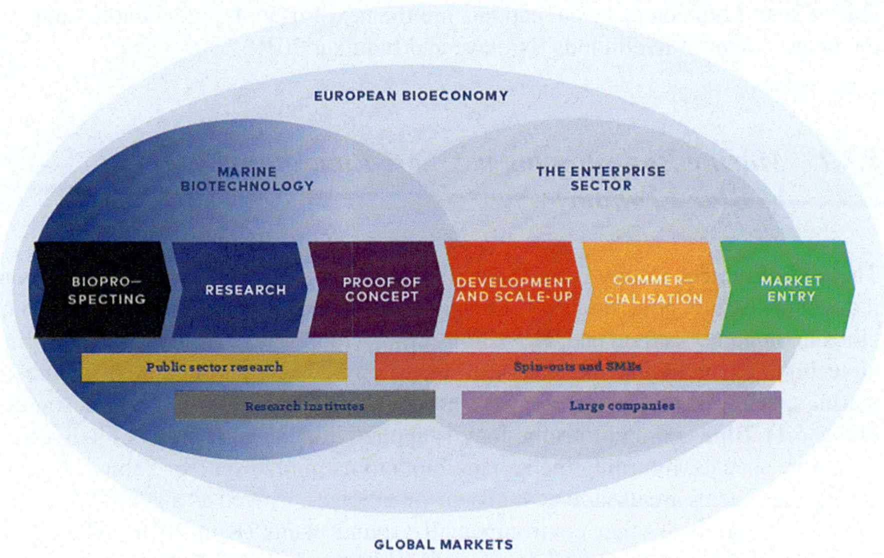

Fig. 5.1 Contribution of marine biotechnology to business areas along the value chain. (Source: Hurst et al. 2016)

economy actions is mostly dependent on marine resources including marine energy, aquaculture, tourism, marine mining, and offshore gas and oil studies. Recently, global enterprises are being taken on the blue economy by key nations like Australia, China, and European Union (EU) (Bennett et al. 2019). The focus on coastal safety and economic and environmental actions is specified in association with the bordering countries. This is to take benefit of our unused marine and coastal properties for increasing the manufacture and productivity of the ocean and water ecosystems. Blue economy idea 2025 document projected the connecting business possible for India and international associates on the sustainability of sea properties for economic profits. Blue economy as a schema for increasing monetary growth can pay to the sustainable development of natural properties for job generation, inspire innovation, and provide openings for knowledge-based productions. Biotechnology has been a very accomplished sector during the last 03 decades and has enhanced the speed of postponement of the biotechnology business in India. Moreover, the bioeconomy has been expected at 100 billion USD in India by 2025 to be proficient by associating with the state and central interventions with an effort on the blue economy in indicating the best policies and regulations stirring investments and entrepreneurship with the active collaboration and joint corporations between the academe and business organizations. This could also perform as a podium for the sustainable use of marine bioresources to support the progress of India's bioeconomy. The other important parts include nutraceuticals, nutrition, fisheries by-product, cosmetics, and the advanced marine biomolecule segment. Therefore, marine-based biotechnological capitals are the next forthcoming valuable trade for the nourishment of livelihoods (Ninawe and Indulkar 2019).

5.3.1 Marine Biotechnology: A Fast-Emerging Sector of Blue Economy

The role of biotechnology is designated as different color codes, e.g., blue, green, red, yellow, white, gold, and gray. Each color code represents the particular area of biotechnological activities. The blue color represents coastal and marine biotechnology including the structure, form, physiology, and chemical nature of marine species. Blue biotechnology is a ground that makes use of ocean bioresources (Table 5.1). Blue coding biotechnology is applied in the conservation of a diversity of marine species, returning the aquatic biota to its unique state of habitat. Ocean-associated species are used to advance new medicines, as well as a genetic learning of plants to produce other environmentally sound plants (Kim 2019). Thus, blue biotechnology is measured as the usage of important marine biological resources as the foundation of biotechnological applications (Fig. 5.2).

From the last century, the concept of "blue economy" has become progressively popular due to its economic significance. Blue economies are founded on the acceptance that coasts, seas, and oceans are playing a major character in addressing

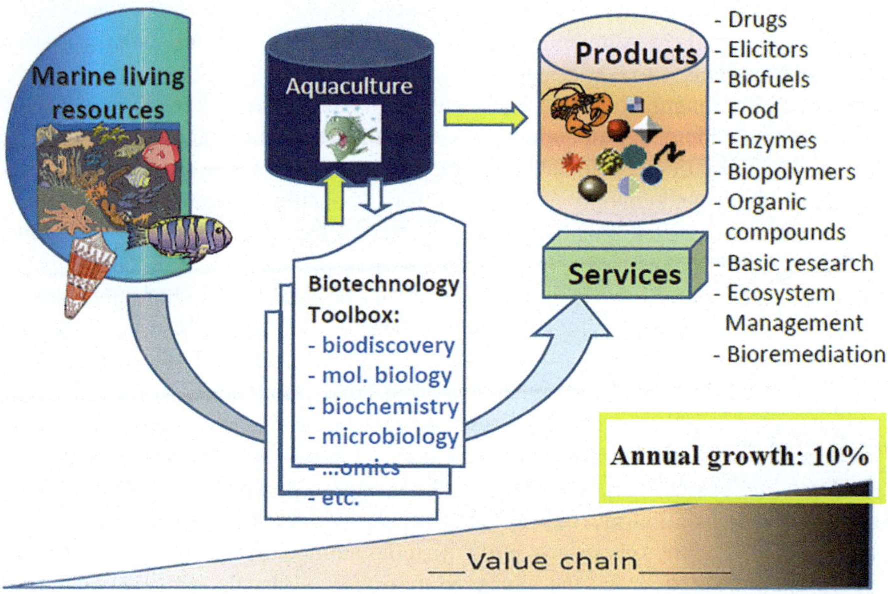

Fig. 5.2 Marine bioresources as the source of biotechnological applications. (Source: Collins et al. 2018)

the several long-term encounters of the global budget. These challenges will be central to unparalleled strains between the present methods of production and feeding and the availability of properties for food, energy, health, water, and raw materials. Other tasks rise in the extents of investment, trade, and industrial affordability of the nations. The marine economy includes numerous sectoral doings, such as those accompanying coastal protection and coastal tourism as well as aquaculture yields and marine fisheries. There are also several emerging actions. Moreover, marine biotechnology reports exceptional encounters of the twenty-first era. More precisely, for the safeguard and managing of the marine environment, the applications of marine biotechnology goal are to improve industrial biotechnology, agriculture, the environment, health, and the biomedical sector (Badri et al. 2019). The present research also emphasizes a diverse variety of biological applications, e.g., anti-inflammatory, antioxidants, anti-dirt, artificial blood, bio-based plastics, and biofuels. Apart from the benefit of blue biotechnology, there is the threat of the unplanned extraction of marine plant and animal because this field is not well-defined and poorly regulated till now, so that planned activities are necessary for this emerging field. The precautionary norm should be useful to biomaterials, bio-nanotechnology, and the introduction of shellfish, genetically modified fish, and microorganisms. An exclusively new group of businesses is being designed to trade credits for sequestration in coastal sediments and carbon storage and marine alive biomass. Concern has been elevated about the commodification of the sea, but it may be balanced by the value of protecting and restoring environments that provide paybacks such as better productivity in coastal zones (Spalding 2016). Our understanding of the ocean economy is changing day by day. Industry segments are being evolving to reflect demands for sustainability and efficiency. Thus, this examination and categorization of the blue economy are necessary. The blue economy must confirm sustainable economic growth. At the center of the blue economy, the concept must be decoupling with socioeconomic development and environmental perspective (Wenhai et al. 2019). Moreover, the human connection with the ocean is varying day by day and the previous ocean economy is acquainting to return to additional burdens, environmental needs, and climate change. Goal 14, "Conserve and Sustainably use the Oceans, Seas and Marine Resources for Sustainable Development," of the just implemented UN Sustainable Development Goals (SDG) for the global ocean frameworks 07 targets and 03 means of enactment connecting to the sustainable usage of the ocean (Table 5.2).

5.3.2 Ocean Bioprospecting: Mining the Untouched Prospective of Living Marine Assets

A crucial part of biotechnology is bioprospecting, which can be designated as a methodical exploration in marine beings from the ocean, ocean bed, and coast including all types of organisms, e.g., micro-organisms and larger organisms. The purpose of marine bioprospecting is to search components, compounds, or genetic resources. There is a wide range of applications that extend well beyond the pharma sector, e.g., functional foods, cosmetic products, human food, animal and fish feed, and biofuels (Abida et al. 2013). The technological expedition involved in bioprospecting such as finding an organism with a biotechnological perspective to having a marketable product (Fig. 5.3).

The market of the world is apprehending the sole medicinal perspective of ocean resources and the linked economic profits. The condition can certainly become more advantageous to countries like India, China, and Africa that have huge coastlines gifted with a varied diversity of marine life and related traditional understanding influenced by the native communities. For example, in China, dried seahorses (Hippocampus spp.) have been used as an old-style medicine for curing respiratory problems and sexual dysfunction for the last 500 years (Hunt and Vincent 2006). Similarly, in Indian Ayurvedic medicine, pearls and oysters have been used comprehensively for centuries as a therapy for cardiac problems, bone development, tuberculosis, etc. In 1950, Werner Bergmann documented commercial use of marine resources with the discovery of spongouridine from the sponge, Tethya crypta, on the coasts of Florida. This product is used to develop antiviral, anticancer, anti-HIV drugs. In the present day, countries like the USA, Canada, Japan, Europe, Fiji, India, China, Australia, Philippines, and Hawaii are exploring the marine bioresources for novel drug development (Demunshi and Chugh 2010). Also, the traditional knowledge remaining in several parts of the world enhances the understanding that

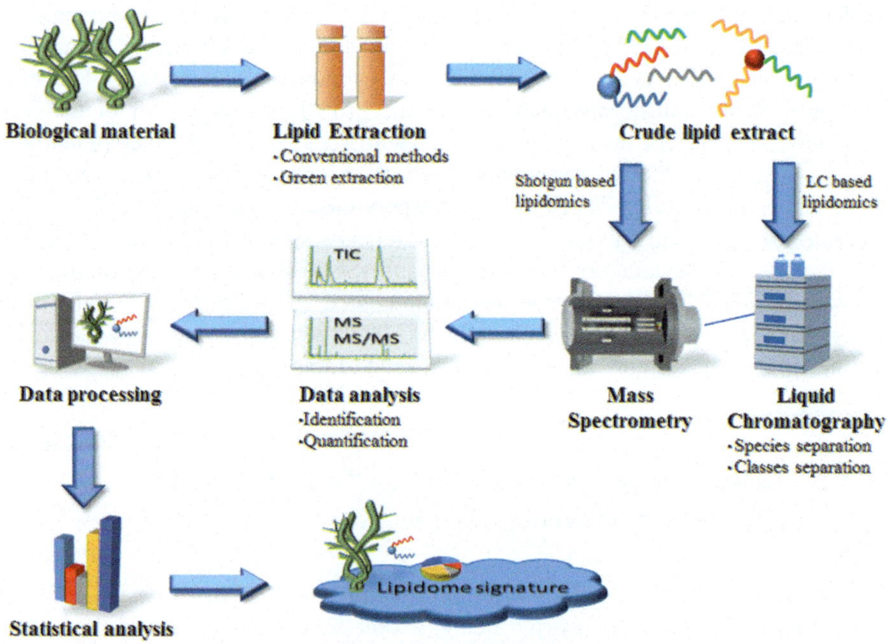

Fig. 5.3 Technical stages of development in marine bioprospecting (source: Maciel et al. 2016)

may help as a key model to pursue research on marine entities for novel drug development. In the last two or three decades, studies on ocean natural products largely involved the collection of organisms from the sea, their extraction, and the analysis of these extracts. Several new compounds have been isolated, and many were established with exciting biological activities. However, the use of many promising materials was troubled by difficulties regarding reproduction and scale-up. Therefore, only a few marine natural products entered the clinical trials, although a large number has been described from marine biota (Imhoff et al. 2011). There is a massive possibility of bioprospecting ocean life prevailing in the seas and oceans through the globe. The seaside populations, e.g., China and India, have settled traditional marine information that uses these bioresources in a varied range of cosmeceutical and pharmaceutical applications. The marketable world is conscious of the consequence of maritime bioresources and related traditional information to discovery drugs for the still hopeless diseases. Moreover, it is important to remember that the sustainable, rational, conservation, and reasonable use of marine capitals may have specific necessities, representing the need for distinct regulatory and judicial frameworks precisely on oceanic biodiversity (Demunshi and Chugh 2010).

5.3.3 Marine Metagenomics for Bioprospecting and Drug Discovery

Bioprospecting for bioactive particles has been dedicated to the capability of an effective microorganism to yield compounds of attention under confined optimal circumstances at the industrial level. This approach has been conventionally practiced due to ancient plant-based drug discovery using ethnobotanical evidence about biodiversity and its nature. Moreover, drug detection from microorganisms has been carried out from the target microbes and their capacity. On an old-style base, microorganisms have been relocated from their natural environments to laboratory plates and bioreactors with various labors to mimic the natural environment to yield compounds of interest (Fig. 5.4). Challenges arising from this exercise have mounted immense interest in understanding the dynamics of laboratory environments and natural environmental parameters regarding metabolites produced by organisms of interest (Maghembe et al. 2020). Metagenomics involves decoding information interlocked into the DNA of the entire microbial community in a target. Whole metagenome sequencing provides more detailed insights into community diversity and function. The generation of protein and genome databases signifies a notable advance in microbial community categorization. High-throughput sequencing of metagenomes is agreeable to downstream analysis, giving insight into the entire community structure, relative differences among ecosystems, and accurate descriptions of strains of biotechnological significance (Chaudhary et al. 2013). Through meta-genomic selection, studies have established the molecular

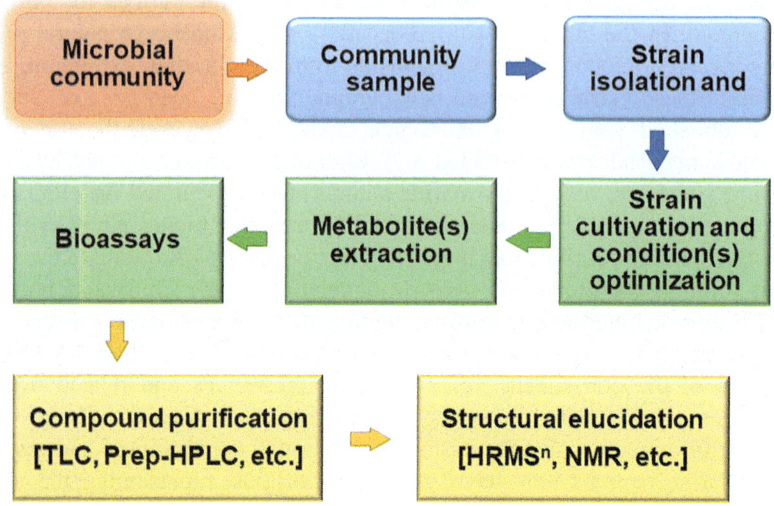

Fig. 5.4 A typical workflow for conventional bioprospecting and drug discovery (source: Maghembe et al. 2020)

adaptation of the microorganisms to their environs, through metabolic linkage and cluster analysis. Present metagenomic studies aim at all domains of life in the environment. Meta-proteomics have been effectively applied to the functioning of microbial communities (Das and Patel 2020). DNA sequencing technology is also playing a major role in forensic medicine, medical diagnostics, and medical biotechnology.

5.4 Market Trends of Blue Biotechnology

In the present time, market-linked use of ocean resource biotechnology has been grown sharply. For several commercialized goods, the marketplace topped some billion USD per annum by 2020, with a compound yearly growth rate of 4–5% (Leary et al. 2009). But owing to the lack of a universally putative description for the segment, it is challenging to assess its structure, possibility, and socioeconomic representation (ECORYS 2014). The estimated value of the blue biotechnology segment will reach EUR 3.5 billion by the year 2025 (Global Industry Analysts 2015). ECORYS (2014) intended that blue biotechnology presently subsidizes nearly 2%–5% of the entire biotechnology business. In the year 2012, the European blue biotechnology segment was between EUR three hundred and two million and seven hundred and fifty-four million (in terms of income). Moreover, health-related biotechnology is measured as the major and most fast-growing segment for ocean biotechnology (Global Industry Analysts 2011).

There are several prospective uses of biotechnology in the marine environment as follows.

5.4.1 Polyunsaturated Fatty Acids (PUFAs)

The discovery of PUFAs (i.e., omega-6 and omega-3) and their significance for social well-being have long been recognized. The removal of PUFAs from fish has permitted its conventional usage in everyday life. It is identified that fish accrue PUFAs through the feeding of algae. Nowadays, PUFAs are extracting straight from algae, and the efficacy of extraction has improved (Medina et al. 1998). Moreover, algae (PUFA-rich) also benefit the growing and existence of shellfish (Reis Batista et al. 2013). Applying this knowledge, aquaculture production might be increased.

5.4.2 Microbiomes

The word microbiome initiates from gene sequencing knowledge in microbiology and mentions a whole microbial population within a definite environmental niche. Microbiomes in diverse environments have been shown to variation in population diversity and density as a function of changes in environmental conditions. Portraying microbiomes and their varieties in and around ships (i.e., outer surfaces and tanks) can lead to new observing systems to form the appearance of environment-damaging organisms and may also prime to bioremediation to decrease organic pollutants (Briand 2011). The same procedure can be used to assess fish health and retort in rearing in aquaculture.

5.4.3 Coatings

Coatings with anticorrosive belongings are presently being established and verified (Eduok et al. 2015). A study of an anticorrosive biocoating comprising encapsulated bacteria from a Saudi hot spring has been established to prevent corrosion. This type of bio-coating may have prospective uses for ship hull safety and defense of off-shore fixings of any type of instrument.

Moreover, the international blue biotechnology marketplace has been divided based on application, invention, and end user. The marketplace has been categorized as drug discovery, bioengineering, vaccine development, genomics, and others (based on application). Based on products, the market has been classified as pharma products, enzymes, biopolymers, bulk chemicals, and others. Based on the product, the market has been characterized as biotechnology companies, hospitals, research institutes, pharmaceutical companies, laboratories, and others. The blue biotechnology market is also classified based on regions, i.e., Americas, Europe, Asia-Pacific, and the Middle East and Africa. The European blue biotechnology marketplace has been classified into Eastern and Western Europe. Western Europe has further been segmented as the UK, Spain, Germany, Italy, France, and the rest of Western Europe. America has also been segmented into South and North America, with the North American market divided into the Canada and USA. The blue biotechnology marketplace in Asia-Pacific has been classified into India, Japan, China, South Korea, Australia, and the rest of Asia-Pacific.

In the global blue biotechnology market, North America has the largest market share and will continue to dominate its top position over the period due to the growing awareness and the increasing demand for blue biotechnology products in vaccine development. There are a huge number of biotechnology companies involved in research and development activities in blue biotechnology, rising demand for aquatic biotechnology in the cosmetic manufacturing industry, and swelling end-users applications like cosmetics, pharmaceuticals, and others drive the development of this market. The European marketplace is the second largest share in the World. The growing focus on marine technology for the research and development of innovative drugs has amplified the growth of the blue biotechnology market in the European region, whereas the market share in the Asia-Pacific region displays a substantial growth in this market but blue biotechnology is not much used in various fields yet. The lack of awareness about the practice of marine microorganisms and their applications slows down the market in this region. The market in the Africa and Middle East is the lowest portion of the global blue biotechnology marketplace due to an immature healthcare sector, deprived medical services, and absence of technical information (https://www.marketresearchfuture.com).

5.5 Conclusion

Although the blue biotechnology is an emerging sector nowadays, still there is lot of scope to emerge this sector to its true prospective. Due to number of the following barricades in the worldwide blue biotechnology segment (ECORYS 2014), i.e., struggle in sample collection, the high price of sampling, the struggle of property rights under marine governance, problem of cost-effective data obtainability, and fragile management system between investors and public research.

Blue biotechnology sector needs to provide a basic research, because ocean biotechnology is a comparatively new region and seeing the present little info on marine biodiversity. Incentives are desired for all significant stack holders to safeguard that the whole invention and growth channel are proven (OECD 2013). ECORYS (2014) showed that EU effectiveness in the blue biotechnology lies in support of research and development actions. The EU is strong in economic support for enterprises involved in research, developing significant infrastructure, and original new ways to entrance in marine resources. The capability for researchers and businesses to access new marine resources is vital and may presently be limiting the EU blue biotechnology segment. Presently, the competition between nations rises, and as a result, the material (mostly from precarious surroundings) will become more demanding. Admittance will also be prejudiced by the development of regulation in coastal states regarding the protection of genetic assets within their EEZs.

Numerous barriers currently exist concerning the growth of the blue biotechnology segment. One of the most important obstacles is blue biotechnology sector that has been funded mainly by policy bodies and rather ignored by "the sector" (i.e., large companies). Other worries are related to profit sharing from the recognition of new marine biological assets, both on oceans and between nations. The absence of simplicity can cause legal insecurity and risks blue biotechnology markets. These worries also have significances for policy required to overwhelmed barricades and to support the global market touch its complete blue biotechnology prospective.

References

Abida H, Ruchaud S, Rios L, et al (2013) Bioprospecting Marine Plankton. Marine Drugs 11:4594–4611. doi: https://doi.org/10.3390/md11114594

Badri MA, Cartier S, Girard J-M, et al (2019) Marine Biotechnology in Québec: Science and Innovation Highlights. Industrial Biotechnology 15:152–156. doi: https://doi.org/10.1089/ind.2019.29171.mab

Bennett NJ, Cisneros-Montemayor AM, Blythe J, et al (2019) Towards a sustainable and equitable blue economy. Nat Sustain 2:991–993. doi: https://doi.org/10.1038/s41893-019-0404-1

Chaudhary AK, Dhakal D, Sohng JK (2013) An Insight into the "-Omics" Based Engineering of Streptomycetes for Secondary Metabolite Overproduction. BioMed Research International 2013:968518. doi: https://doi.org/10.1155/2013/968518

Das S, Patel B (2020) Marine resources and animals in modern biotechnology. In: Animal Biotechnology. Elsevier, pp. 567–591

Demunshi Y, Chugh A (2010) Role of traditional knowledge in marine bioprospecting. Biodivers Conserv 19:3015–3033. doi: https://doi.org/10.1007/s10531-010-9879-9

Hunt B, Vincent ACJ (2006) Scale and Sustainability of Marine Bioprospecting for Pharmaceuticals. AMBIO: A Journal of the Human Environment 35:57–64. doi: https://doi.org/10.1579/0044-7447(2006)35[57:SASOMB]2.0.CO;2

Imhoff JF, Labes A, Wiese J (2011) Bio-mining the microbial treasures of the ocean: New natural products. Biotechnology Advances 29:468–482. doi: https://doi.org/10.1016/j.biotechadv.2011.03.001

Kim S-K (2019) Essentials of Marine Biotechnology. Springer International Publishing, Cham

Maciel E, Leal M, Lillebø A, et al (2016) Bioprospecting of Marine Macrophytes Using MS-Based Lipidomics as a New Approach. Marine Drugs 14:49. doi: https://doi.org/10.3390/md14030049

Maghembe R, Damian D, Makaranga A, et al (2020) Omics for Bioprospecting and Drug Discovery from Bacteria and Microalgae. Antibiotics 9:229. doi: https://doi.org/10.3390/antibiotics9050229

Ninawe AS, Indulkar ST (2019) Blue Economy Mission: India's Focus. AAF 3:1–4. https://doi.org/10.24966/AAF-5523/100017

Rampelotto PH, Trincone A (eds) (2018) Grand Challenges in Marine Biotechnology. Springer International Publishing, Cham

Spalding MJ (2016) The New Blue Economy: the Future of Sustainability. Journal of Ocean and Coastal Economics 2. https://doi.org/10.15351/2373-8456.1052

Trincone A (2014) Increasing knowledge: the grand challenge in marine biotechnology. Front Mar Sci 1. https://doi.org/10.3389/fmars.2014.00002

Vieira H, Leal MC, Calado R (2020) Fifty Shades of Blue: How Blue Biotechnology is Shaping the Bioeconomy. Trends in Biotechnology 38:940–943. https://doi.org/10.1016/j.tibtech.2020.03.011

Wenhai L, Cusack C, Baker M, et al (2019) Successful Blue Economy Examples With an Emphasis on International Perspectives. Front Mar Sci 6:261. https://doi.org/10.3389/fmars.2019.00261

https://www.marketresearchfuture.com/

FAO, 2017. Blue Growth Initiative. Rome.

Leary, D., Vierros, M., Hamon, G., Arico, S., Monagle, C., (2009). Marine genetic resources: A review of scientific and commercial interest. Marine Policy 33(2): 183–194.

ECORYS, (2014). Study in support of Impact Assessment work on Blue Biotechnology, FWC MARE/2012706 – SC C1/2013/03 – 13 June 2014

Global Industry Analysts Inc. (2011) "Marine Biotechnology: A Global Strategic Business Report"

Global Industry Analysts Inc. (2015) "Marine Biotechnology: A Global Strategic Business Report" http://www.strategyr.com/pressMCP-1612.asp

Medina, A. R., Grima, E. M., Gim'enez, A. G. and Gonz'alez, M. I., (1998). Downstream processing of algal polyunsaturated fatty acids. *Biotechnology Advances*, *16*(3), pp. 517–580.

Reis Batista, I., Kamermans, P., Verdegem, M. C. J., Smaal, A. C., (2013) Growth and fatty acid composition of juvenile Cerastoderma edule (L.) fed live microalgae diets with different fatty acid profiles. Aquaculture Nutrition 20(2): 132–142. https://doi.org/10.1111/anu.12059 http://www.marinebiotech.eu/sites/marinebiotech.eu/files/public/library/CSA%20project%20reports/Marine%20Biotechnology%20RTDI%20in%20Europe.pdf

Briand F (Ed.) (2011). New Partnerships for Blue Biotechnology Development: innovative solutions from the sea. Proceedings of the CIESM International Workshop, Monaco, 11–12 Nov 2010. http://www.ciesm.org/WKBIOTECH REPORT 2010.pdf

Eduok, U., Suleiman, R., Gittens, J., Khaled, M., Smith, T. J., Akid, R., El Ali, B., Khalil, A., (2015) Anticorrosion/antifouling properties of bacterial spore-loaded sol–gel type coating for mild steel in saline marine condition: a case of thermophilic strain of Bacillus licheniformis. RSC Advances 5: 93818–93830. https://doi.org/10.1039/c5ra16494j

OECD, (2013). Marine Biotechnology: Enabling Solutions for Ocean Productivity and Sustainability, OECD Publishing. https://doi.org/10.1787/9789264194243-en. This project has received funding from the European Union's Horizon 2020 research and innovation programme under grant agreement No 652629.

EIU (2015) The Blue Economy: Growth, Opportunity and a Sustainable Ocean Economy, Briefing Paper prepared for the World Ocean Summit 2015.

Collins, Jane & Broggiato, Arianna & Vanagt, Thomas. (2018). Blue Biotechnology.

Hurst D, Børresen T, Almesjö L, De Raedemaecker F, Bergseth S (2016) Marine biotechnology strategic research and innovation roadmap: insights to the future direction of European marine biotechnology, 1st edn. Marine Biotechnology ERA-NET, Oostende

Mohanty SK, Dash P, Gupta A, and Gaur P (2015) Prospects of Blue Economy in the Indian Ocean, Research and Information System for Developing Countries, New Delhi.

Chapter 6
Powering the Blue Economy: An Assessment of Marine Renewable Energies

Sameer Guduru and Kapil Narula

6.1 Climate Change and Clean Energy Transition: An Opportunity for Blue Economy

Climate change has emerged as one of the primary challenges for humanity in this millennium. The world has rapidly moved toward consumerism in the post liberalization and globalization period and as a consequence has witnessed large demand for energy. The disastrous effects of climate change such as sea-level rise (SLR), ocean acidification, erratic rainfall patterns, and increased intensity of natural disasters from tropical storms are commonplace today. Linkages with climate change can be drawn for various critical challenges such as food security, energy security, water security, human migration patterns, loss of territorial sovereignty, etc. In short, climate change is intricately and explicitly linked with human security, a nontraditional aspect of security. Consumer driven demand has revolutionized access to goods and services across borders toward the end of the last century; however, it also has resulted in unsustainable practices in energy supply and consumption. This has resulted in large greenhouse gas (GHG) emissions into the atmosphere such as carbon dioxide (CO_2), methane (CH_4), nitrous oxide (N_2O), etc. which are primarily emitted as a result of use of fossil fuels (Singh and Singh 2012).

S. Guduru (✉)
National Maritime Foundation, New Delhi, India

K. Narula
Institute for Environmental Sciences (ISE), University of Geneva, Geneva, Switzerland

S. Hazra, A. Bhukta (eds.), *The Blue Economy*,
https://doi.org/10.1007/978-3-030-96519-8_6

6.1.1 Climate Change and Mitigation Strategies

Given the nature of this existential threat, there is a need to simultaneously reduce the impact of climate change and also put in place strategies for its mitigation. This requires careful, well-tailored solutions that address the root cause for climate change. Firstly, the way energy is produced and consumed in the world has to radically undergo a transformation toward more sustainable practices that have a lower impact on climate. And secondly, it is important to identify the most vulnerable areas and people who will face the maximum impact of climate change, such as in developing countries having low per capita income and human development indices. Global efforts toward mitigating climate change constitute several strategies. Some countries are preparing themselves to prevent the risk of coastal area inundation as a result of SLR caused by the melting of glacial ice as well as polar ice caps due to increasing temperatures. This is being carried out by developing coastal climate resilient infrastructure which is able to withstand intense weather events and natural disasters triggered by climate change. However, other countries, particularly the small island developing states (SIDS) are less fortunate and face an existential crisis threatening the end of cultures. These issues raise several pertinent questions that the global community has to ponder over. These include legal aspects related to the territorial sovereignty of such SIDS, which can quickly escalate into geopolitical contestations. Such scenarios are not entirely restricted to SIDS but also to countries littoral to coastlines. This is especially true in the case of developing countries in the Indo-Pacific Region (IPR). These countries concentrated especially in East, South, and Southeast Asia have significant coastal populations and are still in their development phase. Such countries have to strike a fine balance between their development agendas and climate action and this is highly challenging considering their low per capita GDPs and lack of financial resources (United Nations 2015a).

Climate change and its mitigation have over the last few years driven global governance agendas both nationally and internationally in examining clean energy sources. Energy consumption is intricately linked to economic growth of a country and its demand is predicted to grow substantially over the coming decades due to the growth in global population as well as increased per capita consumption. In fact, the global population is estimated to breach the 10 billion mark by 2050. China, India, the Philippines, Vietnam, etc. have over the last few decades displayed high economic growth rates which were only possible due to increased fossil fuel consumption. Thus, both India and China are amongst the top three global emitters of GHG. Given the production of a significant proportion of energy from fossil fuels, global efforts toward mitigating climate change can only be successful if climate action occurs in the countries of the Indo-Pacific region, specifically India and China. Keeping these aspects in mind, global initiatives on climate change have led to the 2015 United Nations Climate Change Conference (UNCCC) or COP 21 held in Paris (United Nations Framework Convention on Climate Change 2015). COP 21 mandates countries to set their own targets in order to achieve the clean energy transition via a mechanism called "nationally determined contributions", which includes targets for reduction in GHG emissions as well as adopting clean and renewable sources of energy. In addition, UN sustainable development goals (SDGs) enable countries to work collectively toward sustainable development. Specifically, SDG #7 and SDG #13 focus on "Affordable and Clean Energy" and "Climate Action" respectively (United Nations 2015b).

According to the U.S. Energy Information Administration's (EIA) assessment for international energy markets, it is estimated that by the year 2050, global energy usage will increase by 50% (U.S. Energy Information Administration (EIA) 2019). This massive demand for energy is primarily driven by rapidly growing developing countries in the Indo-Pacific region, which is home to more than 4.3 billion people (close to 60% of the global population). Without additional GHG emissions' reduction policies, the global carbon budget to keep the world below 2 °C increase in temperature will get exhausted over the next couple of decades. And over the next decade or so if a 1.5 °C target is considered. Therefore, it is abundantly clear that global efforts need to be progressed collectively and seriously for reducing GHG emissions. In order to keep global temperature increase within 2 °C, it is essential to reduce global carbon emissions to 760 giga tons. This in turn, requires that the contribution of renewable energy resources must increase from 15% in 2015 to 66% in 2050. In addition, improved energy efficiency and reduction in transmission losses also have to be addressed for accelerating the process of decarbonization, while at the same time maintaining decent economic growth trajectories in developing countries (International Renewable Energy Agency 2018).

6.1.2 Clean Energy Transition and Electricity Generation

Electricity is a versatile energy carrier which can be used to provide various forms of energy services. In the case of electricity production, there are myriad options for undergoing a renewable energy transition. In the past two decades, there has been an exponential growth in generation of onshore wind energy (OnWE), followed more recently by solar photovoltaics (SPV). Electricity generated from these sources has already reached cost parity with electricity generated from fossil fuels even without accounting for externalities. This has resulted in widespread diffusion of these technologies. The drive for adopting SPV has been witnessed in China and India, thanks to incentives from the government and lower cost of solar modules. The International Solar Alliance (ISA), a multilateral forum for accelerating the adoption of SPV was jointly set up by France and India, targeting the countries of the tropical regions which experience long summers and have good solar resource potential. Similarly, the installed capacity of wind energy generation has grown considerably. The case of India throws light on the accelerated adoption of SPV and OnWE. The Indian government's targets for electricity from renewable energy sources are highly ambitious with projections of up to 450 gigawatts (GW) to be installed by the end of 2030. Of this, 300 GW and 140 GW are the projected installation capacities from SPV and OnWE, respectively (The Energy and Resources Institute 2020).

However, several issues with SPV and OnWE still exist that may hinder such ambitious goals. Both SPV and OnWE need large swathes of land for installation. This is a major bottleneck for their wider adoption in a country like India where population density is very high. Apart from competing with alternate uses of land, expensive land can contribute to higher cost of electricity generation. It has already been reported that such projects have affected grazing patterns of domesticated animals which can eventually affect the produce from the farm sector. Moreover, the intermittency and low capacity factor of SPV and OnWE are another challenges. SPV cannot produce electricity during the nights and under inclement weather conditions. Similarly, OnWE suffers from intermittency when wind patterns become erratic. Moreover, OnWE also suffers from obstructions on land, which reduces its efficiency. Finally, the lack of effective grid level energy storage mechanisms including battery storage, have to be taken into account in order to address intermittency. Therefore, relying excessively on SPV and OnWE may not be entirely pragmatic and there is a need to look at electricity production from other sources of renewable energy (Fares 2015). It is in this context that alternate forms of renewable energy with large untapped potentials need to be explored.

6.1.3 Linking Renewable Energy Resources and the Blue Economy

Ideas related to sustainable development have led to the evolution of novel concepts such as the "Blue Economy". According to a 2017 joint report by the World Bank and the UN entitled *The Potential of the Blue Economy*, (World Bank and United Nations Department of Economic and Social Affairs 2017).

> *The "blue economy" concept seeks to promote economic growth, social inclusion, and the preservation or improvement of livelihoods while at the same time ensuring environmental sustainability of the oceans and coastal areas.*

Therefore, furthering the concept of blue economy requires novel schemes which encompass ideas related to protecting the ocean environment while at the same time maximizing their output for the benefit of coastal communities. Such schemes can address several issues such as cleaner shipping fuels, renewable energy resources from the oceans, promoting offshore farming, sustainable aquaculture, etc. One of the potential areas for furthering the blue economy is to utilize renewable energy resources from oceans to reduce fossil fuel use. Offshore electricity production can power coastal communities and impact several other coastal industries including shipbuilding and allied industries, coastal tourism, fisheries, and aquaculture leading to enhanced opportunities for local employment. Moreover, these technologies do not produce any GHG emissions while generating clean electricity and therefore contribute to sustainable development. They also have the potential to provide electricity in isolated coastal communities, remote islands, archipelagic states, and even continental states that are littoral to the oceans. The following section briefly touches on various forms of ocean renewable energy resources (ORERs).

6.2 Ocean Renewable Energy Resources

Oceans play several important functions that are beneficial for life on earth. Close to half of all the oxygen on the planet is produced from the oceans. They also act as huge sinks for carbon dioxide in the atmosphere, thereby regulating the impact of global warming. Ocean currents moving from equatorial regions to higher latitudes regulate global weather and climate patterns. However, often, it goes unnoticed that oceans can also be major sources of energy. ORERs can primarily be classified into three categories (International Renewable Energy Agency 2020a):

1. Kinetic Resources.
2. Thermal Gradient-based Resources.
3. Salinity Gradient-based Resources.

Kinetic resources, as the name suggests utilize kinetic energy in the form of ocean waves, tides, currents, and offshore winds to drive turbines and converters to produce electricity. Thermal gradient, i.e., the temperature difference between warm surface water and deep seawater can be used to produce electricity. Similarly, salinity gradient, i.e., the difference in salt concentrations in the water can also be exploited for electricity generation. Estimated global potential of various forms of ORERs is given in Table 6.1.

This section briefly discusses all forms of ORERs except Offshore Wind Energy (OffWE) and Ocean Thermal Energy Conversion (OTEC). These two forms of ORERs, as can be noticed from Table 6.1, have huge potential and can accelerate the clean energy transition quite dramatically. These are therefore dealt with in subsequent sections.

Table 6.1 Estimated global potential of ocean energy resources (International Renewable Energy Agency 2020b; International Energy Agency 2019)

Form of ocean energy	Estimated global potential (TWh/year)
Ocean thermal energy conversion (OTEC)	44,000
Wave energy	29,500
Salinity gradient energy	1650
Tidal energy	1200
Offshore wind energy	420,000
Total	496,350

6.2.1 *Kinetic ORERs*

Kinetic ORERs include the following forms of energy:

1. Ocean Wave Energy Conversion (OWEC).
2. Ocean Tidal Energy Conversion.
3. Ocean Current Energy Conversion.

The principle behind their functioning is the conversion of kinetic energy in tides, waves, and currents to electricity. Each of these forms of energy is briefly discussed below.

6.2.1.1 Ocean Wave Energy Conversion (OWEC)

Waves in the ocean are primarily generated due to the interaction of surface water with wind. This results in the transfer of kinetic energy from wind to water and leads to generation of waves. Wind blowing across the water surface results in friction and creation of a pressure differential, which compels water to rise above and the resultant stresses make the waves grow in size. This kinetic energy carried by moving wave can then be harnessed by a variety of technologies to produce electricity. These include, Oscillating Water Column (OWC), the Oscillating Body Converter (OBC), and the Overtopping Converter (OC). Apart from these, there are many other technologies for OWEC, but the above mentioned ones are more common as compared to others.

6.2.1.2 Ocean Tidal Energy Conversion

Tides are periodic waves that occur due to the gravitational pull of the moon and the sun acting on earth's oceans. Such waves have a long wavelength and are perceived as the rise and fall in the surface level of the ocean. The crest of these waves reaching a coastline is termed as "high tide", while the trough on hitting the coastline is termed as "low tide". These periodic movements of the seawater bring along with them large quantities of water which can be utilized to produce electricity, primarily in two different ways, viz. tidal barrages and tidal streams. Tidal barrages are created such that water at high tide is allowed to enter toward land. This water is then stored in a reservoir and is released back during low tide. The water rushing back runs the turbines and electricity is produced. The minimum requirement of the tidal range, i.e., the difference in the level of the surface at high tide and low tide should be 5 meters for harnessing tidal energy. In the case of tidal streams, turbine structures are usually installed below the water surface. The mechanism of operation exploits geographical restrictions to the flow of water along narrow channels. Water in such locations during high tide rushes along narrow channels carrying large amounts of kinetic energy. This energy is harnessed in the form of electricity by deploying tidal wave energy converters underwater, which look similar to wind turbines, except that the turbines in this case are run by seawater during the tides.

6.2.1.3 Ocean Current Energy Conversion

Ocean currents also can be used to produce electricity from the oceans. Ocean currents are created due to a variety of factors including temperature differential, salinity differential, density differential, etc. Such differentials lead to the movement of seawater from high density, high temperature, and higher salinity toward regions where these are lower. For ocean current energy to be converted into electricity in a profitable manner, the minimum speed of the current should be 2.5 m/s. Ocean current energy can be converted into useful electrical energy by use of submerged converters. Some of the most common types are underwater horizontal axis turbines, underwater kites, etc. Fast-flowing currents with a minimum speed of 2.5 m/s are usually uncommon in the vicinity of coastal territories, which is a challenge for utilization of ocean currents.

6.2.2 Ocean Salinity Gradient

This form of marine energy exploits the salinity gradient in water to produce electricity. When a semipermeable membrane separates water with a higher salt density compared to the other side, the lower salt density water automatically moves toward the higher side. This results in the creation of hydrostatic pressure through the principle of osmotic pressure. This pressure can then be utilized to run turbines for producing electricity. Even though the principle is simple to understand, the practicalities involved have not allowed the full exploitation of this energy source. The semipermeable membranes are highly expensive and require frequent maintenance, which leads to an increase in the cost of electricity production. With improvised membranes employing nanotechnology as demonstrated in 2016, these technologies have again gained attention and are in their nascent stages of adoption. Maturity of such technologies is likely to happen over the next few decades.

Even though these technologies theoretically offer many possibilities, these have still not matured, limiting their wider adoption. Moreover, technologies for harnessing tidal energy and ocean currents are economically viable only in a few geographic locations in this world. Further, due to the nature of their rotational underwater component, there is a possibility of their interference with marine fauna and their impact on underwater ecosystems is not fully understood. Lastly, given the dynamic nature of the marine environment, these systems need frequent maintenance. Nevertheless, given their immense potential, faster adoption is expected which may reduce the cost of electricity from these technologies in the near future. Table 6.2 below provides the current global total installed capacity of various ORERs.

Table 6.2 Current global total installed capacity of ocean renewable energy resources (IRENA 2020)

Forms of ocean energy	Current global installed capacity (MW)
OTEC	0.23
Ocean wave	2.31
Salinity gradient	0.05
Tidal barrage + Tidal stream	521.5 + 20.6
Offshore wind	2900
Total	3444.69

Two forms of ORERs that have the highest potential to be exploited are OTEC and OffWE. The following section discusses these two forms of ORERs and their potential contribution to furthering the blue economy especially in tropical regions such as South Asia, Southeast Asia, Africa, Latin America, Caribbean, and in the SIDS of the Pacific Ocean.

6.3 Ocean Renewable Energy Resources and Blue Economy

With growing global discourse around the blue economy and sustainable use of ocean resources, there is a large scope for exploration and innovation. While novel solutions for furthering the blue economy are always proposed, it is important to have a realistic perspective regarding the possibility of their scalability. The selected solutions should be globally replicable and so that economies of scale can lead to cost reduction. Two such solutions in the context of ORERs that have the highest energy potential are OTEC and OffWE. This section discusses how these forms of ORERs can play a role in creating robust blue economies in the Indian Ocean and the Indo-Pacific Region.

6.3.1 Powering Blue Economy with Ocean Thermal Energy Conversion (OTEC)

As mentioned earlier, OTEC has a global energy potential of 44,000 TWh/year which makes it a promising renewable energy resource. This is especially true for countries in the tropical regions where ocean surface temperature remains high throughout the year, as compared to deep seawater, fulfilling the requirement of a 20 degree celsius temperature differential. These plants can benefit the Indian Ocean Region and more broadly the Indo-Pacific Region by producing clean renewable energy with other additional advantages. These include supplying freshwater to coastal communities in continental countries as well as SIDS, air conditioning in coastal communities, and even opportunities for aquaculture. OTEC plants in conjunction with Low Temperature Thermal Desalination (LTTD) can be self-sustaining in nature. This means that the electricity generated in an OTEC plant can power its own processes such as pumping deep seawater, etc. while at the same time producing electricity and freshwater for coastal towns and island communities. This is extremely beneficial for SIDS as they are often limited in their land resources and therefore cannot afford to have power plants located in sensitive ecosystems. Having such plants offshore leads to limited use of onshore land which can be utilized for alternative purposes. In addition, deep seawater pumped from depths over 600 meters is typically around 4 degree celsius. This water can be pumped into cities in coastal areas to maintain air conditioning thereby providing sustainable cooling services. Moreover, the water supplied by LTTD systems can address water shortages faced by SIDS due to small catchment areas and erratic rainfall patterns due to climate change. Ironically, SIDS contribute the least to GHG emissions but have to face the maximum brunt of climate change including SLR leading to loss of sovereign territory. LTTD has already been demonstrated and the technology has been implemented by India's National Institute of Ocean Technology (NIOT), Chennai. LTTD plants are currently operational and are providing desalinated water to the inhabitants of the islands in India's Lakshadweep archipelago (Guduru et al. 2020). Another advantage of pumping out deep seawater is the fact that it is rich in nutrients and is an ideal resource for aquaculture. Cold water species such as lobsters and salmon dwell in nutrient-rich waters and they can be cultured to grow seafood. This benefits fishing communities along coastal areas and can reduce their dependence on high seas. In addition, by mixing cold seawater with warm surface water, optimal temperatures for the aquaculture of non-native species can be achieved, which can lead to diversification of fish farming, greatly benefiting fishing communities. In addition, algal aquaculture farming as well as production of health supplementing microalgae such as spirulina can further the prospects of a blue economy. OTEC-based aquaculture in Japan and Europe has produced a variety of fish and vegetable products including Flounder, Flatfish, Butter Fish, Trout, Sandfish, Snow Crab, Shrimp, Oyster, Microalgae, and Sea Trumpet among others (World Ocean Initiative 2020; European Commission 2020).

While OTEC and LTTD promise a lot of potential, they are still plagued by a few disadvantages. A self-sustaining OTEC cum LTTD plant is not yet a reality. Deep seawater for LTTD plants is primarily pumped using fossil fuels. Adopting solar or other forms of renewable energy for producing electricity for plant operations is a possible way ahead. Another major disadvantage of OTEC is the use of very long submerged pipes often longer than 600 meters, which is an engineering challenge. Disposal of brine solution in the waters close to the vicinity of such plants can also impact marine ecosystems. Instead the brine solution can be used for restoring mangrove forests as some species of mangroves thrive in highly saline environments. For example, studies point out that black mangrove or *A. Germanis* has a high tolerance to salinity and mangrove reforestation can be achieved using brine solutions. Apart from this, brine solution can be utilized to produce different salts that are utilized in various applications (Chan et al. 2020).

Even though the principle behind OTEC has been predicted a long time ago and the technologies have already been demonstrated, its wider adoption has not yet taken off. This is partly because the technology has a low efficiency of less than 10%. This makes OTEC plants economically unviable. Since the technology is not mature, it is difficult to predict the levelized cost of electricity (LCOE) generation in the future. In 2018, the LCOE was estimated to be around 0.04–0.29 USD/kWh from a large-scale 100 MW OTEC plant (Langer et al. 2020). Improving the technology further will make it a more attractive option and efforts are currently underway to achieve this. Given that OTEC technology is extremely suitable for providing energy to remote and isolated island communities, its higher cost of electricity production should not be a major challenge. In the case of the Indo-Pacific region, there are a number of SIDS as well as archipelagic states such as Indonesia, India, and the Philippines, with several remote islands. OTEC can be exploited for the development of local communities as well as for furthering tourism by ensuring uninterrupted power and water supply. A whole gamut of opportunities is possible and development of shore-based ancillary industries is a co-benefit that policymakers have to keep in mind when framing policies toward adoption of such innovative technologies. As of today, OTEC plants are operational in the United States and Japan. Such plants have previously been deployed for experimental purposes and have been decommissioned in India, Japan, and South Korea. New projects are being planned by Japan, China, Korea, India, France, the Netherlands, Singapore, Monaco, Maldives, and Iran. Current global installed capacity of OTEC as of today is just 0.23 MW. However, companies like US-based Makai Ocean Engineering, Lockheed Martin, and several other public and private funded institutions across Japan, India, Brazil, and Malaysia are dedicating efforts toward improving the technology. OTEC can therefore emerge as an important source of renewable energy in the next few decades.

6.3.2 *Powering Blue Economy with Offshore Wind Energy (OffWE)*

OffWE can be perceived as an extension of wind energy in a marine environment. Even though wind farms on land have been in place for decades, OffWE has over the last decade and a half spurred a lot of interest in terms of technological development as well as deployment. The first discernable and obvious difference between OffWE and Onshore Wind Energy (OnWE) is the size of the turbines. OffWE turbines are much larger compared to their OnWE counterparts. This is done to achieve maximum exploitation of wind power at sea, where there are no obstructions such as those on land. Even though a large proportion of OffWE is still produced from wind turbines within continental shelves and up to a sea floor depth of about 60 meters, new manufacturers with improvised turbine technology and larger turbine size are venturing deeper into the sea. This is because wind speeds are more uniformly distributed and the output is much more consistent. In addition, technological advancement has now made it possible to install floating wind turbines that are capable of handling and withstanding tropical cyclones, which make them an attractive option compared to any rival ORER.

The global OffWE potential is estimated to be about 420,000 TWh/year which is several times the current global demand for electricity. 2019 was an important year in the growth of OffWE with 6.1 GW of capacity addition, taking the cumulative global installed capacity up to 29.1 GW. From being just 1% of global wind capacity in 2009, OffWE has grown to over 10% in 2019. Table 6.3 displays countries with the largest OffWE installations. Europe, unsurprisingly, is the largest region for installations due to large investments, incentives, and policy initiatives as far as market penetration is concerned. However, the biggest growth of OffWE is expected to be in the Indo-Pacific region in the near future as more and more countries in the region look at OffWE as a viable resource for producing electricity. China has emerged as a global leader in new capacity augmentation and markets like India, Taiwan, Vietnam, Japan, Indonesia, and South Korea are set to grow rapidly over the coming decades. It is estimated that the global OffWE market will be valued at 1 trillion USD by the year 2040 and more than 205 GW of new offshore installed wind capacity is forecasted by 2030. Given the global drive toward adopting renewable energy solutions, the adoption of OffWE as a viable alternative to provide grid connected power is likely to increase further. Moreover, emergence of floating offshore wind provides an opportunity for exploration and is expected to add a capacity of 6 GW globally, by the year 2030. Improvements in turbine technology will improve both efficiency and will reduce maintenance costs, resulting in LCOE reduction and increased adoption. The LCOE of OffWE is predicted to decrease from USD 0.13/kWh in 2018 to between USD 0.05 and 0.09/kWh by 2030 and to USD 0.03 and 0.07/kWh by 2050 (IEA 2019).

OffWE is intricately linked to the shipbuilding sector, rare earth metal extraction and processing, and other ancillary industries. Moreover, most OffWE companies

Table 6.3 Offshore wind installed capacity by country (IEA 2019)

Country	Installed capacity (GW)
United Kingdom	9.723
Germany	7.493
China	6.838
Denmark	1.703
Belgium	1.556
Netherlands	1.118
South Korea	0.073
United States	0.03
Others	0.6

are located either in the vicinity of ports or within the ports itself for logistical reasons. The use of permanent magnets inside the wind turbines is crucial for the production of electricity and these are made from rare earth metals. As OffWE is getting integrated with the electricity grid, it will have a positive impact on coastal communities via developmental opportunities which will contribute to the blue economy of a country. Coastal tourism is another sector that can benefit from OffWE given the possibility of having turbines in the proximity of remote and isolated islands that are away from the main grid. Fishing sector is another important sector that can experience a positive impact due to the uninterrupted power from OffWE that can be supplied to cold storage facilities. Employment opportunities for skilled manpower due to OffWE and allied industries are plenty and can boost economic growth rates. For instance, in the European Union (EU), it is predicted that by 2030, almost 210,000 people will be employed in the OffWE sector (European Union 2020).

Floating wind technologies due to its flexibility of adoption even in tropical regions is likely to play a significant role in the near future. They can withstand intense weather events including storms and therefore provide uninterrupted power to coastal communities in developing countries. Given the immense potential of these technologies, major European offshore oil and gas companies have already ventured into this arena to utilize their experience in offshore projects to further OffWE. This is also a part of a wider diversification strategy away from fossil fuels in order to combat climate change (Deign 2020).

6.4 Challenges and Future Scope for ORER Adoption

In general, ORERs are much more expensive compared to their other renewable energy counterparts such as SPV and OnWE. Even though ORERs offer a lot of scope for adoption, their development is still limited due to the following challenges (IRENA 2020; Rinaldi 2020):

- Lack of accessibility to offshore areas leading to difficulty in deployment of technologies.
- The farther the location of offshore ORER power plants, the longer is the submarine cables for bringing electricity to the coastline for feeding it into the grid. Such cables are highly expensive and elevate the cost of offshore installations dramatically, which eventually impacts the LCOE.
- The demand for highly skilled manpower.
- Existing technologies have low efficiencies.
- ORER converters are economically viable only in certain locations under specific conditions.

- By virtue of being present in a dynamic and corrosive environment like the ocean means the need for frequent maintenance operations which can further add to the costs.
- So far only few countries have ventured into these forms of ORERs. Hence, the technological capabilities are only relegated to a few technologically advanced countries when their real need for deployment is in the developing world.
- Impact on marine fauna and ecosystems is another threat that ORERs may pose. This is especially significant in the case of technologies that have an associated underwater pumping mechanism or rotating blades which may harm marine fauna. Also, the noise generated by these technologies can alter the patterns that marine fauna usually follow, resulting in the loss of their habitats.
- Finally, offshore installations are highly expensive and need to be safeguarded from manmade threats thereby needing the adoption of new security and patrolling measures.

Since, these technologies except for OffWE are not mature yet; their adoption can be accelerated by combining them with the production of other form of energy carriers such as hydrogen. Green hydrogen is an emerging form of renewable energy that is produced via electrolysis of water. Traditionally, the production of hydrogen has been via techniques like methane reformation wherein organic molecules containing hydrogen are reacted with water to produce hydrogen and carbon dioxide as a by-product. Therefore, such processes inherently have a carbon footprint associated with them. This is being bypassed by producing hydrogen, instead via electrolysis of water, which is broken down into its constituent elements oxygen and hydrogen. The electrolysis is carried out with electricity generated by ORERs thereby eliminating the carbon footprint associated with traditional methods of hydrogen production from fossil fuels. Simultaneously, since hydrogen itself is seen as a clean energy carrier, ORERs can play a significant role in realizing economies of scale. Hydrogen can be stored as a fuel to be utilized on demand, which makes it attractive alongside battery-operated electric vehicles. Off late, several such hybrid approaches by integrating offshore floating SPV with OTEC, OffWE with hydrogen production, etc. are being tested. With improved power generation efficiencies and larger OffWE turbines, offshore OTEC and OWEC plants, the futuristic possibility of offshore refueling stations for ships is possible. Moreover, plans are afoot to create offshore cities that are entirely powered by ORERs. In addition, the emergence of long endurance autonomous underwater vehicles for a variety of applications ranging from scientific research to antisubmarine warfare offers prospects for hydrogen as well as for other battery storage technologies. These can benefit significantly when combined with ORERs in creating offshore recharging stations (Guduru and Chauhan 2020). Offshore solar energy and marine biomass and algae are also potential energy sources which have the potential to be developed as future renewable energy resources (Narula 2019).

The other possibility to hasten ORER adoption is by creating both bilateral and multilateral mechanisms in order to jointly develop technologies, or by transferring technology from advanced countries to developing countries. If carried out effectively, this will not lead to creation of technology silos. Moreover, looking at climate change through the prism of human and energy security offers countries an opportunity to arrive at a common global and regional consensus. Such mechanisms can be setup at multilateral forums like the UN General Assembly, G20, BRICS, QUAD, ASEAN, etc. The European Union has already initiated multilateral mechanisms in incentivizing and furthering ORERs with special focus on OffWE.

Table 6.4 Blue economy sectors benefiting from ORERs

Sector	Role played by ORERs
Aquaculture	Nutrient-rich deep seawater used in OTEC/LTTD can be harnessed to raise a variety of marine species for human consumption away from marine-protected areas and sensitive ecosystems reducing the possibility of illegal, unreported, and unregulated fishing
Offshore farming	Offshoring farming of healthy food and medicinal supplements such as spirulina, blue green algae, and fish food
Tourism coastal community development	Coastal and island tourism can immensely benefit from ORERs especially in remote off grid islands where electricity is currently produced by diesel generators
Powering ports and naval establishments	ORERs can be utilized to supply power to port and naval establishments where usually space is limited and can be expensive for installing solar or wind plants. This offers the possibility of providing auxiliary power to berthed ships for carrying out their activities on board as well as loading and offloading. Shipping pollution at ports is a major contributor to GHG emissions in coastal cities. In addition, ORERs can power navigation systems, IoT-enabled sensors, radar installations, communication systems, etc. for transit of vessels and guidance along narrow approach channels and in the vicinity of obstacles
Shore-based industry	Shore-based ancillary industry in and around ports can be powered by ORERs. This can lead to employment opportunities and investment boosting local economies, development, and urbanization.
Hydrogen production	Green hydrogen production from ORERs via electrolysis offers prospects related to creating hydrogen-powered economies of scale. Not only hydrogen can play the role of a fuel in shipping, powering naval assets, movement of logistics, it can also be stored, transported, and be used in other industries as a source of energy or as feedstock. In addition, it can play the role of achieving net carbon neutrality by reacting with atmospheric CO_2
Offshore refueling/ powering	Offshore refueling using green hydrogen produced by ORERs will allow ships to refuel along the way instead of arriving at ports for the same. This is beneficial from an economic point of view as well as reduces emissions. In addition with the emergence of high endurance autonomous underwater vehicles and drones, their range can be further extended for the purposes of underwater domain awareness, antisubmarine warfare, etc.
Ocean observation and research	Wave-powered sensor buoys can be deployed for collecting and transmitting information and observations for scientific research such as salinity, hydrology, chemical composition, ocean dynamics, etc.

Such efforts are encouraging and pave way for ORER adoption in the near future. Table 6.4 lists out the various other sectors related to blue economy that can benefit from the adoption of ORERs.

6.5 Conclusion

In conclusion, climate change mitigation efforts require a variety of mechanisms in order to achieve COP 21 targets as well as to realize SDG #7 and #13. The electricity sector requires new technological advancements for undergoing a clean energy transition and ORERs offer a significant opportunity. While SPV and OnWE have become viable options for electricity generation, at par with fossil fuels, scarcity of land and lack of effective storage mechanism makes them unviable in the long run, requiring alternate solutions to harness energy for achieving sustained economic growth. It is in this context that ORERs can play a major role in aiding a clean energy transition. Even though these technologies are not yet mature for large-scale deployment, except OffWE, their wider adoption is possible in the near future by way of financial incentives as well as technological advancement. This will eventually bring down the LCOE and make them economically viable. In addition, creating focused multilateral mechanisms and transfer of technology can help economically backward countries to gain access to these technologies for a faster renewable energy transition. More importantly, ORERs have a larger role to play in furthering the blue economy and to sustainably harness the ocean resources. Sectors including tourism, coastal community development, aquaculture, desalination, powering ports and naval establishments, green hydrogen production, etc. can also be achieved. Off these, OTEC and OffWE offer the biggest prospects with huge theoretical renewable energy potential as well as in furthering various sectors of the blue economy. Framing of policies to incentivize their adoption can lead to wider adoption of ORERs with improved technologies. This will further bring down the costs of electricity generated from ORERs and make them a financially viable alternative to fossil fuel.

References

Singh, B.R. and Singh, O. (2012). Study of Impacts of Global Warming on Climate Change: Rise in Sea Level and Disaster Frequency, Global Warming - Impacts and Future Perspective. B.R. Singh, *IntechOpen*, https://doi.org/10.5772/50464. Retrieved from https://www.intechopen.com/books/global-warming-impacts-and-future-perspective/study-of-impacts-of-global-warming-on-climate-change-rise-in-sea-level-and-disaster-frequency

United Nations (2015a). *Issues Facing Small Island Developing States 'Global Challenges' Demanding Collective Responsibility, Secretary-General Tells Security Council.* Retrieved from https://www.un.org/press/en/2015/sc11991.doc.htm#:~:text=On%20the%20%E2%80%9Cfront%20lines%20of, Ban%20said%20in%20his%20presentation.

United Nations Framework Convention on Climate Change (2015). *COP 21*. Retrieved from https://unfccc.int/process-and-meetings/conferences/past-conferences/paris-climate-change-conference-november-2015/cop-21
United Nations (2015b). *The 17 Goals*. Retrieved from https://sdgs.un.org/goals
U.S. Energy Information Administration (EIA) (2019). *EIA projects nearly 50% increase in world energy usage by 2050, led by growth in Asia*. Retrieved from https://www.eia.gov/todayinenergy/detail.php?id=41433#:~:text=In%20its%20newly%20released%20International,50%25%20between%202018%20and%202050.
International Renewable Energy Agency (2018). *Global Energy Transformation: A roadmap to 2050*, International Renewable Energy Agency, Abu Dhabi. Retrieved from https://www.irena.org/-/media/Files/IRENA/Agency/Publication/2018/Apr/IRENA_Report_GET_2018.pdf
The Energy and Resources Institute. (2020). *India to have 60% power capacity from renewable energy by 2030: RK Singh*. Retrieved from https://www.teriin.org/press-release/india-have-60-power-capacity-renewable-energy-2030-rk-singh
Fares, R. (2015). *Renewable Energy Intermittency Explained: Challenges, Solutions, and Opportunities*. Retrieved from https://blogs.scientificamerican.com/plugged-in/renewable-energy-intermittency-explained-challenges-solutions-and-opportunities/
World Bank and United Nations Department of Economic and Social Affairs. (2017). *The Potential of the Blue Economy: Increasing Long-term Benefits of the Sustainable Use of Marine Resources for Small Island Developing States and Coastal Least Developed Countries*. World Bank, Washington DC. Retrieved from https://sustainabledevelopment.un.org/content/documents/2446blueeconomy.pdf
International Renewable Energy Agency (2020a). *Innovation Outlook: Ocean Energy Technologies*, IRENA, Abu Dhabi. Retrieved from https://www.irena.org/publications/2020/Dec/Innovation-Outlook-Ocean-Energy-Technologies
International Renewable Energy Agency (2020b). *Fostering a blue economy: Offshore Renewable Energy*, IRENA, Abu Dhabi. ISBN 978-92-9260-288-8. Retrieved from https://www.irena.org//media/Files/IRENA/Agency/Publication/2020/Dec/IRENA_Fostering_Blue_Economy_2020.pdf
International Energy Agency (2019). *Offshore Wind Outlook 2019*, IEA, Paris. Retrieved from https://www.iea.org/reports/offshore-wind-outlook-2019
IRENA (2020). *Innovation Outlook: Ocean Energy Technologies*. Retrieved from https://www.irena.org/publications/2020/Dec/Innovation-Outlook-Ocean-Energy-Technologies
Guduru, S., Bajaj, P. and Gonsalves, O.N., India's low temperature thermal desalination technology: Water diplomacy with Small Island Developing States in the Indo-Pacific Region, Maritime Affairs: Journal of the National Maritime Foundation of India, 1–16, 2020, Routledge, https://doi.org/10.1080/09733159.2020.1855816. Retrieved from https://www.tandfonline.com/doi/full/10.1080/09733159.2020.1855816?src=
World Ocean Initiative (2020). *Accelerating Energy Innovation for the Blue Economy*, The Economist Group Limited. Retrieved from https://www.woi.economist.com/wp-content/uploads/2020/11/AcceleratingEnergyInnovationfortheBlueEconomy.pdf
European Commission (2020). The EU Blue Economy Report. 2020. Publications Office of the European Union. Luxembourg. Retrieved from https://ec.europa.eu/maritimeaffairs/sites/maritimeaffairs/files/2020_06_blueeconomy-2020-ld_final.pdf
Chan, E.C.C et al. (2020). *Environmental Impact Assessment of the Operation of an Open Cycle OTEC 1MWe Power Plant in the Cozumel Island, Mexico, Ocean Thermal Energy Conversion (OTEC) - Past, Present, and Progress*, Albert S. Kim and Hyeon-Ju Kim, IntechOpen, https://doi.org/10.5772/intechopen.91179. Retrieved from: https://www.intechopen.com/books/ocean-thermal-energy-conversion-otec-past-present-and-progress/environmental-impact-assessment-of-the-operation-of-an-open-cycle-otec-1mwe-power-plant-in-the-cozum
Langer J., Quist, J., Blok, K. (2020). *Recent progress in the economics of ocean thermal energy conversion: Critical review and research agenda*, Renewable and Sustainable Energy Reviews,

Volume 130, 2020, 109960, ISSN 1364-0321, https://doi.org/10.1016/j.rser.2020.109960. Retrieved from http://www.sciencedirect.com/science/article/pii/S1364032120302513

IEA (2019). *Offshore Wind Outlook 2019*, Retrieved from https://www.iea.org/reports/offshore-wind-outlook-2019

European Union (2020). *Briefing to the European Parliament: Offshore Wind Energy in Europe*. EU, Brussels, Retrieved from https://www.europarl.europa.eu/RegData/etudes/BRIE/2020/659313/EPRS_BRI(2020)659313_EN.pdf

Deign, J. (2020). So, What Exactly is Floating Offshore Wind? Greentechmedia.com, Retrieved from https://www.greentechmedia.com/articles/read/so-what-exactly-floating-offshore-wind

Rinaldi, G. (2020). Offshore Renewable Energy, Renewable Energy - Resources, Challenges and Applications, Mansour Al Qubeissi, Ahmad El-kharouf and Hakan Serhad Soyhan, IntechOpen, https://doi.org/10.5772/intechopen.91662. Retrieved from https://www.intechopen.com/books/renewable-energy-resources-challenges-and-applications/offshore-renewable-energy

Guduru, S. Chauhan, P. (2020) *Hydrogen-Fuel Adoption: An Ocean Renewable Energy Approach Part 5: Hydrogen-Fuel- The Option-of-Choice for India.* Retrieved from https://maritimeindia.org/part-5-hydrogen-fuel-the-option-of-choice-for-india/

Narula, K. (2019). *Renewable energy from oceans*. Lecture Notes in Energy, Springer, Singapore, pp. 163-186, https://doi.org/10.1007/978-981-13-1589-3_8.

Chapter 7
Transitioning Regional Fisheries and Aquaculture into the Blue Economy Framework

Shekar Bose

7.1 Introduction

The notion of the 'blue economy' was launched at the Rio +20 United Nations Conference on Sustainable Development (UNCSD), held in Rio de Janeiro, June 20–22, 2012—an exceptional illustration of successful ocean diplomacy by the small island developing states (SIDS). The initiative has socio-economic, political and cultural relevance to all coastal states and countries with vested interests in waters beyond the exclusive economic zones (EEZs). The concept signifies the integration of traditional ocean industries (for instance, fisheries, marine tourism, maritime transport and security, oil and gas, etc.) and emerging activities, such as offshore renewable energy, aquaculture, seabed extractive activities and marine biotechnology and bioprospecting to ensure sustainable management of ocean wealth (Techera 2018; WB and UNDESA 2017). The operational definition of the 'blue economy' has not been widely agreed upon as yet (WB and UNDESA 2017) and it is rational to think that the concept will evolve in the future with progress in technology.

The socio-economic importance of oceans and seas is also reflected in the 2016 report of the Organisation for Economic Co-operation and Development (OECD). Based on a 'business-as-usual' model, the ocean economy contribution to global value added is predicted to increase from USD 1.5 trillion in 2010 to over USD 3 trillion in 2030 (OECD 2016). In 2030, strong growth is predicted to occur in marine aquaculture, offshore wind energy, fish processing and shipbuilding and repair sectors rather than the offshore oil and gas sectors that were dominant (accounted for one-third of the total contribution) in 2010 (OECD 2016). In 2010, the global ocean economy accounted for about 31 million direct full-time jobs and industrial capture

S. Bose (✉)
Independent Researcher, Brisbane, QLD, Australia

S. Hazra, A. Bhukta (eds.), *The Blue Economy*,
https://doi.org/10.1007/978-3-030-96519-8_7

fisheries was the largest employer followed by the maritime and coastal tourism sector. Under the 'business-as-usual' model, ocean industries are predicted to create approximately 40 million full-time jobs by 2030 and offshore wind energy, marine aquaculture, fish processing and port activities are expected to take the lead (OECD 2016).

The 'blue growth' initiative launched by the Food and Agriculture Organisation (FAO) is nested in the 'blue economy' initiative and is particularly aimed at the long-term sustainability of fisheries and aquaculture (FAO 2017). While both initiatives are aligned with the Sustainable Development Goals (SDGs) and envisioned for ocean-based economic growth that is environmentally sustainable, socially inclusive and equitable (FAO 2017), a balance between the conflicting objectives of environmental sustainability and economic development needs to be maintained (Al-Masroori and Bose 2011).

Fisheries and aquaculture—an important sector of the ocean economy- provides significant benefits to human welfare by supporting food and nutritional security as well as the livelihood of millions of people around the world (FAO 2020; Kelleher et al. 2012). According to the 2020 report by the FAO, in 2018 about 59.51 million people were engaged in the primary sector of fisheries and aquaculture, of which about 38.98 million and 20.53 million people were engaged in capture fisheries and aquaculture, respectively. The highest proportion of primary sector workers were found in Asia (85 per cent), followed by Africa (9 per cent), the Americas (4 per cent) and Europe and Oceania (1 per cent each). In 2018, women made up 14 per cent of the total fishers and fish farmers (FAO 2020). Furthermore, with regard to capture fisheries and aquaculture production, many of the top performing countries in the world belong to the east, south and southeast regions (hereafter, labelled as sub-regions) of Asia (FAO 2020).

Therefore, to combat poverty, enhance food and nutrition security and strengthen economies of Asia, particularly the above-mentioned sub-regions, harnessing the economic potential of fisheries and aquaculture in a sustainable manner is crucial. To ensure the flow of socio-economic benefits from coastal and marine living resources to local, regional and global communities, sustainable use of such resources must be affirmed (FAO 2020). Otherwise, the supporting function of fisheries and aquaculture in providing food and nutritional security and livelihoods to millions of people in the region would be at stake. The problem will be more acute with the rise in global population that is expected to reach 9 billion by 2050 (WB 2013). This concern has raised awareness among local and regional actors. From a regional perspective, the following initiatives, among others, affirm such change: (a) the commencement of a four-year (2015–2018) project on blue economy measurement for Southeast Asian economies by the Partnerships in Environmental Management for the Seas of East Asia (PEMSEA) in mid-2015 (McIlgorm 2016), (b) the adoption of the 'Blue Economy Declaration' at the first Indian Ocean Rim Association (IORA) Ministerial Blue Economy Conference in 2015 that aimed at harnessing oceans and maritime resources to drive economic growth, job creation and innovation without undermining the long-term sustainability of natural resources and the protection of ocean environments (Bose 2021), (c) the Action

Plan for 'Healthy Oceans and Sustainable Blue Economies' launched by the Asian Development Bank (ADB) in May 2019 to scale up investments and technical assistance to $5 billion over 5 years (ADB 2019).

With particular attention to the aforementioned sub-regions of Asia, the main objectives of the present chapter are threefold. First, to describe the current outlook, socio-economic contributions and untapped potential of the fisheries and aquaculture sector in the sub-regions. Second, to provide a brief account of both internal and external challenges for the fisheries and aquaculture sector in the sub-regions that inhibit progress towards the sustainable management of the sector and its transition to the 'blue economy.' Third, to suggest the way forward to effectively manage the transition of the regional fisheries and aquaculture to the 'blue economy.'

7.2 The Current Outlook and Potentials of Fisheries and Aquaculture in the Region

As the world's largest contributor to capture fisheries and aquaculture production, Asia plays a vital role in both regional and global food and nutritional security (FAO 2020). From the sub-region's perspective, the significant contribution of fisheries and aquaculture to the sub-region's food and nutritional security, foreign exchange earnings and job creation has been emphasised in various studies (Pomeroy et al. 2019; Pomeroy 2012; Abdullah and Kuperan 1997).

Table 7.1 presents the contribution of fisheries and aquaculture of Asia and the top performing countries belonging to east, south and southeast regions covering the period 2000–2018. The results show that during 2000–2018, Asia and the top producers from the sub-regions of Asia accounted for, on average, about 51.1 per cent (with an average growth rate of 0.55 per cent and 48.8 per cent (with an average growth rate of 0.47 per cent) of the world total capture (inland and marine) production of fish by quantity, respectively. In 2018, three of the East Asian countries (China, Japan and South Korea), six of the Southeast Asian countries (Indonesia, Vietnam, Philippines, Myanmar, Thailand and Malaysia), and two of the South Asian countries (India and Bangladesh) were in the world's top 25 capture fisheries countries by quantity (FAO 2020).

The same pattern was also observed in global aquaculture production of fish. During the same period, Asia accounted for on average about 88.4 per cent (with an average growth rate of 0.07 per cent) and 80.7 per cent (with an average growth rate of 0.31 per cent) of the world total aquaculture production of fish by quantity and value, respectively. The top producers from the subregions of Asia accounted for on average about 87.0 per cent (with an average growth rate of 0.02 per cent) and 78.9 per cent in value (with an average growth rate of 0.30 per cent) of the world total aquaculture production of fish by quantity and value, respectively.

In 2018, four of the East Asian countries (China, Japan, South Korea and Taiwan), seven of the Southeast Asian countries (Indonesia, Vietnam, Myanmar, Thailand,

Table 7.1 Fisheries and aquaculture production (Live weight): 2000–2018

	Capture production			Aquaculture production (excluding aquatic plants)					
	World	Asia's share in world production	Top sub-regional producers' share in world production[a]	World		Asia's share in world production		Top sub-regional producers' share in world production[b]	
	Quantity	Quantity	Quantity	Quantity	Value	Quantity	Value	Quantity	Value
Year	(million tonnes)	(%)	(%)	(million tonnes)	(USD billion)	(%)	(%)	(%)	(%)
2000	93.6	46.6	44.6	32.4	47.8	87.7	79.5	86.7	77.9
2001	90.8	47.7	45.6	34.6	48.6	87.5	79.0	86.4	77.4
2002	91.1	47.4	45.2	36.8	49.9	88.0	79.4	86.9	77.8
2003	88.3	50.0	47.8	38.9	54.3	87.8	78.3	86.6	76.3
2004	92.9	47.4	45.1	41.9	60.1	88.0	78.0	86.8	75.8
2005	92.5	48.2	46.0	44.3	66.1	88.5	77.1	87.2	75.0
2006	90.2	50.6	48.2	47.3	75.2	88.4	75.8	87.1	73.5
2007	90.5	50.9	48.6	49.9	91.8	88.5	77.8	87.2	75.7
2008	89.5	51.2	49.1	52.9	106.7	88.8	80.0	87.5	78.1
2009	89.1	51.4	49.3	55.2	113.8	88.8	80.7	87.5	78.9
2010	87.1	53.6	51.4	57.7	131.2	88.7	81.5	87.3	79.7
2011	91.6	50.8	48.6	59.8	154.8	88.3	81.3	86.8	79.7
2012	88.6	53.4	51.1	63.5	169.7	88.2	83.3	86.7	81.6
2013	89.7	53.3	51.0	67.0	191.9	88.8	82.6	87.2	81.1
2014	90.4	54.9	52.6	70.5	210.5	88.4	82.5	86.9	80.9
2015	91.7	54.1	51.7	72.8	206.2	88.8	85.1	87.2	83.4
2016	89.6	54.5	52.0	76.5	222.9	88.9	84.4	87.3	82.7
2017	93.1	53.1	50.5	79.5	237.8	88.8	83.8	87.2	82.1
2018	96.4	51.4	48.6	82.1	250.1	88.7	84.0	87.0	82.2
Average	**90.9**	**51.1**	**48.8**	**56.0**	**131.0**	**88.4**	**80.7**	**87.0**	**78.9**
Average growth (%)	**0.17**	**0.55**	**0.47**	**5.30**	**9.63**	**0.07**	**0.31**	**0.02**	**0.30**

Source: FAO Fishery and Aquaculture Statistics 2018 (FAO 2020)

[a] China, Indonesia, India, Vietnam, Japan, Philippines, Myanmar, Bangladesh, Thailand, Malaysia, Korea Rep. were in the world's top 25 capture fisheries countries

[b] China, India, Indonesia, Vietnam, Bangladesh, Myanmar, Thailand, Philippines, Japan, Korea Rep, Taiwan, Cambodia, Malaysia were the top 25 aquaculture producers in the world

Philippines, Malaysia and Cambodia) and two of the South Asian countries (India, Bangladesh) were in the world top 25 aquaculture producers who produced more than 160,000 tonnes (FAO 2020). In the same year, five of the East Asian countries (China, Japan, South Korea, North Korea and Taiwan), six of the Southeast Asian countries (Indonesia, Philippines, Malaysia, Vietnam, Cambodia and Myanmar) and one of the South Asian countries (India) were in the world's top 25 producers of aquatic plants by quantity (FAO 2020). Overall, the current situation indicates that the decline of marine capture gives rise to the development of aquaculture – a substitute to common-property- in the region.

Table 7.2 presents the marine and inland contribution of fisheries and aquaculture in the world and in Asia covering the period 2000–2018. The results indicate that the share of inland waters in the total global capture production increased from 9.2 per cent in 2000 to 12.5 per cent in 2018 (with an average growth rate of 1.71 per cent), while the share of marine captures declined from 90.8 per cent to 87.5 per cent (with an average growth rate of −0.20 per cent). Similar trends were also observed in the case of Asia. The share of inland waters in the total Asian captures increased from 12.5 per cent in 2000 to 16.0 per cent in 2018 (with an average growth rate of 1.39 per cent), while the share of marine captures declined from 87.5 per cent to 84.0 per cent (with an average growth rate of −0.23 per cent).

With regard to aquaculture, the share of marine waters in the total global production increased from 57.7 per cent in 2000 to 62.5 per cent in 2018 (with an average growth rate of 0.45 per cent), while the share of inland production declined from 42.3 per cent to 37.5 per cent (with an average growth rate of −0.68 per cent). In the global aquaculture production, the share of Asia's marine waters increased from 86.7 per cent in 2000 to 87.0 per cent in 2018 with an average growth rate of 0.02 per cent, while the share of inland waters increased from 77.9 per cent in 2010 to 82.2 per cent in 2018 with an average growth rate of 0.30 per cent.

Similar to the global trend, the share of marine waters (inland waters) in the total Asian production increased (decreased) from 60.9 per cent (39.1 per cent) in 2000 to 65.5 per cent (34.5 per cent) in 2018 with an average growth rate of 0.40 per cent (−0.69 per cent). During 2000–2018, inland fisheries capture production (including aquaculture) in Asia was comparatively higher in quantity (on average about 86.4 per cent of the world total) than that of marine capture (on average about 56.0 per cent of the world total) and hence is crucial for their magnitude of socio-economic contributions in the region.

In Asia, the number of fishers and fish farmers increased from about 40.43 million people in 2000 to about 50.38 million people in 2018, an annual growth rate of about 1.28 per cent. In 2018, Asia accounted for the highest number of fish workers and farmers (about 78.9 per cent and 95.5 per cent in fisheries and aquaculture, respectively) of the world total for fisheries (about 38.98 million) and aquaculture (about 20.53 million) (FAO 2020).

Table 7.3 presents the total value of trade in fishery products for the period 2000–2018. The results suggest that during 2000–2018, Asia's export (import) share in total value of world fish exports (imports) ranged from 34.9 per cent (39.6 per cent) in 2010 to 37.3 per cent (34.3 per cent) in 2018. Of which, the export (import) share of the top trading countries in the sub-regions ranged from 94.4 per cent (96.8 per cent) in 2010 to 94.0 per cent (91.3 per cent) in 2018. In the same period, the total value of world fish exports (imports) increased at an annual growth rate of 6.2 per cent (5.6 per cent). Over the same period, the total value of Asia's fish exports (imports) increased at an annual growth rate of 6.6 per cent (4.7 per cent),

Table 7.2 Marine versus inland fisheries and aquaculture production (Live weight): 2000–2018

	Capture production						Aquaculture production (excluding aquatic plants)					
	In World's production		In Asia's production		Asia's share in world production		In world's production		In Asia's production		Asia's share in world production	
	Marine	Inland	Marine	Inland	Marine	Inland	Marine	Inland	Marine	Inland	Marine	Inland
Year	(%)	(%)	(%)	(%)	(%)	(%)	(%)	(%)	(%)	(%)	(%)	(%)
2000	90.8	9.2	87.5	12.5	44.9	63.4	57.7	42.3	60.9	39.1	86.7	77.9
2001	90.6	9.4	87.2	12.8	45.9	64.7	57.1	42.9	60.6	39.4	86.4	77.4
2002	90.8	9.2	87.6	12.4	45.8	63.4	57.2	42.8	60.4	39.6	86.9	77.8
2003	90.2	9.8	87.6	12.4	48.5	63.4	57.3	42.7	60.7	39.3	86.6	76.3
2004	90.7	9.3	87.8	12.2	45.9	62.0	58.2	41.8	61.8	38.2	86.8	75.8
2005	89.8	10.2	86.5	13.5	46.4	64.0	58.6	41.4	62.0	38.0	87.2	75.0
2006	89.1	10.9	85.8	14.2	48.7	65.9	58.8	41.2	62.3	37.7	87.1	73.5
2007	88.9	11.1	85.7	14.3	49.1	65.2	59.5	40.5	63.0	37.0	87.2	75.7
2008	88.6	11.4	85.3	14.7	49.2	66.2	60.9	39.1	64.2	35.8	87.5	78.1
2009	88.4	11.6	85.1	14.9	49.5	66.1	61.2	38.8	64.5	35.5	87.5	78.9
2010	87.5	12.5	84.4	15.6	51.7	67.0	62.2	37.8	65.2	34.8	87.3	79.7
2011	88.5	11.5	85.3	14.7	48.9	65.3	62.1	37.9	65.3	34.7	86.8	79.7
2012	87.7	12.3	84.7	15.3	51.6	66.6	62.3	37.7	65.7	34.3	86.7	81.6
2013	87.8	12.2	85.1	14.9	51.6	65.3	62.9	37.1	65.9	34.1	87.2	81.1
2014	87.8	12.2	85.4	14.6	53.4	65.6	62.9	37.1	66.0	34.0	86.9	80.9
2015	87.8	12.2	85.3	14.7	52.5	65.5	62.9	37.1	66.0	34.0	87.2	83.4
2016	87.3	12.7	84.8	15.2	52.9	65.4	62.7	37.3	65.5	34.5	87.3	82.7
2017	87.2	12.8	84.0	16.0	51.1	66.4	62.3	37.7	65.2	34.8	87.2	82.1

	Capture production						Aquaculture production (excluding aquatic plants)					
	In World's production		In Asia's production		Asia's share in world production		In world's production		In Asia's production		Asia's share in world production	
	Marine	Inland	Marine	Inland	Marine	Inland	Marine	Inland	Marine	Inland	Marine	Inland
Year	(%)	(%)	(%)	(%)	(%)	(%)	(%)	(%)	(%)	(%)	(%)	(%)
2018	87.5	12.5	84.0	16.0	49.3	66.2	62.5	37.5	65.5	34.5	87.0	82.2
Average	**88.8**	**11.2**	**85.7**	**14.3**	**49.3**	**65.1**	**60.5**	**39.5**	**63.7**	**36.3**	**87.0**	**78.9**
Average growth (%)	**−0.20**	**1.71**	**−0.23**	**1.39**	**0.53**	**0.23**	**0.45**	**−0.68**	**0.40**	**−0.69**	**0.02**	**0.30**

Source: FAO Fishery and Aquaculture Statistics 2018 (FAO 2020)

Table 7.3 Total value of trade in fishery commodities, by world, Asia and top sub-regional countries:2000–2018

	World		Continent: Asia		Top sub-regional producers	
	Export	Import	Export	Import	Export	Import
Year	(USD billion)	(USD billion)	(USD billion)	(USD billion)	(USD billion)	(USD billion)
2000	55.8	60.1	19.5	23.8	18.4	23.0
2001	56.5	59.7	19.3	22.0	18.3	21.2
2002	58.5	62.2	19.7	23.0	18.5	22.1
2003	64.1	67.4	21.0	22.3	19.5	21.3
2004	71.9	75.7	24.1	26.3	22.6	25.2
2005	78.9	82.5	26.5	27.5	24.7	26.3
2006	86.3	90.9	29.1	28.3	27.6	26.8
2007	93.7	98.9	31.4	29.1	29.8	27.3
2008	102.4	108.1	35.0	32.9	33.1	30.8
2009	96.5	99.9	34.1	30.7	32.3	28.5
2010	110.7	111.1	41.5	35.6	39.4	33.0
2011	129.7	130.0	50.3	42.5	47.9	39.5
2012	130.6	129.0	52.0	44.0	49.7	40.5
2013	139.5	133.5	54.8	42.3	52.3	38.6
2014	148.6	141.3	58.0	43.5	55.3	39.3
2015	133.3	127.6	51.8	41.6	49.3	37.5
2016	142.7	135.0	54.6	43.9	51.8	39.5
2017	156.0	146.3	59.2	48.7	56.0	44.1
2018	165.4	159.7	61.7	54.8	58.0	50.0
Average	**106.4**	**106.3**	**39.1**	**34.9**	**37.1**	**32.4**
Average growth (%)	**6.2**	**5.6**	**6.6**	**4.7**	**6.6**	**4.4**

Source: FAO Fishery and Aquaculture Statistics 2018 (FAO 2020)

while for the top trading countries in the sub-regions, the annual growth rate of exports (imports) was 6.6 per cent (4.4 per cent).

In 2018, about 66.3 million tonnes (live weight equivalent) of the total fishery and aquaculture production (about 178.5 million tonnes) was exported with a recorded value of about USD 165.4 billion and China, Vietnam and India were in the top four exporting countries (FAO 2020). In the same year, about 88 per cent (about 156.4 million tonnes) of the total fishery and aquaculture production (about 178.5 million tonnes) was used for direct human consumption and live and fresh forms accounted for about 44 per cent. The remaining 12 per cent (22 million tonnes) was utilised mainly for the production of fishmeal and fish oil (FAO 2020).

In 2017, global per capita consumption of fish was estimated at 20.3 kg, with fish accounting for about 17.3 per cent of the global population's intake of animal proteins and 6.8 per cent of all proteins consumed. Preliminary estimates for 2018 indicate a further growth in per capita consumption to about 20.5 kg, of which aquaculture production in total available food fish supply was estimated at 10.8 kg compared to 9.7 kg of capture fisheries (FAO 2020).

In 2017, the per capita consumption of fish in Asia was estimated at 24.1 kg, with fish accounting for about 23.1 per cent of their average per capita intake of animal protein and 8.1 per cent of all proteins consumed. In the same year, the per capita consumption of fish in many Asian countries was higher than that of the world. The following countries (with per capita fish consumption) exemplify the point: Hong Kong (66.5 kg), Malaysia (57.8 kg), South Korea (57.2 kg), Myanmar (45.9 kg), Japan (45.8 kg), Indonesia (44.7 kg), Cambodia (42.7), China (38.8), Vietnam (37.7 kg), Sri Lanka (30.3), Taiwan (29.6 kg), Thailand (29.5), Philippines (26.2 kg), Bangladesh (26.0 kg) and Laos (25.3 kg) (FAO 2020).

Another important characteristic of the sub-region's fisheries is the extent of small-scale fisheries (SSF) and their socio-economic contributions (Pomeroy 2012; Mills et al. 2011; Salayo et al. 2008). However, SSF's contributions to livelihoods and food security are often undermined (Kelleher et al. 2012; Mills et al. 2011; Teh et al. 2011) and are likely to be overshadowed by the higher profile interest in ocean issues (Funge-Smith and Bennett 2019). Therefore, given the SSF's extent and socio-economic contribution in the region (Kelleher et al. 2012), the likelihood of achieving the 'blue economy' goals increases with the adoption of a holistic approach by integrating SSF into the 'blue economy' agenda, changing the existing policy discourse in relation to SSF which is influenced by economic growth paradigms as pointed out by Béné et al. (2010) and providing effective leadership (Sutton and Rudd 2015).

Apart from providing products for direct consumption (e.g. food, harvested through commercial and recreational/sport fishing, medicines, cosmetics, pharmaceutical products, genetic materials, etc.), fisheries and aquaculture (including aquatic plants)—as part of the marine ecosystem—can also generate a wide range of economic benefits. These include: (a) productive use benefits (e.g. pearl industry), (b) non-consumptive benefits (marine tourism activities such as dolphin and turtle watching, snorkelling, diving, etc.), (c) socio-cultural benefits (e.g. aesthetic, artistic, educational, spiritual, religious and/or scientific values) and (d) non-use benefits (e.g. option benefits, existence benefits and bequest benefits).[1]

The case of marine turtles can serve as an example of embracing all categories of values (Busaidi et al. 2019; de Vasconcellos Pegas and Stronza 2010; Campbell and Smith 2006; Troëng and Drews 2004; Tisdell and Wilson 2002). Troëng and Drews (2004) estimated the socio-economic value of marine turtle conservation involving nine developing countries (including three countries from the sub-regions) around the world and the estimated gross revenue of consumptive and non-consumptive use for each case study was on average USD 581,815 per annum (range from USD158 to USD 1,701,328 per annum) and USD1,659,250 (range from USD 41,147 to USD 6,714,483 per annum), respectively. A cross-country study involving five Asian cities by Jin et al. (2010) revealed that the representative households of the study sites

[1] See Barbier (2012) and Grant et al. (2013) for further details on various categories of values.

positively valued the benefits of marine turtle conservation. Based on a case study from Malaysia, Teh et al. (2018) obtained that the total economic value of marine turtles was USD 23 million per year and argued that 1146 tourism jobs (equivalent to annual employment income of USD 469,000) could potentially be generated by protecting such marine species. De Brauwer et al. (2017) estimated that the economic value of muck dive tourism for both Indonesia and the Philippines collectively was more than USD$ 150 million per annum. In addition, over 2200 jobs were created by the dive tourism industry and attracted more than 100,000 divers annually.

However, despite the significant non-consumptive values of the highly valued groups of species such as marine turtles, whales, dolphins, sharks, rays, etc. (Farr et al. 2014; Wilson and Tisdell 2003; Wilson and Tisdell 2001; Marcovaldi and Dei Marcovaldi 1999; Loomis and Larson 1994), the quantification and monetisation of such species and marine ecosystem services[2] remain limited in, species (Teh et al. 2018; Ishizaki et al. 2011; Jin et al. 2010), services (Dang et al. 2021; Brander et al. 2012) and geographical coverage (Olewiler et al. 2016; BOBLME 2014) in the sub-regions. Failure to convey such economic value to policy-makers will potentially hamper the efforts to protect and conserve those valuable marine resources and environmental degradation (Jones-Walters and Mulder 2009). Furthermore, in the 'blue economy' context, McIlgorm (2016) observed that China, Vietnam and Indonesia had a substantially higher marine economy's share in national gross domestic product (GDP) compared to that of Australia, NZ, Canada, USA, France and UK. Despite this, economic assessment of the marine economy is rare (Song et al. 2013; Zhao et al. 2014) and in its early stages in the sub-region (McIlgorm 2016).

While a detailed overview of the available valuation methods along with the underlying theory, their advantages and disadvantages, and pre-conditions for their use is beyond the scope of this chapter, a brief discussion on some of the widely used methods is given below.[3] The provisioning services of fisheries and aquaculture along with some of the tourism and recreation services are offered through well-developed markets and hence, their economic values can be generally estimated following the market price approach based on observed market prices (Schuhmann and Mahon 2015). However, many of the above-mentioned non-exploitative benefits and uses of marine living resources are not priced in the market. In such a case, two complementary methods, namely the 'stated preference' and the 'revealed preference' methods can be used to generate economic value. Under the 'stated preference' methods, *contingent valuation* and *conjoint analysis* can be applied (Ishizaki et al. 2011; Alriksson and Öberg 2008; Veisten 2007; Adamowicz et al. 1998). They are both founded on behavioural economics and are survey-based (Hanemann 1991; Barbier et al. 2009; Carson 2012) and the contingent valuation method has been the most widely used method to measure non-use values (Nunes

[2] For descriptions of ecosystem services, see Werner et al. (2014) and Grant et al. (2013).

[3] Further details on the valuation methods can be found in Kanninen (2007) and Brander et al. (2012), and a review of their application involving marine species can be found in Lew (2015).

and van den Bergh 2001). On the other hand, the revealed preference methods include, among others, the *travel cost method*, where costs incurred travelling to and at a site used as a proxy for price of recreation (Pascoe et al. 2014) and the *hedonic price method* to infer the prices which individuals are willing to pay for recreational goods (Carter and Liese 2010).

Other methods of estimating the economic value include *the replacement cost* approach that involves calculation of replacement cost based on market data (Troëng and Drews 2004) and the cost (damage) avoidance approach, which uses estimates of the expenditures that would be incurred to prevent, diminish or avoid harmful effects associated with degradations of natural resources (Schuhmann and Mahon 2015).

It is difficult to formulate a clear universal statement about the applicability of methods as such applicability, to a great extent, would be subject to particular circumstances. In applying valuation techniques, one needs to consider the following aspects with great caution: (1) limitations of these valuation methods, (2) the use of appropriate econometric techniques, (3) technical skills and capacity needed to perform the valuation tasks, (4) data limitations, (5) the costs of collecting data (Harrison and Lesley 1996), etc., among others. In any case, such exercise could at best produce only a crude estimate of such economic value.

In summary, it is quite clear that the fisheries and aquaculture sector in the sub-regions of Asia plays an important role in the socio-economic development of the region. In addition, the socio-economic potentials of non-consumptive benefits and uses of marine living resources and the importance of valuing such uses are highlighted. Therefore, reflecting on the current state of affairs, one should envisage a holistic approach—that takes into account the values produced by fisheries and aquaculture in the sub-regions—to sustainable management of fisheries and aquaculture resources.

7.3 Challenges to the Transition to the Blue Economy

Before turning to proposals for radical change of the *status quo* situation, it is important to give a brief account of the key challenges faced by the fisheries and aquaculture sector in the sub-regions and motivations for change. Despite the remarkable performance of the fisheries and aquaculture sector in the sub-regions over the period 2000–2018, the overall experience on the progress towards sustainability of fisheries resources has been far from satisfactory.

A number of factors both endogenous and exogenous to the fisheries and aquaculture sector have been inhibiting the sector's progress towards long-term sustainability. Some of these factors are listed, in no particular order, below.[4]

[4] Further details on various issues pertaining to fisheries and aquaculture in the sub-regions can be found in SEAFDEC (2017) and WB and UNDESA (2017).

First, the twin problem of overfishing and overcapacity (Pagkalinawan et al. 2020; Pomeroy 2012; Williams and Staples 2010; FAO-WFC 2008; Salayo et al. 2008; Stobutzki et al. 2006). Second, the extent of Illegal, Unreported and Unregulated (IUU) fishing within national jurisdiction (Kuperan and Jahan 2020) and beyond (Latun et al. 2016; Johns 2013; Williams 2013). The gravity of IUU fishing in the region was exemplified by the joint declaration of the International Day (June 5, 2019) by the United Nation's Agencies, the Food and Agriculture Organisation (FAO), the International Labour Organisation (ILO) and the International Organisation for Migration (IOM) in Bangkok, Thailand for the fight against IUU fishing (www.fao.org/3/ca4937en/ca4937en.pdf. Accessed March 31, 2021). Third, weak enforcement of rules and lack of rule compliance (Kuperan and Jahan 2020; Dang et al. 2017; SEAFDEC 2017; Goldstein 2013; Catedrilla et al. 2012; Boonstra and Dang 2010). Fourth, inadequacy of the current legislative arrangements (Dang et al. 2017; Williams and Staples 2010). Fifth, information deficiency (Pomeroy 2012; de Graaf et al. 2011). Sixth, the countries lack of a coherent all-embracing approach as policies involving ocean industries are generally set at different government agencies (McIlgorm 2016; Williams and Staples 2010). Seventh, government fisheries agencies lack capacity needed to address the big marine fisheries challenges (Williams and Staples 2010). Eighth, the low prioritisation of fisheries issues (Funge-Smith and Bennett 2019; Teh et al. 2011). Ninth, the inadequacy of national and sub-regional plans along with the allocation of adequate funds to effectively address international policy change and subsequent expectations on coastal states (SEAFDEC 2017). For example, the blue economy initiative elevates complexity to a new level through the convergence of ocean-based industries and creates new and unprecedented expectations on coastal states with respect to sustainable management of ocean wealth. Tenth, the inefficient use of economic resources in fisheries (Nga et al. 2020; Larry et al. 2017; Teame 2017; Yang and Lou 2016; Wiyono and Hufiadi 2014; Zen et al. 2002) and aquaculture operations (Anh Ngoc et al. 2018; Zongli et al. 2017; Alam et al. 2012).

On the other hand, some of the notable external challenges facing the fisheries and aquaculture sector in the sub-regions are as follows. First, is the climate change threat to coastal and marine fisheries and aquaculture (Watkiss et al. 2019; Ding et al. 2017; Allison et al. 2009; Cochrane et al. 2009). Second is the existing maritime territorial dispute. For instance, the recent territorial disputes in the South China Sea are considered to be unproductive for sub-regional fisheries (Pornpatimakorn 2012) and a critical factor in China-ASEAN (the Association of Southeast Asian Nations) relations (Acharya 2013). The third is the extent of maritime piracy (Hastings 2020; Dillon 2005). The fourth is the impact of land and sea-based pollution (e.g. oil spill and ocean dumping) on marine living resources (Todd et al. 2010). The fifth is the regional conflict due to disparity of national values and economic and political interests. Shifts in the balance of regional power due to the rise of China and India are claimed to be a greater challenge to unity and cohesion in regional organisations (Acharya 2013; Berlin 2011; Rumley et al. 2012).

The scope of the challenges listed above is not confined within national jurisdictions and therefore, their remedial measures are beyond the scope of the conventional nation-state type organisation. Therefore, it is time to search for arrangements and innovative options to create a mutually beneficial regime (Bose 2021).

7.4 Managing the Transition: The Way Forward

The complexity under the 'blue economy' together with the challenges faced by the fisheries and aquaculture sector in the region necessitate the engagement of nation-states in reorganizing intra-agency cooperation at both national and regional levels guided by the existing international binding instruments, for example, the 1982 UN Convention on the Law of the Sea (UN 1982), Agenda 21 (UNCED 1992), the 1995 UN Fish Stocks Agreement (UN 1995), etc., and innovative approaches (Bose 2021; Pomeroy et al. 2019; Al-Balushi et al. 2016) to managing regional public goods (Barrett 2020). Therefore, in managing the transition of the regional fisheries and aquaculture sector to the 'blue economy' and stressing the strategic significance of regional cooperation as a mechanism, the following strategic steps are proposed:

(a) *Organisational Reform:* A close scrutiny of the objectives of the existing organisations of the Asian sub-regions such as, The Association of Southeast Asian Nations (ASEAN), The South Asian Association of Regional Cooperation (SAARC), The Bay of Bengal Initiative for Multi-Sectoral Technical and Economic Cooperation (BIMSTEC), etc. indicates that individually they are limited either in geographical coverage or sectoral coverage, or both to effectively embrace and reconcile activities of ocean industries in achieving the goals stipulated in the 'blue economy' initiative. Therefore, to realise the vast economic potential of the fisheries and aquaculture sector in the sub-regions, the existing mandate of sub-regional and regional organisations needs to be broadened and aligned with the 'blue economy' goals.

    While the formation of common interests in the 'blue economy' initiative and interest in the creation of a mutually beneficial regime in the sub-regions are preconditions for such an integration of regional organisations, the fulfilment of such preconditions does not serve as a sufficient condition for action unless those common interests and shared goals are reinforced with agreements and implementation. It would not be surprising at all if nation-states lack interest to accomplish such tasks due to diversity among countries in their resource endowments, cultural heritage and uneven political interests and power that often undermine effective reform. While the existing literature on collective action advocates the superiority of regional cooperation and organisational integration to unilateral action (Bose et al. 2017; Acharya 2014; Meinzen-Dick and di Gregorio 2004), such attempts may be overshadowed temporarily by political interests of some members and persistent power rivalry between members. For example, the present geopolitical conditions such as the South China Sea disputes (Clark-Shen et al. 2020) and political upheaval in Myanmar, among others, are not conducive to harnessing broader cooperation possibilities in the sub-regions. However, cooperation with the majority carries with it a means of placing unwilling/non-compliant members with political power at a disadvantageous position and is likely to diminish their influence on the actions of other cooperative members (Barnes 1995). Furthermore, the cooperation should not be *ad hoc* in nature; it must be institutionalised with the establishment of strong compliance mechanisms at the outset that can be expected to alter the behaviour of members.

Mutual trust (Wallis 2011), institutional innovations (Bose 2021) and greater transparency (Clark et al. 2015) will be necessary for achieving more efficient and effective governance that legitimises the inherent process, approach and management actions. Such governance should also value fairness, equality in opportunity and national security; otherwise, it would impair the effectiveness of the governance mechanisms. Inter-governmental organisations such as the Indian Ocean Rim Association (IORA) could take a leading role in the development of such a governance framework.

(b) *Strengthening Regional Organisations*: The 'blue economy' initiative does not necessarily demand for setting up a new regional organisation if the existing sub-regional and regional organisations can be horizontally integrated with an extended scope, an all-embracing governance mechanism and political authority to make collective binding decisions. In addressing the implementation challenge of the ecosystem-based approach to the Baltic Sea fisheries management, Haapasaari et al. (2021) suggested a similar approach based on the highly likely prospect of acceptability and adaptability of such an approach. Furthermore, such reform will minimise a country's involvement in multiple sub-regional and regional organisations and, thereby, increase efficiency through the reduction of time and costs (Linn and Pidufala 2008).

With particular reference to the ecosystem-based approach to fisheries management in the sub-regions, there has been increasing recognition of the need for regional organisations (for example, Indian Ocean Tuna Commission (IOTC), Regional Commission for Fisheries (RECOFI), etc.) to improve their governance of fisheries and conservation and management of fishery resources (Clark et al. 2015; Gilman et al. 2014). The effectiveness of existing Asian organisations in relation to managing security problems and the economic vulnerabilities of their members have also been questioned (Acharya and Johnston 2007). The complexity of issues involved demands for improved and inclusive governance based on political leadership, a common vision, comprehensive legislations, effective partnerships involving member-states, political authority to make decisions and investment in scientific research to help knowledge-based decision making (Acharya and Johnston 2007; Bennett et al. 2019) among others. A sense of mutual trust among member-states and public trust in organisational activities and performance must be restored to achieve organisational effectiveness. To foster necessary collaboration across the nations and integration across sectors, a model of joint leadership involving three well-recognised Asian powers—Japan, China and India would be more promising.

Intervention of international institutions such as ESCAP may be required for resolving any disputes.

(c) *Rule Consistency (or Institutional Synergy)*: The task involves harmonisation of relevant national and regional regulatory legislations and policies with the related binding (e.g. UNCLOS, Agenda 21) and non-binding (e.g. FAO Code of Conduct for Responsible Fisheries (FAO (2011)) international instruments. This type of substantial institutional synergy is not unrealistic or unfeasible. The creation of a single European currency illustrates the case in point. Furthermore, to achieve the 'blue economy' goals, the operations of the ocean-based industries including fisheries and aquaculture need to be brought under an integrated governance framework. The implication of rule consistency is that it gives rise to policy coherence that not only benefits the sustainable management of ocean wealth but can also help with making provisional agreements. For example, with particular reference to the South China Sea territorial disputes, Clark-Shen et al. (2020) pointed out the scope of developing cooperative agreements based on the existing commonalities of national fisheries laws and policies to safeguard the dwindling fisheries resources. In the context of South China Sea, Hsiao (2020) discussed the potential elements of four provisional fisheries enforcement arrangements (involving China, Japan Taiwan, the Philippines, Malaysia and Indonesia) and proposed the establishment of a jointly managed maritime zone with provisional measures that would facilitate the commencement of institutionalised cooperation on maritime law enforcement and fisheries issues.

(d) *Creating Incentives for Member-states*: The proposed organisational integration and institutional harmonisation stated in (a) and (b) above would perform a number of important economic functions of collective action regime such as economies of scale, reduction of transaction costs, reduction of intra-group competition, increase of global competitiveness and risk reduction (Bose 2021; Acharya 2014). However, such steps obviously do not overcome the aforementioned challenges to fisheries unless there are other special mechanisms to raise the self-interest of member-states. The prospective benefits of collaboration in innovation, science and technology, research and development, sharing of country-specific skills and expertise, information sharing, etc. will likely prove to be powerful incentives for enhancing regional cooperation. These initiatives could also help diffuse the political sensitivities that have often held up progress towards South Asian economic integration. India's bilateral science and technology cooperation with 83 countries in the world and development cooperation with ASEAN, SAARC, BIMSTEC, etc. in the region illustrate the point (Sharma and Varshney 2019). Another incentive mechanism is the provision of financial and capacity enhancing technical support to member countries to evaluate domestic regulations and address the implementation challenges (WB and UNDESA 2017). Regional institutions such as the Asian Development Bank (ADB) and the Economic and Social Commission for Asia and Pacific (ESCAP) can play a leading role using their convening power in this matter.

(e) *Creating Public Awareness and Policy Makers' Interest*: As pointed out earlier, while some piecemeal studies on ecosystem services are available, they do not explicitly reflect the overall economic value of marine living resources of the sub-region. It is important to note that a number of studies (Costanza et al. 1997; Millennium Ecosystem Assessment 2005) and initiatives (for instance, the Economics of Ecosystem and Biodiversity (TEEB), the Global Ocean Biodiversity Initiative, etc.) in relation to the economic value of the world's ecosystem services have played a significant role in raising public awareness of the ecosystem services paradigm (Jones-Walters and Mulder 2009) and research interest in estimating the value of ecosystem services at both local (Perez-Verdin et al. 2016) and global levels (Barbier et al. 2009; De groot et al. 2012; Pendleton et al. 2016).

Therefore, despite methodological flaws, pervasive uncertainties involved and the immensity in measuring non-use value and non-market goods and services (Eberle and Hayden 1991; Diamond and Hausman 1994; Carson et al. 2001; Hausman 2012; Lloyd-Smith 2018), such an undertaking will help create public awareness, generate political interest and create a 'level playing field' for fisheries and aquaculture in decision-making (Schuhmann and Mahon 2015; Gómez-Baggethun et al. 2010). For example, the damage assessment of the *Exxon Valdez* oil spill of March 1989 using the CV method (Carson 2012) and the support received from the National Oceanic and Atmospheric Administration (NOAA) Panel (Arrow et al. 1993) illustrate the point.

7.5 Concluding Remarks

Under an ever-evolving concept of the 'blue economy,' the oceans and seas are in a stage of transition. It should be recognised that the sustainability of oceans resources is not optional—it is essential for achieving the 'blue economy' and national developmental goals. In this regard, hard choices have to be made to achieve an appropriate balance between efficiency and equity, and sustainable resource management rather than resource exploitation (Bennett et al. 2019). These choices will test the political will and the creativity of the solutions. Ultimately, successful transition to the 'blue economy' requires a higher degree of regional integration than now exists. While there are no easy solutions in changing the organisational form, achieving rule consistency and remedying their inherent imperfections, it is not infeasible. Under the prevailing circumstances, the only available remedy is to try to make them work better. This is the modest underlying hope of this chapter.

Acknowledgements The author would like to thank Somnath Hazra and Anindya Bhukta for their useful feedback on an earlier outline of the chapter. I also thank Amlan Bose for his help in revising the chapter. Any remaining errors are the author's responsibility.

References

Abdullah, N. M. R., & Kuperan, K. (1997). Fisheries management in Asia: the way forward. *Marine Resource Economics*, *12*(4), 345-353.

Acharya A. (2014) Foundations of Collective Action in Asia: Theory and Practice of Regional Cooperation. In: Capannelli G., Kawai M. (eds) The Political Economy of Asian Regionalism (pp. 19-38). Springer, Tokyo. 193p. https://doi.org/10.1007/978-4-431-54568-2_2

Acharya, A. (2013). ASEAN 2030: Challenges of building a mature political and security community, ADBI Working Paper, No. 441, Asian Development Bank Institute (ADBI), Tokyo.

Acharya, A., & Johnston, A. I. (eds). (2007). *Crafting cooperation: Regional international institutions in comparative perspective*. Cambridge University Press, UK, 318p.

Adamowicz, W., Boxall, P., Williams, M., & Louviere, J. (1998). Stated preference approaches for measuring passive use values: choice experiments and contingent valuation. *American Journal of Agricultural Economics*, *80*(1), 64-75.

ADB (Asian Development Bank). (2019). Action Plan for Healthy Oceans: Investing in Sustainable Marine Economies for Poverty Alleviation in Asia and the Pacific. https://www.adb.org/sites/default/files/related/145036/Action%20Plan%20for%20Healthy%20Oceans%20and%20Sustainable%20Blue%20Economies.pdf (accessed March 29, 2021).

Alam, M. F., Khan, M. A., & Huq, A. A. (2012). Technical efficiency in tilapia farming of Bangladesh: a stochastic frontier production approach. *Aquaculture International*, *20*(4), 619-634.

Al-Balushi, A. H., Bose, S., & Govender, A. (2016). Stakeholders' views on management arrangements: A case of kingfish fishery in the Sultanate of Oman. *Natural Resources*, *7*(05), 251-264.

Allison, E. H., Perry, A. L., Badjeck, M. C., Neil Adger, W., Brown, K., Conway, D., ... & Dulvy, N. K. (2009). Vulnerability of national economies to the impacts of climate change on fisheries. *Fish and fisheries*, *10*(2), 173-196.

Al-Masroori, H. S., & Bose, S. (2011). Fisheries sustainability and sustainable development. Journal of Fisheries and Aquatic Science, 6(1), 1–21.

Alriksson, S., & Öberg, T. (2008). Conjoint analysis for environmental evaluation. *Environmental Science and Pollution Research*, *15*(3), 244-257.

Anh Ngoc, P. T., Gaitán-Cremaschi, D., Meuwissen, M. P., Le, T. C., Bosma, R. H., Verreth, J., & Lansink, A. O. (2018). Technical inefficiency of Vietnamese pangasius farming: A data envelopment analysis. *Aquaculture Economics & Management*, *22*(2), 229-243.

Arrow, K., Solow, R., Portney, P. R., Leamer, E. E., Radner, R., & Schuman, H. (1993). Report of the NOAA panel on contingent valuation. *Federal Register*, *58*(10), 4601-4614.

Barbier, E. B. (2012). Progress and challenges in valuing coastal and marine ecosystem services. *Review of Environmental Economics and Policy*, *6*(1), 1-19.

Barbier, E. B., Baumgärtner, S., Chopra, K., Costello, C., Duraiappah, A., Hassan, R., ... & Perrings, C. (2009). The valuation of ecosystem services. In: Shahid Naeem, Daniel E. Bunker, Andy Hector, Michel Loreau, Charles Perrings (Eds) *Biodiversity, ecosystem functioning, and human wellbeing: An ecological and economic perspective*, 248-262. Oxford University Press, UK, 368p.

Barnes, B. (1995). The Elements of Social Theory, Princeton University Press, Princeton, New Jersey, USA. 263p.

Barrett, S. (2020). Regional Public Goods: Conceptual Foundations. In: Bambang Susantono and Cyn-Young Park (Eds), Future of Regional Cooperation in Asia and the Pacific. Part 1 (pp.3-36), Asian Development Bank, Metro Manila, Philippines, 640p. http://hdl.handle.net/11540/12854.

Béné, C., Hersoug, B., & Allison, E. H. (2010). Not by rent alone: analysing the pro-poor functions of small-scale fisheries in developing countries. *Development Policy Review*, *28*(3), 325-358.

Bennett, N. J., Cisneros-Montemayor, A. M., Blythe, J., Silver, J. J., Singh, G., Andrews, N., ... & Sumaila, U. R. (2019). Towards a sustainable and equitable blue economy. *Nature Sustainability*, *2*(11), 991-993.

Berlin, D. (2011). The rise of India and the Indian Ocean. *Journal of the Indian Ocean Region*, *7*(1), 1-31.
BOBLME (2014). Assessing, demonstrating and capturing the economic value of marine & coastal ecosystem services in the Bay of Bengal Large Marine Ecosystem BOBLME-2014-Socioec-02, Andaman Sea Fisheries Research Development Center, Thailand, 70p.
Boonstra, W. J., & Dang, N. B. (2010). A history of breaking laws—Social dynamics of non-compliance in Vietnamese marine fisheries. *Marine Policy*, *34*(6), 1261-1267.
Bose, S. (2021). Law Enforcement, Compliance, and Fisheries Sustainability. In: Jawad L.A. (ed) The Arabian Seas: Biodiversity, Environmental Challenges and Conservation Measures. Springer, Cham, 829-824, https://doi.org/10.1007/978-3-030-51506-5_35.
Bose, S., Al-Jabri, O., Osman, N. M., Al Habsi, S. H. and Al Shidi, R. (2017). Role of Small and Medium Enterprises and Agricultural Cooperatives in Promoting Sustainable Agriculture for Rural Development and Food Security, Amity Journal of Agribusiness, 2 (2):1-21.
Busaidi, M. A., Bose, S., Claereboudt, M., & Tiwari, M. (2019). Sea turtles tourism in Oman: Current status and future prospects. *Tourism and hospitality research*, *19*(3), 321-336.
Brander, L. M., Wagtendonk, A. J., Hussain, S. S., McVittie, A., Verburg, P. H., de Groot, R. S., & van der Ploeg, S. (2012). Ecosystem service values for mangroves in Southeast Asia: A meta-analysis and value transfer application. *Ecosystem services*, *1*(1), 62-69.
Campbell, L. M., and Smith, C. (2006). What makes them pay? Values of volunteer tourists working for sea turtle conservation. *Environmental Management*, *38*(1), 84–98.
Carson, R. T. (2012). Contingent valuation: A practical alternative when prices aren't available. *Journal of Economic Perspectives*, *26*(4), 27-42.
Carson, R. T., Flores, N. E., & Meade, N. F. (2001). Contingent Valuation: Controversies and Evidence. *Environmental and Resource Economics*, *19*, 173–210.
Carter, D. W., & Liese, C. (2010). Hedonic valuation of sportfishing harvest. *Marine Resource Economics*, *25*(4), 391-407.
Catedrilla, L. C., Espectato, L. N., Serofia, G. D., & Jimenez, C. N. (2012). Fisheries law enforcement and compliance in District 1, Iloilo Province, Philippines. *Ocean & coastal management*, *60*, 31-37.
Clark, N. A., Ardron, J. A., & Pendleton, L. H. (2015). Evaluating the basic elements of transparency of regional fisheries management organizations. *Marine Policy*, *57*, 158-166.
Clark-Shen, N., Hsiao, A., Oei, L., Rao, M., Sun, N. Y., & Douglas, A. (2020). A comparative study of fishery and environment laws in the south China sea: Utilizing existing laws to promote peace. *Marine Policy*, *121*, 103568.
Cochrane, K., De Young, C., Soto, D., & Bahri, T. (2009). Climate change implications for fisheries and aquaculture. *FAO Fisheries and aquaculture technical paper*, *530*, 212.
Costanza, R., d'Arge, R., De Groot, R., Farber, S., Grasso, M., Hannon, B., ... & Van Den Belt, M. (1997). The value of the world's ecosystem services and natural capital. *Nature*, *387*(6630), 253-260.
Dang, A. N., Jackson, B. M., Benavidez, R., & Tomscha, S. A. (2021). Review of ecosystem service assessments: Pathways for policy integration in Southeast Asia. *Ecosystem Services*, *49*, 101266.
Dang, N. B., Momtaz, S., Zimmerman, K., & Nhung, P. T. H. (2017). Effectiveness of formal institutions in managing marine fisheries for sustainable fisheries development: A case study of a coastal commune in Vietnam. *Ocean & Coastal Management*, *137*, 175-184.
De Brauwer, M., Harvey, E. S., McIlwain, J. L., Hobbs, J. P. A., Jompa, J., & Burton, M. (2017). The economic contribution of the muck dive industry to tourism in Southeast Asia. *Marine Policy*, *83*, 92-99.
de Graaf, G. J., Grainger, R. J., Westlund, L., Willmann, R., Mills, D., Kelleher, K., & Koranteng, K. (2011). The status of routine fishery data collection in Southeast Asia, central America, the South Pacific, and West Africa, with special reference to small-scale fisheries. *ICES Journal of Marine Science*, *68*(8), 1743-1750.

De Groot, R., Brander, L., Van Der Ploeg, S., Costanza, R., Bernard, F., Braat, L., ... & Van Beukering, P. (2012). Global estimates of the value of ecosystems and their services in monetary units. *Ecosystem services*, *1*(1), 50-61.

de Vasconcellos Pegas, F., & Stronza, A. (2010). Ecotourism and sea turtle harvesting in a fishing village of Bahia, Brazil. *Conservation and Society*, *8*(1), 15-25.

Diamond, P. and Hausman, J. (1994). Contingent valuation: Is some number better than no number? *Journal of Economic Perspectives*, 8: 45-64.

Dillon, D. R. (2005). Maritime piracy: Defining the problem. *SAIS Review of International Affairs*, *25*(1), 155-165.

Ding, Q., Chen, X., Hilborn, R., & Chen, Y. (2017). Vulnerability to impacts of climate change on marine fisheries and food security. *Marine Policy*, *83*, 55-61.

Eberle, W. D., & Hayden, F. G. (1991). Critique of contingent valuation and travel cost methods for valuing natural resources and ecosystems. *Journal of Economic Issues*, *25*(3), 649-687.

FAO. (2020). *FAO Yearbook. Fishery and Aquaculture Statistics 2018, FAO*. Rome. https://doi.org/10.4060/cb1213t

FAO. (2017). Blue growth initiative: Partnering with countries to achieve the Sustainable Development Goals, FAO, Rome (www.fao.org/policy-support/policy-themes/blue-growth/en, Accessed March 13, 2021).

FAO (2011). Code of Conduct for Responsible Fisheries, Food and Agriculture Organization of the United Nations, Rome, 91p.

FAO-WFC (Food and Agriculture Organization of the United Nations and World Fish Center). (2008). *Small-scale Capture Fisheries: A Global Overview with Emphasis on Developing Countries*. World Bank, Washington, DC.

Farr, M., Stoeckl, N., & Beg, R. A. (2014). The non-consumptive (tourism)'value'of marine species in the Northern section of the Great Barrier Reef. *Marine Policy*, *43*, 89-103.

Funge-Smith, S., & Bennett, A. (2019). A fresh look at inland fisheries and their role in food security and livelihoods. *Fish and Fisheries*, *20*(6), 1176-1195.

Gilman, E., Passfield, K., & Nakamura, K. (2014). Performance of regional fisheries management organizations: ecosystem-based governance of bycatch and discards. *Fish and Fisheries*, *15*(2), 327-351.

Goldstein, L. J. (2013). Chinese fisheries enforcement: Environmental and strategic implications. *Marine Policy*, *40*, 187-193.

Gómez-Baggethun, E., De Groot, R., Lomas, P. L., & Montes, C. (2010). The history of ecosystem services in economic theory and practice: from early notions to markets and payment schemes. *Ecological economics*, *69*(6), 1209-1218.

Grant, S. M., Hill, S. L., Trathan, P. N., & Murphy, E. J. (2013). Ecosystem services of the Southern Ocean: trade-offs in decision-making. *Antarctic Science*, *25*(5), 603-617.

Haapasaari, P., Ignatius, S., Pihlajamäki, M., Bryhn, A., Sarkki, S., Tuomisto, J., ... & Kuikka, S. (2021). Integrated governance for managing multidimensional problems: Potentials, challenges, and arrangements. *Marine Policy*, *123*, 104276.

Hanemann, W. M. (1991). Willingness to pay and willingness to accept: how much can they differ? *The American Economic Review*, *81*(3), 635-647.

Harrison, G. W., & Lesley, J. C. (1996). Must contingent valuation surveys cost so much? *Journal of Environmental Economics and Management*, *31*(1), 79-95.

Hastings, J. V. (2020). The Return of Sophisticated Maritime Piracy to Southeast Asia. *Pacific Affairs*, *93*(1), 5-30.

Hausman, J. (2012). Contingent valuation: from dubious to hopeless. *Journal of Economic Perspectives*, *26*(4), 43-56.

Hsiao, A. (2020). Opportunities for fisheries enforcement cooperation in the South China Sea. *Marine Policy*, *121*, 103569.

Ishizaki, A., Teel, T. L., & Yamaguchi, M. (2011). Contextual factors influencing support for sea turtle management actions in Ogasawara Islands, Japan: An application of conjoint analysis. *Human Dimensions of Wildlife*, *16*(5), 287-298.

Jin, J., Indab, A., Nabangchang, O., Thuy, T. D., Harder, D., & Subade, R. F. (2010). Valuing marine turtle conservation: A cross-country study in Asian cities. *Ecological Economics*, *69*(10), 2020-2026.

Johns, M. (2013). Enhancing responsible fishing practices in South East Asia to combat Illegal, Unreported and Unregulated (IUU) fishing. *Australian Journal of Maritime & Ocean Affairs*, *5*(3), 112-119.

Jones-Walters, L., & Mulder, I. (2009). Valuing nature: The economics of biodiversity. *Journal for Nature Conservation*, *17*(4), 245-247.

Kanninen, B. J. (ed). (2007). *Valuing environmental amenities using stated choice studies: a common sense approach to theory and practice* (Vol. 8). Springer, Dordrecht, 337p. https://doi.org/10.1007/1-4020-5313-4

Kelleher, K., Westlund, L., Hoshino, E., Mills, D., Willmann, R., de Graaf, G., & Brummett, R. (2012). *Hidden harvest: The global contribution of capture fisheries*. World bank and WorldFish, 96p.

Kuperan, K. V., & Jahan, K. M. (2020). Noncompliance a major threat in fisheries management–experience from the Artisanal Coastal Fisheries of Bangladesh. *Journal of International Studies*, *6*, 97-113.

Larry, N. D., Astronomo, I. J. T., & Shemaiah, G. P. (2017). Technical Efficiency of Handline Fishers in Region 12, Philippines: Application of Data Envelopment Analysis. *Asian Fisheries Science*, *30*, 215-226.

Latun, A. R., Ali, M., Ahmad, M. T. A., & Katoh, M. (2016). Boosting national mechanisms to combat IUU fishing: Dynamism of the Southeast Asian fisheries sector. *Fish for the People*, *14*(1), 36-43.

Lew, D. K. (2015). Willingness to pay for threatened and endangered marine species: a review of the literature and prospects for policy use. *Frontiers in Marine Science*, *2*, 96.

Linn, J. F., & Pidufala, O. (2008). The experience with regional economic cooperation organizations: Lessons for Central Asia. *Wolfensohn Center for Development Working Paper*, (4), The Brooking Institution, Massachusetts Avenue, Washington, DC, USA, 30p.

Lloyd-Smith, P. (2018). A Note on the Robustness of Aggregate Ecosystem Service Values. *Ecological Economics*, *146*, 778-780.

Loomis, J. B., & Larson, D. M. (1994). Total economic values of increasing gray whale populations: results from a contingent valuation survey of visitors and households. *Marine Resource Economics*, *9*(3), 275-286.

Marcovaldi, M. A., & Dei Marcovaldi, G. G. (1999). Marine turtles of Brazil: the history and structure of Projeto TAMAR-IBAMA. *Biological conservation*, *91*(1), 35-41.

McIlgorm, A. (2016). Ocean economy valuation studies in the Asia-Pacific Region: Lessons for the future international use of national accounts in the Blue Economy. *Journal of Ocean and Coastal Economics*, *2*(2), 6.

Meinzen-Dick, R., & di Gregorio, M. (eds). (2004). Collective Action and Property Rights for Sustainable Development: Understanding Property Rights, 2020 Focus Brief 11. International Food Policy Research Institute (IFPRI), Washington, DC, 36p.

Millennium Ecosystem Assessment (2005). Ecosystems and Human Well-being: Biodiversity Synthesis Report. World Resources Institute, Washington, DC.

Mills, D. J., Westlund, L., de Graaf, G., Kura, Y., Willman, R., & Kelleher, K. (2011). Underreported and undervalued: small-scale fisheries in the developing world, Chapter 1 (pp. 1-15), Robert S. Pomeroy and Neil Andrew (Eds), In Small-scale Fisheries Management: Frameworks and Approaches for the Developing World, CAB international, Cambridge, USA, 247p.

Nga, C. T. H., Eide, A., Armstrong, C. W., & Long, L. K. (2020). Economic performance and capacity utilisation in Vietnamese purse seine fishery. *Asian Fisheries Science*, 33:58–64.

Nunes, P. A., & van den Bergh, J. C. (2001). Economic valuation of biodiversity: sense or nonsense? *Ecological Economics*, *39*(2), 203-222.

OECD (2016), *The Ocean Economy to 2030*, OECD Publishing, Paris, 252p. https://doi.org/10.1787/9789264251724-en.

Olewiler, N., Francisco, H. A., & Ferrer, A. J. G. (Eds.). (2016). *Marine and coastal ecosystem valuation, institutions, and policy in Southeast Asia*. Springer, Singapore. 391p.

Pagkalinawan, L. K., Oposa, A., & McGovern, E. (2020). Regional Cooperation and Integration for Ocean Health and a Sustainable Blue Economy. In: Bambang Susantono and Cyn-Young Park (eds), Future of Regional Cooperation in Asia and the Pacific. Part 4, (pp. 535-590), Asian Development Bank, Metro Manila, Philippines, 640p. http://hdl.handle.net/11540/12854.

Pascoe, S., Doshi, A., Thébaud, O., Thomas, C. R., Schuttenberg, H. Z., Heron, S. F., ... & Calgaro, E. (2014). Estimating the potential impact of entry fees for marine parks on dive tourism in South East Asia. *Marine Policy*, *47*, 147-152.

Pendleton, L. H., Thébaud, O., Mongruel, R. C., & Levrel, H. (2016). Has the value of global marine and coastal ecosystem services changed? *Marine Policy*, *64*, 156-158.

Perez-Verdin, G., Sanjurjo-Rivera, E., Galicia, L., Hernandez-Diaz, J. C., Hernandez-Trejo, V., & Marquez-Linares, M. A. (2016). Economic valuation of ecosystem services in Mexico: Current status and trends. *Ecosystem Services*, *21*, 6-19.

Pomeroy, R.S. (2012). Managing overcapacity in small-scale fisheries in Southeast Asia. Marine Policy 36:520–527.

Pomeroy, R. S., Garces, L. R., Pido, M. D., Parks, J. E., & Silvestre, G. (2019). The role of scale within an Ecosystem Approach to fisheries management: Policy and practice in Southeast Asian seas. *Marine Policy*, *106*, 103531.

Pornpatimakorn, S. (2012). Approaches for the Sustainable Development and Management of Fisheries in the Southeast Asian Region. Secretariat, Southeast Asian Fisheries Development Center (SEAFDEC), Bangkok, Thailand, 12p.

Rumley, D., Doyle, T., & Chaturvedi, S. (2012). 'Securing' the Indian Ocean? Competing regional security constructions. *Journal of the Indian Ocean Region*, *8*(1), 1-20.

Salayo, N., Garces, L., Pido, M., Viswanathan, K., Pomeroy, R., Ahmed, M., ... & Masae, A. (2008). Managing excess capacity in small-scale fisheries: Perspectives from stakeholders in three Southeast Asian countries. *Marine Policy*, *32*(4), 692-700.

Schuhmann, P. W., & Mahon, R. (2015). The valuation of marine ecosystem goods and services in the Caribbean: A literature review and framework for future valuation efforts. *Ecosystem Services*, *11*, 56-66.

SEAFDEC (Southeast Asian Fisheries Development Center). (2017). The Southeast Asian State of Fisheries and Aquaculture 2017, Southeast Asian Fisheries Development Center Bangkok, Thailand. 160 p.

Sharma, J., & Varshney, S. (2019). Science Diplomacy and Cooperation in Science and Technology in India. *Science Diplomacy Review*, *1*(2), 11-22.

Song, W. L., He, G. S., & McIlgorm, A. (2013). From behind the Great Wall: The development of statistics on the marine economy in China. *Marine Policy*, *39*, 120-127.

Stobutzki, I. C., Silvestre, G. T., & Garces, L. R. (2006). Key issues in coastal fisheries in South and Southeast Asia, outcomes of a regional initiative. *Fisheries Research*, *78*(2-3), 109-118.

Sutton, A. M., & Rudd, M. A. (2015). The effect of leadership and other contextual conditions on the ecological and socio-economic success of small-scale fisheries in Southeast Asia. *Ocean & Coastal Management*, *114*, 102-115.

Teame, G.T. (2017). Efficiency and Capacity Utilization of India's Marine Fisheries. Management and Organizational Studies 4 (3):14-24.

Techera, E. J. (2018). Supporting blue economy agenda: fisheries, food security and climate change in the Indian Ocean. Journal of the Indian Ocean Region, 14(1), 7-27.

Teh, L. S., Teh, L. C., & Jolis, G. (2018). An economic approach to marine megafauna conservation in the coral triangle: Marine turtles in Sabah, Malaysia. *Marine Policy*, *89*, 1-10.

Teh, L. S., Teh, L. C., & Sumaila, U. R. (2011). Quantifying the overlooked socio-economic contribution of small-scale fisheries in Sabah, Malaysia. *Fisheries Research*, *110*(3), 450-458.

Tisdell, C., & Wilson, C. (2002). Ecotourism for the survival of sea turtles and other wildlife. *Biodiversity & Conservation*, *11*(9), 1521-1538.

Todd, P. A., Ong, X., & Chou, L. M. (2010). Impacts of pollution on marine life in Southeast Asia. *Biodiversity and conservation*, *19*(4), 1063-1082.

Troëng, S., & Drews, C. (2004). Money talks: economic aspects of marine turtle use and conservation. World Wildlife Fund (WWF) International, Gland, Switzerland (www.panda.org), 62p.

UN (United Nations). (1982). The Law of the Sea. Official text of the United Nations Convention on the Law of the Sea with Annexes and Tables. United Nations, New York, p 224.

UN (United Nations) (1995). Agreement for the implementation of the provisions of the United Nations Convention on the Law of the Sea of 10 December 1982 relating to the conservation and management of straddling fish stocks and highly migratory fish stocks. United Nations, New York.

UNCED (United Nations Conference on Environment and Development) (1992). Agenda 21, Rio Declaration, Forest Principles. United Nations, New York.

Veisten, K. (2007). Contingent valuation controversies: Philosophic debates about economic theory. *The Journal of Socio-Economics*, *36*(2), 204-232.

Wallis, J. J. (2011). Institutions, organizations, impersonality, and interests: The dynamics of institutions. *Journal of Economic Behavior & Organization*, *79*(1-2), 48-64.

Watkiss, P., Ventura, A. & Poulain, F. (2019). *Decision-making and economics of adaptation to climate change in the fisheries and aquaculture sector.* FAO Fisheries and Aquaculture Technical Paper No. 650. Rome, FAO.

WB (World Bank). (2013). Fish to 2030: Prospects for Fisheries and Aquaculture, Agriculture and Environmental Services Discussion Paper-03, World Bank Report Number 83177-GLB, World Back, Washington DC, USA, 80p.

WB and UNDESA (World Bank and United Nations Department of Economic and Social Affairs). 2017. The Potential of the Blue Economy: Increasing Long-term Benefits of the Sustainable Use of Marine Resources for Small Island Developing States and Coastal Least Developed Countries. World Bank, Washington, DC. 36p.

Werner, S. R., Spurgeon, J. P., Isaksen, G. H., Smith, J. P., Springer, N. K., Gettleson, D. A., & Dupont, J. M. (2014). Rapid prioritization of marine ecosystem services and ecosystem indicators. *Marine Policy*, *50*, 178-189.

Williams, M. J. (2013). Will new multilateral arrangements help Southeast Asian states solve illegal fishing? *Contemporary Southeast Asia*, 35(2), 258-283.

Williams, M. J., & Staples, D. (2010). Southeast Asian Fisheries. In: R. Quentin Grafton, Ray Hilborn, Dale Squires, Meryl Williams, Maree Tait (eds) Handbook of Marine Fisheries Conservation and Management (pp. 243-257). *Oxford University Press, New York*, 770p.

Wilson, C., & Tisdell, C. (2001). Sea turtles as a non-consumptive tourism resource especially in Australia. *Tourism Management*, *22*(3), 279–288.

Wilson, C., & Tisdell, C. (2003). Conservation and economic benefits of wildlife-based marine tourism: Sea turtles and whales as case studies. *Human Dimensions of Wildlife*, *8*(1), 49–58.

Wiyono, E.S. and Hufiadi. (2014). Measuring the technical efficiency of purse seine in tropical small-scale fisheries in Indonesia. *Asian Fisheries Science*, 27: 297–308.

Yang, C., & Lou, X. (2016). Technical efficiency study on Japanese marine fisheries applying stochastic frontier analysis and data envelopment analysis approaches. *International Journal of Fisheries and Aquatic Studies,* 4(4): 135-141.

Zen, L. W., Abdullah, N. M. R., & Yew, T. S. (2002). Technical efficiency of the driftnet and payang seine (Lampara) fisheries in West Sumatra, Indonesia. *Asian Fisheries Science*, *15*(2), 97-106.

Zhao, R., Hynes, S., & He, G. S. (2014). Defining and quantifying China's ocean economy. *Marine Policy*, *43*, 164-173.

Zongli, Z., Yanan, Z., Feifan, L., Hui, Y., Yongming, Y., & Xinhua, Y. (2017). Economic efficiency of small-scale tilapia farms in Guangxi, China. *Aquaculture Economics & Management*, *21*(2), 283-294.

Chapter 8
Desalination of Ocean Water: How Far Does It Contribute to the Blue Economy?

Somnath Hazra, Suvajit Banerjee, and Sourav Kumar Das

8.1 Introduction

Day by day, the frown is becoming more transparent, and the danger line is closely approaching the ultimate and unfortunate climax of severe water shortages. Even today, water is a scarce resource insufficiently available for meeting the enormous demand in all world regions. The situation is near critical in some parts of the globe as far as freshwater availability is concerned. It is quite pronounced for the dry regions with deficient annual average rainfall, which are predominantly the living spaces of the inhabitants. Those who are economically weak and vulnerable are in the face of a severe threat to the survival of a large number of population. Many countries have already addressed this issue and complemented the deficit of water by using the water sources, which are nonconventional and strategic adoption of this essential resource for the sustenance of the human race. Globally, the most acceptable methodology of water conversion is the adoption of desalination. Some economies have also committed to the reuse of wastewater effluents. For instance, in the Kingdom of Saudi Arabia (KSA), this reuse strategy is known to be a socially acceptable and economically viable method to draw benefit from this water for all purposes. However, the application of re-usable water is very limited to date and only for groundwater recharge and irrigation purposes (KAUST 2011). To resolve the water shortage problem for the coastal countries, ocean water desalination is a

S. Hazra (✉)
School of Oceanographic Studies, Jadavpur University, Kolkata, West Bengal, India

S. Banerjee
Department of Economics and Politics, Vidya Bhavana, Visva Bharati University, Bolpur, West Bengal, India

S. K. Das
Lalbaba College, Howrah, West Bengal, India

S. Hazra, A. Bhukta (eds.), *The Blue Economy*,
https://doi.org/10.1007/978-3-030-96519-8_8

well-recognized contrivance to eradicate the drinking water problem (Abderrahman 2000).

Another school of thought considers that the desalination process can also increase to meet the future water demand caused by population expansion. The statistics of demographics have estimated that the growth of population in the world will increase by 33% from 7 to 9.3 billion between 2011 and 2050, while the water of natural resources at large will remain constant. Consequently, by 2030 the global water demand is expected to grow from 4500 to 6900 billion cubic-m/year, portraying a 53% increase (Mauter and Fiske 2020). Today, water shortages are a global phenomenon due to the impacts of accelerated climate change, greenhouse gas emissions induced by economic development, and population growth (Oki and Kanae 2006).

Many scientifically conducted evidence-based studies predicted that the countries that are significantly exposed to the high risks of shortage of water and drought conditions are principally found around western North America, the Middle East, Mediterranean, eastern Australia, northern China, Chile, and western Asia. Nowadays, this problem can be solved effectively through seawater desalination, which can be implemented in countries, located near the oceans and seas (Sadeghi et al. 2020). There are lots of benefits of seawater desalination. The desalination of seawater is the natural hydro-cycle, which has been used in various coastal areas. Due to some technological up-gradation, the unit cost of desalinated water production has significantly declined. As a result, this is now available at a competitive price with the other conventional water resources (Gao et al. 2017).

This chapter tries to examine the economic assessment of the desalination of seawater. The preheating of feed water of RO systems is a valuable and best method for reduction in the desalination power consumption and, as a result, decreasing the freshwater total cost. The seawater permeability in the membranes is inversely proportional to the feed water viscosity, whereas, by temperature increase, the viscosity will decrease, and resultantly, the RO membrane will become permeable more (Humphries et al. 1993). According to Dsaldata Report, RO technology has achieved remarkable gains in desalinated water production from seawater (6.9 km^3/year) (Dsaldata 2015). Various previous research studies have estimated the development of seawater desalination in future, which has been based on various scales. In 2016, to identify potential areas of desalination, Hanasaki et al. proposed a model on seawater desalination. Kirshen (2007) estimated the requirement of desalinated water for commercial needs or growing domestic by 2025 in 10 water-scarce economies. Kim et al. (2016) have used an economic model to estimate the total desalinated water growth as the year 2100 at a basin scale.

Globally, there are 15,906 plants of operational desalination with a capacity of about 95.37 million m^3 per day (34.81 billion m^3 per year). The desalination plants were based on thermal technologies, mainly located in water-scarce oil-rich regions, particularly in the Middle East. From the post-1980, membrane technologies were introduced, specifically, RO, gradually dominating other desalination technologies. In the year 2000, the desalinated water produced by RO was approximately 11.4 m^3/day. Since 2000, the number and capacity of RO plants both are increasing exponentially; on the other hand, the thermal technologies have received a marginal increment. Now, the production of desalinated water from RO is supporting 69% of the total desalinated water produced globally (Jones et al. 2019).

It has been observed that the maximum desalination plants are situated in China, United States and Australia, Europe, North Africa, and Middle East. A relatively less number of plants have been seen in South America and Africa. Globally, the desalination plants are majorly concentrated around the coastline. Worldwide, the plants are supplying municipal water, but it is dominant in the Middle East and North Africa (Table 8.1).

It has been seen that Saudi Arabia is relatively successful in desalinating seawater to reduce the water demand-supply deficit. Previously, the countries are not accepting this technique to overcome the water-deficit issues because the production cost of water from these desalination plants is very high, but nowadays the production cost has declined due to technological upgradation like the discovery of the device

Table 8.1 Geographical region-wise number, capacity and global share of desalination plants

	Number of desalination plants	Desalination capacity (million m^3 per day)	Percentage
Global scenario	15,906	95.37	100
Geographic region			
The Middle East and North Africa	4826	45.32	47.5
East Asia and Pacific	3505	17.52	18.4
North America	2341	11.34	11.9
Western Europe	2337	8.75	9.2
Latin America and the Caribbean	1373	5.46	5.7
Southern Asia	655	2.94	3.1
Eastern Europe and Central Asia	566	2.26	2.4
Other	303	1.78	1.9
Income level			
High income	10,684	67.24	70.5
Upper middle income	3075	19.16	20.1
Lower middle income	2056	8.88	9.3
Low income	53	0.04	0.0
Sector uses			
Municipal uses	6055	59.39	62.3
Industry uses	7757	28.80	30.2
Power uses	1096	4.56	4.8
Irrigation uses	395	1.69	1.8
Military uses	412	0.59	0.6
Other uses	191	0.90	0.4

Source: Jones et al. (Jones et al. 2019)

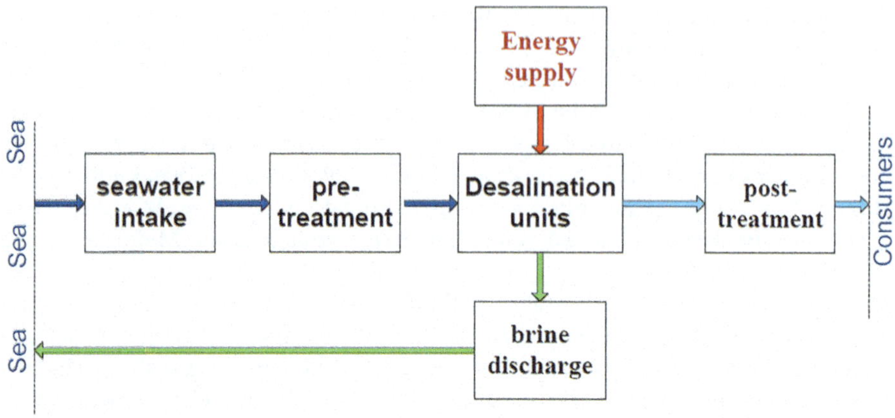

Fig. 8.1 Flowchart of seawater desalination plant

for energy recovery in reverse osmosis (RO) membrane process (Ettouney and Wilf 2009; Kamal 2008). Some other applications also can reduce the production cost of desalted water from the ocean water, and these are hybrid desalination, desalination plants, and the cogeneration principle for energy use (Buros 2000; Al-Mutaz and Al-Namlah 2004; Al-Karaghouli and Kazmerski 2013). In spite of these technological developments, the production costs of desalted water are still significantly more than conventional water. It has been seen that the cost of production of 1 m^3 cube of water by a plant is five times higher than groundwater (Ghaffour et al. 2013; Al-Zubari 2003; Al-Karaghouli and Kazmerski 2013) (Fig. 8.1).

The high capital and operational cost of desalination are the main economic barrier to implementing large-scale seawater desalination plants (Ziolkowska 2015). For desalination, water reuse and recycling have been considered and applied increasingly for providing extra usable water. Combining desalination technology and reuse of wastewater strategies can convert wastewater to high-quality water, which suits various users in the agriculture and industrial sector. Therefore, it is imperative to consider the water use efficiency improvement wherever and whenever there is water stress. However, to adopt any prudent water use strategy, the precondition of economic viability must ensure that the marginal costs of adapting to the new strategy should not higher the marginal costs of the water supply through desalination (Zhou and Tol 2005).

It has also been observed that the too much costs of desalted water production are restricting the extensive adoption of desalination despite the plentiful availability of seawater resources. Against this backdrop, one must know the relationship between the cost and factors contributing to these costs. Much research has been conducted to understand the relationship between these factors and the ultimate cost of desalted water. Most of the researchers have developed several models to calculate the cost of desalted water to reduce the problem of water scarcity and make it affordable for the population. It is seen that most of the cost estimation models have been based on the construction cost (capital cost) only. Our review of previous researches deduced that in the earlier attempts, the authors did not undertake a comprehensive evaluation process where the operational costs also have a significant share of the total cost of production.

This study proposes a methodology to evaluate seawater desalination and assess its economic feasibility for Asian countries. In this chapter, we have developed a detailed methodology to compare the price of water with the production cost of desalination to know the attainability of seawater desalination. This assessment will also help to identify the potential countries where economically, seawater desalination is possible. Here, we have taken help from two statistical models to estimate the production cost and water price.

8.2 Materials and Methods

These established models are purely based on empirical study, and most of the models relied on the collected country-specific data. Different countries have different socioeconomic and other situations, so the same models may not be applicable for all the nations. For example, the labor cost, cost of energy consumption, and the salinity of seawater may not be the same for all the countries, so empirical models may not be compatible between all the nations.

Until now, no such comprehensive economic study has been conducted to know the cost-effective implementation of the desalination plant and ocean water desalination plant drivers. Consequently, decision-makers and planners are not clear about the performance of the desalination industry to reduce water scarcity. Therefore, in this chapter, we have attempted to develop a methodology to understand the cost of water production in desalination plants and the drivers of the cost component.

It is necessary to understand the significant factors of the overall production water cost of the desalination plants. These factors are used to develop the cost model of desalted water produced from these desalination plants. Perfect information on the future cost of the plant, always a chance of cost reduction through improvement in the management performance (Kaplan and Cooper 1998). As an example, we can say that valuable information on the cost of the desalination plant and the management can avoid the delay in operational and maintenance activities.

8.2.1 Process of Data Collection

To analyze the present desalination capacity, desalinization data of different desalinization plants can be used primarily. These desalinization data give us information on capital cost, plant capacity, and location. We can use socioeconomic data from the government database or any other international database like World Bank. We will collect the gross domestic product (GDP) and population data also, and we will use it for historical simulation. The data on the price of electricity can be collected from the ministry of energy resource or the International Energy Association. The Shared Socioeconomic Pathway (SSP) database will get from IIASA.

8.2.2 Assessment of Economic Feasibility

The ions are separated from the water during desalination, which can be used as an essential treatment step for reusing many water sources. Furthermore, since desalination is associated with efficient freshwater production from the sea, the desalination process is also necessary for municipal wastewater recycling. Again, since the municipalities are the leading receivers of desalinated water, then the municipal water price can be used to indicate the economical of water for consumers.

The feasibility index (Fi) can be used as an indicator of desalinization plant implementation potential. The feasibility index can be measured as follows:

$$\mathrm{Fi} = \mathrm{Wp} / \mathrm{Cp} \tag{8.1}$$

where Wp implies water price (price per m^3) in a given region, and Cp indicates the desalinization plant's unit production cost (price per m^3) in a given time. Therefore, if the value of Fi is greater than 1, this implies a high potential for implementing a desalinization plant in the said region.

Thus, to know the future potential areas of desalination, the value of Fi should be greater than equal to 1. To calculate the value of Fi, the value of Cp and Wp has to be developed separately.

The water production cost from desalination plants also includes capital and operation costs (Fryer 2010; Frioui and Oumeddour 2008; Kim et al. 2009; Khayet 2013; Huehmer et al. 2011).

8.2.3 Estimation of Capital Cost

The cost of water production from a desalination plant is based on the two major components: cost of capital costs and the cost of operation and management (O&M). The categories of capital cost and operation cost are as follows (Fig. 8.2 and 8.3):

Capital cost (Ca) can be calculated as follows:

$$\mathrm{Ca} = \mathrm{IISY}\,\mathrm{x}\left\{ r \times (1+r)n \right\} / \left\{ (1+r)n - 1 \right\} \tag{8.2}$$

where Ca implies the annual amortized capital cost, the IISY implies the initial investment in the starting year, and r and n are the annual discount rates and the

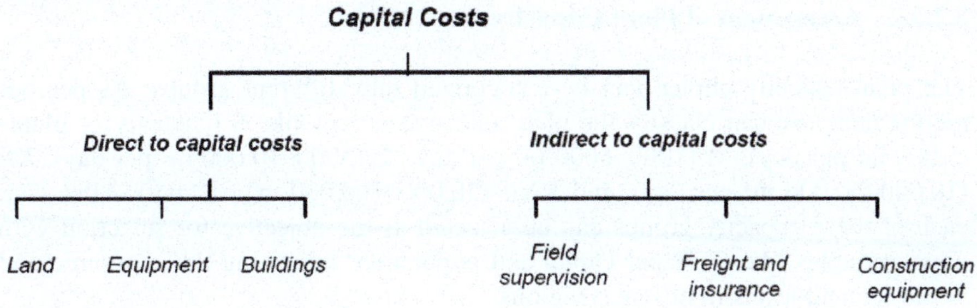

Fig. 8.2 Categories of capital costs

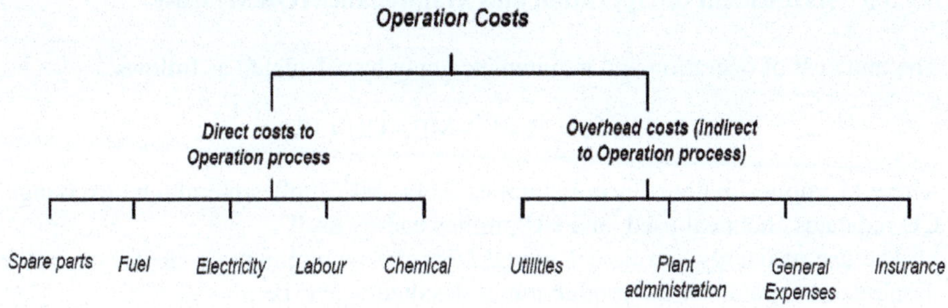

Fig. 8.3 Categories of operation costs

desalination plant life. Therefore, the yearly output of the plant can be used to shift the annual amortized capital cost and the annual operation and maintenance (O&M) cost (Ca/O&M) to a unit annual amortized capital cost (UAACC) and a unit annual O&M cost (UAOMC), respectively. So, the unit production cost (Cp, price per m^3) for producing one m^3 of desalinized water has been calculated as follows:

$$\mathrm{Cp} = \mathrm{UAACC} + \mathrm{UAOMC} = \mathrm{Ca} / \mathrm{Capacity} + \left(\mathrm{Ca} / \mathrm{O}\,\&\,\mathrm{M}\right) / \mathrm{Capacity} \quad (8.3)$$

8.2.3.1 Estimation of Drivers of Capital Cost

The desalination plant's capital cost is correlated with plant capacity (CAP), total installed capacity (TIC), GDP. per capita, distance from seashore, and one dummy variable may be oil exporting. The functional form of this can be as follows:

$$\text{Capital cost} = \mathrm{f}\left(\text{CAP,TIC,GDP per capita,Distance,Oil} - \text{exporting}\right) \quad (8.4)$$

The above function can be estimated by an ordinary least-squares (OLS) method.

8.2.3.2 Assessment of Plant Capacity

The plant capacity choice can be categorized into different groups: As per our requirement, we can classify the plant into five groups like S (capacity of plants <1000 m^3 per day), M (1000–5000 m^3 per day), L (5000–10,000 m^3 per day), XL (10,000–50,000 m^3 per day), and XXL (50,000–100,000 m^3 per day). After that, each of these capacity groups can be selected as the objective for selection with other variables like GDP per capita and population, indicating the economic and water demand levels in different regions.

8.2.3.3 Assessment of Operation and Maintenance (O&M) Cost

The unit cost of operation and maintenance may be calculated as follows:

$$\mathrm{O\&M} = \mathrm{M} + \mathrm{L} + \mathrm{ME} + \mathrm{CU} + \mathrm{EU} \quad (8.5)$$

where M implies maintenance, L implies labor, ME implies membrane exchange, CU indicates chemical used, and EU implies energy used.

The use of energy can be calculated from the requirement of electricity in the desalinization plant for the production of desalted water; i.e.,

$$\mathrm{EU} = \mathrm{PE} \times \mathrm{EC} \quad (8.6)$$

where EU is the unit energy cost in price/m^3, PE implies the electricity price in the region (price/kWh), and EC implies electricity consumption for one unit seawater desalination (kWh/m^3).

8.2.4 Assessment of Water Price (WP)

The water price can be a function of four independent variables namely GDP per capita, energy price (electricity price, PE), population density (PD), and water withdrawal per capita (Wpc).

This can be written as follows:

$$\mathrm{Wp} = \mathrm{f}\left(\mathrm{GDP,PE,PD,Wpc}\right)$$

The equation can be written as follows:

$$\mathrm{Wp} = b0 + b1\mathrm{Xi1} + ei \quad (8.7)$$

where Wp is the water price of a region, X implies the independent variables (i.e., GDP per capita, PE, PD, and Wpc), *bis* are the coefficient, and e is the error terms.

8.2.5 Methodology of Future Simulations with Developed PC and WP Models

The estimation of PC and WP can also be able to estimate the production cost of a past unit of desalinization plants and the water price, respectively. After assessment of these, two can be used for future projections of unit production cost and water price across different regions.

Based upon the linear regression analysis, we will use the functional equations across the different capacity groups. A separate equation can be developed for a specific category. For each country, the plant capacity will be selected based on the results of the decision trees.

For both periods (past and future), the operation and maintenance costs can be estimated using constant data on labor, membrane exchange, chemical costs, and the equation of electricity consumption.

Based on the above methodology, we can estimate the water price to know the socioeconomic condition of the region, the shared socioeconomic pathways (SSPs) can develop the socioeconomic scenarios for the global climate policy development, use in international climate policy studies. According to the methodology, it has been observed that the cost of energy consumption is the essential component for the estimation of the unit production cost. Since the different policies are based on various energy prices, future climate policies will be affected.

8.3 Influence of Different Factors

As we know, 96.5% is the earth's oceans and seas and 1.7% are under the ice, glaciers, etc., and 1.03% is the saline groundwater, soil moisture, saline lakes, etc. Only 0.77% is freshwater, which is usable for human consumption (Gleick 1996). Desalination means the treatment of saline water by removing salt. Desalination is used to reduce the shortage of freshwater through the water in the oceans and seas. A lot of research has been done on the desalination process in many countries, but the percentage of success is very low (A-Sofi 2001). During the Second World War (in 1940), a significant development was seen in the desalination process, when some countries agreed to supply their groups in the arid regions. In the 1960s, the USA has created an office on saline water (OSW) to support research and development activities to improve desalination technologies (Buros 2000). Recently, most of the countries have been very much dependent on desalted water through desalination plants. It has been observed that globally the capacity of desalted water has increased from 5.09 × 106 m^3 per day in 1980 to 74.83 × 106 m^3 per day in 2012 (Pankratz 2013). Globally, various desalination methods exist and produced water can be used for drinking and other domestic purposes. Desalination techniques can be applied to generate freshwater through seawater and brackish water.

It has been observed that in most of the desalination processes, the costs of production are inversely related to the capacity of production (Avlonitis 2002). In some cases, irrespective of the plant size, some expenditures are the same to run the desalination plant like administration costs and partly labor costs. Consequently, to meet the water demand, if the plant capacity has increased, it will dramatically reduce the cost of production (Pankratz 2013). Quality and salinity are essential factors in the selection of desalination technology. If we consider the operation cost, the satisfactory freshwater generation is directly proportional to the salinity; consequently, the energy consumption is also increasing, resulting in greater energy consumption higher for seawater desalination (TDS > 10,000 ppm) for brackish water desalination (TDS < 10,000 ppm) (Greenlee et al. 2009; Farooque et al. 2008; Buros 2000).

Energy consumption is highly dependent on the techniques used in the desalination process and other required activities like pre-treatment processes. It has been seen that if MSF desalination is used then 15–18 (kWh) energy can be consumed for generation one metric cube of freshwater from Seawater, and if MED is applied, then 5.7–15 kWh energy can be consumed (Al-Sahali and Ettouney 2007; Ettouney and Wilf 2009). Plant locations, type of techniques, and human resources are other factors on which production costs depend. Based on the previous desalination cost, the future fee was assessed. The following table shows the future price of desalination plants:

Advancement in desalination technology may not cut down the cost of seawater desalination in the upcoming years. But due to the regulatory process with this reduction in production costs, people start believing the ocean as a source of water that can reduce water scarcity. Moreover, globally, the coastal communities believe that ocean water desalination is a drought-proof alternative for them.

It has been observed that technological progress can cut the cost of production of desalinated water by 20% in the upcoming 5 years and up to 60% in the upcoming 20 years (Table 8.2). It is also observed that the produced freshwater from the desalination plant will be emerging as a feasible and cost-effective potable water production in future. The high cost of seawater desalination is the cost of energy associated with the process. The below figure shows the seawater desalination costs for different segments.

Table 8.2 Desalination costs forecasting for medium and large size projects

Parameter for best in class desalination plants	2016	Coming 5 years	Coming 20 years
Cost of water (US$ per m^3)	0.8–1.2	0.6–1.0	0.3–0.5
Construction cost (US$ per MLD)	1.2–2.2	1.0–1.8	0.5–0.9
Electrical energy use (kWh per m^3)	3.5–4.0	2.8–3.2	2.1–2.4
Membrane productivity (m^3 per membrane)	28–47	35–55	95–120

Source: https://iwa-network.org/desalination-past-present-future/

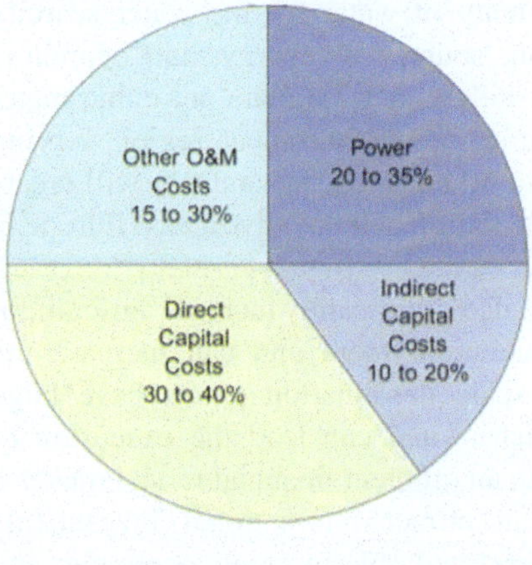

Source: https://iwa-network.org/desalination-past-present-future/

8.4 Challenges and Opportunities

Desalination has become more critical for consumable water as the issue of climate change eventually gravitated and created new patterns of water distribution across the different regions of the globe. Although this has been effective, the process of desalination may be complex, energy-intensive, and expensive. The mission of sustainable mega-scale desalination (fulfilling the demand of several millions of population) is a process of mass production of edible water:

1. Using less energy as possible, and ambitiously with 100% renewable energies;
2. With minimal waste production; and.
3. With the least positive outfalls of water made available.

Recognition of various devices of energy recovery and better performance of membranes and desalination of seawater through reverse osmosis may require 2–6 kWh per cubic meter of water produced (Buonomenna and Bae 2015). All this energy has come from sources of renewable energy. So, for continuous energy supply, energy storage in batteries or any other capacities are required. In the reverse osmosis process, the production of 1 m^3 of freshwater will produce near about 1 m^3 of residual as brine with a reasonably high mineral concentration. Nowadays, retrieval process of minerals is also a very hard task.

Desalinated water can use in cities for domestic uses, but the quality of the water cannot be suitable for all uses. Especially, the mineral content of desalinated water is not suitable for drinking and needs some more treatment to make it useful for drinking. Due to high boron content, untreated urban desalinated wastewater is also not suitable for irrigation purposes. Even the implementation of renewable energy for desalination activity has also some significant technical, financial, and organizational challenges. If all of these can be able to address properly, then the desalination process may allow accessing plenty of water for the water scarcity region. Depending on the plant size, region, sources of energy, and economic scenario, the desalinated water will cost 1.3 dollars to 1.7 dollars per cubic meter (World Bank 2019). It has been observed that in the Mediterranean region, serving 100 million people will cost 15 billion euro to 30 billion euros and this will create huge employment that may inspire the implementation of innovative ideas (European Commission 2019).

Furthermore, treated desalinated water may also significantly increase agricultural activity, which can also generate additional employment and can increase the contribution of agriculture to GDP. Large-scale desalination plants have huge potential to develop the economy of the district and can lead the execution of innovative ideas. A proper execution plan with an efficient financial model makes a sustainable development project, which can attract international investments (including international aid and other transfers) and able to create economic and employment opportunities in the region, which leads to reducing rural–urban migration. Most of the desalination plants employ RO technology. The suppliers of desalination also play a major role in exporting their expertise from one region to another. Most of the large seawater desalination plants worldwide have received design–build or design–build–operate contracts based on long-term water costs and incorporated innovations to the benefit of everybody involved. As new technologies are efficient and reduce the cost of production, traditional business models will hopefully shift to adopt new technologies earlier.

8.5 Conclusions

Freshwater availability is expected to be impacted by the worldwide event of a change in climate, and many European regions are expected to face severe scarcity of water within 2050. Estimation of Water Exploitation Index for 2050 indicated that the coastal Mediterranean regions and also regions in Hungary, France, Northern Italy, Germany, Bulgaria, and Romania may face water scarcity critically. This problem can be solved through desalination in European regions. The desalination process is an energy-intensive technology, and it currently provides 4.2% of the EU public supply water, it accounts for 16% of the energy used by the EU water system. According to the International Energy Agency estimation, it has been observed that the demand for freshwater production through desalination is increasing every day. As a consequence, it is estimated that worldwide, by the year 2040 the consumption of energy for desalination will increase eightfold. It has also been seen that maximum demand for the new desalination plants is coming from the Middle East and Northern African regions. If the demand for desalination plants increases at the same rate in other regions of the world, then we should estimate the desalination capacity based on the water requirement per person per day (150 L/person/day). It is also necessary to estimate the cost of new investments can able to meet the new desalination capacity or not. However, increased desalination capacity will increase energy requirements, carbon emission, and environmental impacts. Hence, before large-scale implementation of desalination, a cost–benefit analysis is also required. To reduce the incremental environmental impact and energy requirement for the implementation of desalination technologies, we should employ some other technologies, especially reuse of water and zero leakage. To reduce the carbon emission for the desalination process and technology, green technology (renewable energy-driven) can be adopted for the desalination plants installation. Based on the estimation of the future growth of the desalination market, we should identify a viable solution that can be able to tackle the increasing water demand with minimum negative impacts and be able to transform this process into policy.

References

Abderrahman, W. A. (2000). Water Demand Management in Saudi Arabia. The International Development Research Centre (IDRC), 16(4), 465–473. https://doi.org/10.1080/713672529

Al-Karaghouli, A., & Kazmerski, L. L. (2013). Energy consumption and water production cost of conventional and renewable-energy-powered desalination processes. *Renewable and Sustainable Energy Reviews*, *24*, 343–356. https://doi.org/10.1016/j.rser.2012.12.064

Al-Mutaz, I. S., & Al-Namlah, A. M. (2004). Characteristics of dual purpose M.S.F. desalination plants. *Desalination*, *166*, 287–294. https://doi.org/10.1016/j.desal.2004.06.083

Al-Sahali, M., & Ettouney, H. (2007). Developments in thermal desalination processes: Design, energy, and costing aspects. *Desalination*, *214*(1-3), 227–240. https://doi.org/10.1016/j.desal.2006.08.020

Al-Zubari, W. K. (2003). Exploiting Natural Resources. In *Exploiting Natural Resources*.

A-Sofi, M. A. K. (2001). Seawater desalination SWCC experience and vision, *135*(November 2000), 121–139.

Avlonitis, S. A (2002). Operational water cost and productivity improvements for small-size R.O. desalination plants. *Desalination*, *142*(3), 295–304. https://doi.org/10.1016/S0011-9164(02)00210-2

Buonomenna, M.G.; Bae, J. (2015) Membrane processes and renewable energies. Renew. Sustain. Energy Rev., 43, 1343–1398.

Buros, O. (2000). *The A.B.C.s of desalting* (Second Edi.). Topsfield, Massachusetts, U.S.A.: International Desalination Association. Retrieved from http://www.miliarium.es/Desalinizador/The A.B.C.s of Desalting.pdf

DesalData (2015). Available online: http://www.desaldata.com/.

Ettouney, H., & Wilf, M. (2009). Commercial Desalination Technologies, An Overview of the Current Status of Applications of Commercial Seawater Desalination Processes. In G. Micale, L. Rizzuti, & A. Cipollina (Eds.), *Seawater Desalination* (pp. 77–107). Berlin, Heidelberg: Springer Berlin Heidelberg. https://doi.org/10.1007/978-3-642-01150-4

European Commission (2019). The EU Blue Economy Report. 2019. Publications Office of the European Union. Luxembourg.

Farooque, A. M., Jamaluddin, A. T. M., Al-Reweli, A. R., Jalaluddin, P. A. M., Al-Marwani, S. M., Al-Mobayed, A. A., & Qasim, A. H. (2008). Parametric analyses of energy consumption and losses in SWCC SWRO plants utilizing energy recovery devices. *Desalination*, *219*(1-3), 137–159. https://doi.org/10.1016/j.desal.2007.06.004

Frioui, S., & Oumeddour, R. (2008). Investment and production costs of desalination plants by semi-empirical method. *Desalination*, *223*(1-3), 457–463. https://doi.org/10.1016/j.desal.2007.01.180

Fryer, J. (2010). An Investigation of the Marginal Cost of Seawater Desalination in California.

Ghaffour, N., Missimer, T. M., & Amy, G. L. (2013). Technical review and evaluation of the economics of water desalination: Current and future challenges for better water supply sustainability. *Desalination*, *309*, 197–207. https://doi.org/10.1016/j.desal.2012.10.015

Gleick, P. H. (1996). Basic Water Requirements for Human Activities : Meeting Basic Needs Basic Water Requirements for Human Activities : Meeting Basic Needs, *21*(2), 83–92.

Greenlee, L. F., Lawler, D. F., Freeman, B. D., Marrot, B., & Moulin, P. (2009). Reverse osmosis desalination: water sources, technology, and today's challenges. *Water Research*, *43*(9), 2317–48. https://doi.org/10.1016/j.watres.2009.03.010

Huehmer, R., Gomez, J., Curl, J., Moore, K., & Huehmer, P. R. (2011). Cost modeling of desalination systems. In *World Congress*. Perth- Western Australia: International Desalination Association I.D.A.

Humphries, J.R., et al., "The CANDESAL nuclear desalination system", paper presented at the IAEA Technical Committee Meeting and Workshop on Coupling Aspects of Nuclear Reactors with Seawater Desalination processes, Vienna, 1993.

Jones, E., Qadir, M., van Vliet, M. T., Smakhtin, V., & Kang, S. M. (2019). The state of desalination and brine production: A global outlook. Science of the Total Environment, 657, 1343-1356.

Kamal, I. (2008). Myth and reality of the hybrid desalination process. *Desalination*, *230*(1-3), 269–280. https://doi.org/10.1016/j.desal.2007.11.030

Kaplan, R., & Cooper, R. (1998). Cost & Effect: Using Integrated Cost Systems to Drive Profitability and Performance (1st ed.). Harvard Business Review Press.

KAUST. (2011). The KICP Annual Strategic Study, Promoting Wastewater Reclamation and Reuse in the Kingdom of Saudi Arabia : Technology Trends, Innovation Needs, and Business Opportunities. Rabigh Saudi Arabia.

KH. Sadeghi, S. H. Ghazaie, E. D. Fedorovich, E. A. Sokolova, and A. S. Shirani, Economic Assessment of the Possible Desalination Processes for the First Unit of Bushehr Nuclear Power Plant, *Thermal Engineering, 2020, 67(5), 271–281.*

Khayet, M. (2013). Solar desalination by membrane distillation: Dispersion in energy consumption analysis and water production costs (a review). *Desalination*, *308*, 89–101. https://doi.org/10.1016/j.desal.2012.07.010

Kim, S.H.; Hejazi, M.; Liu, L.; Calvin, K.; Clarke, L.; Edmonds, J.; Kyle, P.; Patel, P.; Wise, M.; Davies, E. Balancing global water availability and use at basin scale in an integrated assessment model. Clim. Chang. 2016, 136, 217–231.

Kim, Y. M., Lee, Y. S., Lee, Y. G., Kim, S. J., Yang, D. R., Kim, I. S., & Kim, J. H. (2009). Development of a package model for process simulation and cost estimation of seawater reverse osmosis desalination plant. *Desalination*, *247*(1-3), 326–335. https://doi.org/10.1016/j.desal.2008.12.035

Kirshen, P. Adaptation Options and Cost in Water Supply; Tufts University: Medford, MA, U.S.A., 2007.

Lu Gao, Sayaka Yoshikawa, Yoshihiko Iseri, Shinichiro FujimoriD and Shinjiro Kanae, An Economic Assessment of the Global Potential for Seawater Desalination to 2050, Water 2017, https://doi.org/10.3390/w9100763

M. S. Mauter and P. Fiske, Energy & Environmental Science. 2020, https://doi.org/10.1039/D0EE01653E.

Oki, T.; Kanae, S. Global Hydrological Cycles and World Water Resources. Science 2006, 313, 1068–1072.

Pankratz, T. (2013). *IDA desalination Year book "2012-2013", Water Desalination Report.* (R. Owen, Ed.). Oxford, U.K.: Media Analytics Ltd, The Jam Factory.

World Bank. (2019). The Role of Desalination in an Increasingly Water-Scarce World. World Bank.

Zhou, Y.; Tol, R.S.J. Evaluating the costs of desalination and water transport. Water Resour. Res. 2005, 41.

Ziolkowska, J.R. Is Desalination Affordable?—Regional Cost and Price Analysis. Water Resour. Manag. 2015, 29, 1385–1397.

Part III
Regulating Services

Chapter 9
Ocean as a Repository for Waste: An Economic Assessment of Chennai City

Sukanya Das

9.1 Introduction

Coastal zones are unique ecosystems different from the oceanic or terrestrial and are essential for socio-economic development. Coastal ecosystems support life on our planet and affect human societies' present and future well-being (Halkos and Matsiori 2018). India has 13 states and territories along its coastline of India which generates about 33,215 million liters per day (MLD) of sewage while the treatment capacity is much lower. The huge gap of 20,542 MLD of sewage treatment capacity aggravated marine pollution. The Central Pollution Control Board (CPCB 2015) identified over 302 polluted river stretches throughout the country.[1] Given the severity of its impacts have influenced researchers to derive economic value and to suggest for stringent and developing management strategies. The literature in the context of degradation of coastal ecosystem and water quality due to wastewater is limited. Over the years several valuation studies have been conducted for estimating willingness to pay for wastewater treatment. Stated preference techniques have been widely used in the valuation of water quality due to the inflow of sewage. Bouzit et al. (2018) conducted a systematic and comprehensive review of available empirical studies that assessed individual willingness-to-pay (WTP) estimates of recycled wastewater. About 84 WTP estimates from 22 international studies covering 12 countries have been compiled. From meta-regression model, the mean WTP for recycled wastewater was estimated of an amount US$ 52.62 per household/year. It was observed that WTP can vary in a systematic and predictable way in regard to

S. Das (✉)
TERI School of Advanced Studies, New Delhi, Delhi, India
e-mail: sukanya.das@terisas.ac.in

[1] https://geographyandyou.com/coastal-pollution/ accessed on 10th Jan 2021

S. Hazra, A. Bhukta (eds.), *The Blue Economy*,
https://doi.org/10.1007/978-3-030-96519-8_9

the key determinants influencing it, such as socioeconomic and contextual characteristics, individuals' attitudes, and perceptions concerning recycled water.

Breaux et al. (1995) evaluated the impact of effluent discharge on the coastal wetland in the context of Louisiana. The estimated value for wetlands per acre for wastewater treatment ranges from $US 785 per acre to $34,700 per acre in savings over conventional wastewater treatment technologies. In a similar line, another study in Sri Lanka by Wattage and Mardle (2008) on Muthurajawela Marsh estimates the economic benefits of around $eight million per year or $2631/hectare/year. Another CVM study by Kontogianni et al. (2001) on Thermaikos Bay estimated WTP to make a partially operational wastewater treatment plant fully operational. About 480 respondents were interviewed face-to-face. The mean willingness to pay was 5189 Greek Drachmas (1999) per four-monthly water bill for 5 years.

MacDonald et al. (2015) used a choice experiment (CE) to estimate society's willingness to pay (WTP) for improved coastal water quality in Adelaide, Australia. The estimated amount is $12.4 million (Australian dollars) for a 25-day improvement in water clarity, $18.9 million for a 10% increase in seagrass, and $35.8 million for restoring five other reefs to good health. Another CE study estimated the benefits from improved water quality and marine life in Göcek Bay, Turkey. Results showed that residents were willing to pay 18 TL/month for improvements in water quality and 14.8 TL/month for improvements in marine life, while tourists were willing to pay 16.6 TL/tour and 11.2 TL/tour for water quality (Can and Alp 2012).

Perni et al. (2011) estimated the value of improved water quality for a coastal lagoon in Spain. The mean annual willingness to pay was 20.11 Euro for the moderate scenario and 35.34 for the perfect scenario. The nonuse values for moderate and reasonable scenarios were 15.81 and 24.27, respectively.

Indab et al. (2003), in the context of the wastewater intrusion into the marine water, estimated cost estimates and concluded that the use of effluent charge scheme as an efficient management tool for protecting and maintaining good water quality in Sarangani Bay. Another study in Galvenson Bay by Whittington and Swarna (1994) estimated the economic value of improving the environmental quality. A CVM survey was conducted using a split sample of over 1700 respondents and had an average response rate of about 45 percent. The aggregate annual economic value to residents of the area varies from $100 to $125 million.

Given the limited number of studies in developing countries like India, the current study tries to explore citizens WTP for improved wastewater treatment through Contingent Valuation Method (CVM) in coastal city Chennai, where the sea water quality of Bay of Bengal is degraded by wastewater intrusion from Adyar River.

9.2 Background

Chennai is one of the largest metropolises in India and the coastal city which suffers from uncontrolled disposal of wastewater and profound pollution level due to industrialization and rapid urbanization (Giridhar 2001; Shanmugam and Babu 2017).

The massive scale of anthropogenic activities has a severe impact on the marine ecosystem. The Adyar River, which originates from the Chembarambakkam Lake in Kanchipuram district, joins the Bay of Bengal at the Adyar estuary. The river, which stretches across 42.5 km, has a significant contribution to the estuarine ecosystem of Chennai.

Over the years, the Tamil Nadu government has initiated several restoration programs, though the pace has been relatively low. In 2006, the Tamil Nadu government, to address the concern, set up the Adyar Poonga Trust to protect and restore the three rivers, the Buckingham Canal, and other water bodies in the city. The trust was renamed, as CRRT[2] coordinates the work between various government departments, such as the Public Works Department (PWD), the Greater Chennai Corporation (GCC), the Chennai Metropolitan Water Supply and Sewerage Board (Chennai Metro Water), Tamil Nadu Slum Clearance Board (TNSCB), NGOs The Adyar River which stretches across 42 km starts at Adhanur flows through Thiruneermalai, Tambaram, Manapakkam, Alandur, Saidapet. Finally, it terminates at the Bay of Bengal between Santhome beach in the north and Elliot's within Chennai.

The poor condition of the Adyar River aggravated the challenges for the Chennai River Restoration Trust, a nodal agency responsible for its restoration. The total project cost of Adyar River Restoration was about Rs. 2047.02 crore, as per the record of the Municipal Administration and Water Supply department. For medium-term development from 2019 to 2026, the budget is about Rs. 77.40 crore. For the long-term phase (2027–2031) involving riverfront development, river channel improvement, and biodiversity management, the amount sanctioned was about 32.64 crore (Chennai Rivers Restoration Trust (CRRT)).[3]

9.3 Survey Design and Administration

The CVM survey was implemented in February and March 2016 with face-to-face interviews with 150 randomly selected residents covering 15 zones across Chennai. Other than that, an online survey method, LimeSurvey, was implemented with over 50 respondents covering residents in the same zones. The survey was administered to represent the sample population regarding income, social status, proximity to the Adyar River and Buckingham Canal.

Following an extensive review of the literature on wastewater treatment in general, we conducted focus group discussions with the households and consultations with several experts comprising managers and employees of the STPs. They are civil and chemical engineers and hydrologists employed by the Chennai Metro

[2] https://science.thewire.in/environment/chennai-adyar-river-recovery-pollution-investment/ accessed on 25th Jan 20201

[3] https://chennai.citizenmatters.in/chennai-adyar-river-restoration-progress-and-challenges-11786 accessed on 25th Jan 2021

Water Supply and Sewerage Board (CMWSSB) and Chennai Municipal Corporation. Through the focus group discussions, increased water and sewerage tax was chosen as the payment vehicle. The focus group discussants felt this payment vehicle could ensure that everyone contributes, though they strongly felt it was the public authorities' role to improve the existing water and wastewater treatment facilities in Chennai.

Respondents were informed that additional funds would be needed to improve river water quality and the canal in the future. In order for the project to be completed and maintained, they were introduced to a hypothetical situation where the households are willing to contribute to these expenditures. During the pretest of the questionnaire, we observed that the scenario was well understood by the respondents (as per Arrow et al. 1993). Respondents were reminded that there were no right or wrong answers and that we were only interested in their opinions. They were also told that the municipality did not have sufficient funds to improve the wastewater treatment facilities of the Sewage Treatment Plant (STP), and therefore it may be necessary to increase the monthly water and sewerage taxes paid by the households. The respondents were also reminded of their budget constraints and other local public goods that could be funded through their taxes. The payment vehicle as monthly water and sewerage tax was preferred as the most appropriate as it is plausible and familiar to the population investigated. It is similar to the existing supply system and is also harmonious with the local institutional structure of the particular community (Jakobsson and Dragun 1996).

9.3.1 CV Scenario

The households were presented an overview of the existing system of wastewater treatment located in and around Chennai. All the Municipal Wastewater Treatment Plants have more or less similar technology though there is a variation in performance and their geographical locations. Most of the treatment plants have a conventional sewerage system and involve appreciable running costs. The final discharge of the treated water is disposed chiefly to the Adyar River and Buckingham Canal. The wastewater going in these water bodies is partly and insufficiently treated. The respondents were further motivated to accrue the benefits from the upgradation of the existing STPs (quantitative aspects) and new technologies (quantitative aspects). The benefits of upgrading wastewater treatment involve eliminating or reducing foul odor, insects and rodents, overflowing cesspits, and groundwater contamination. For the CV scenario of our study, respondents were presented with the current situation and the progress expected after restoration.

The households were asked to state the maximum increase in their water and sewerage bill that they could afford to pay. They were told that the money would be used to modernize wastewater collection and treatment facilities by the CMWSSB and that the current environmental problems would thus be eliminated.

Four distinct wastewater programs are characterized in terms of the quantity and quality of water. The four wastewater treatment programs and the present situation, i.e., status quo, were defined as follows:

Scenario D (mostly present situation): Not suitable for drinking, swimming, aquaculture, irrigation.
Scenario C: Not suitable for drinking, swimming, aquaculture, irrigation for eatable crops.
Scenario B: Suitable for swimming, aquaculture, irrigation. Not suitable for drinking.
Scenario A: Suitable for drinking, swimming, aquaculture, irrigation.

In addition to the CV scenario questions, data on the household's social, economic, and demographic characteristics were collected. Information was also collected regarding their dwelling characteristics, drainage and sewerage services in their households, and perception of wastewater.

The logistic regression model was estimated with STATA 11.0. In addition, an Ordered Logit Model and a simple Tobit model (see appendix) were also fitted, including a proxy variable of the awareness level of the existence of the Sewage Treatment Plant.

9.4 Empirical Results

Descriptive statistics of the sample are reported in Table 9.1. The survey reveals that, on average, the households have been residents for over 28 years in the locality where the number of members in the family are 4. The average age of the household age is 44 years. About 15% of them have completed primary education, while 34.5% have completed secondary education. 11.7% are in service, 22.4% are self-employed, whereas 46.3% are manual workers. In addition, 74% of the households have an average monthly income from Rs. 5000 to 10,000, whereas 49% falls between Rs. 10,000 and 15,000 with an average monthly income of Rs. 14087.5 for the entire sample.

Almost half of the respondents own the houses they live in, 46.5% rent a house, and 46.5% have individual homes out of all respondents. The average number of years for a respondent residing in his/her home is 13 years. About 42.5% of respondents have a piped sewer system, 36% use a septic tank, and 12.5% have no toilet facilities. Knowledge about the existence of Sewage Treatment Plant is another factor affecting awareness level. About 78.5% of the respondents did not know of its existence.

People use various sources of water. Under potable water sources, most of the respondents use piped and bottled water; a substantial 23.5% still rely on tube wells, and a mere 5.5% are dependent on tanker facilities. Under nonpotable water sources, 55.5% use piped, 46% rely on tube wells, 15.5% on public taps, 9.5% are dependent on tanker facilities, and none of the respondents use bottled water.

Table 9.1 Social, economic, and demographic characteristics of the sampled households

Characteristics	Mean (SE)
Age (in years) of the respondent	44.7 (10.9)
Number of years lived in the area	28.99 (16.98)
Number of members in the household	4.26 (1.39)
Number of children less than 10 years of age	0.34 (0.78)
Monthly household income	14087.50 (9484.77)
Monthly household spending on water	340.32 (405.15)
Number of years residing in the house	13.58 (11.38)
Distance from nearest canal/river	2.39 (3.25)
	Percentage
Household head completed primary education = 1,0 otherwise	15
Household head completed secondary education = 1,0 otherwise	34.5
Employment in service sector = 1,0 otherwise	11.7
Self-employed = 1,0 otherwise	22.4
Manual worker = 1,0 otherwise	46.3
Monthly income 5000–10,000	74
Monthly income 10,000–15,000	49

Source: Wastewater treatment household survey (2016)

Regarding the household's perception of wastewater, the respondents were asked about the water quality of the river/canal in their area, and 74% reported it to be inferior, and only a mere 2.5% thought it was of good quality. About 78.5% of the respondents thought that the water quality had been highly degraded in the past 25 years, and 55.5% considered it was polluted. More than 64% used the river/canal water for recreational purposes. About 53.5% thought wastewater treatment was necessary regardless of the cost, and 42.5% thought it was necessary while holding the current cost of treating wastewater.

It was observed that as the scenario improves, people are slightly less willing to pay higher sewage taxes since an improvement from Scenario D to A would cost a lot higher than that from D to B. This can be supported by the fact that almost 42.5% of respondents thought wastewater treatment was necessary while holding current costs.

For those who said yes, the reasons were reported. Most people strongly agreed with the importance of the treatment, enjoying contributing to environmental conservation, need for river restoration, and that it is a priority as the protection of future generations is necessary. When the people willing to pay higher taxes were asked whether they would contribute the same amount for other water bodies, 17% agreed to pay the same amount; however, 26% said they would pay a lesser amount, and the rest refused to pay.

Of the people who were not willing to pay in the first place, they agreed that the municipality should use existing funds and that it may unwisely use the funds.

In order to incorporate all the factors that may cause respondents to agree to pay the specified bid amount, a logit regression equation was estimated. The logistic

Table 9.2 Logistic regression on determinants of WTP for wastewater treatment programs

	Scenario D to C	Scenario D to B	Scenario D to A
Variables	Coeff (SE)	Coeff (SE)	Coeff (SE)
Constant	−4.98*** (1.43)	−4.91*** (1.43)	−5.66*** (1.50)
Age	0.0064(0.019)	0.0065 (0.019)	0.019 (0.02)
No. of children	−1.04*** (0.41)	−0.96*** (0.39)	−0.59 (0.38)
Income	0.000*** (0.00)	0.000*** (0.00)	0.000*** (0.00)
Female	−0.54(0.56)	−0.61 (0.556)	−0.92 (0.61)
Water quality	1.18** (0.613)	1.22*** (0.621)	1.62** (0.738)
Education	0.79 (0.499)	0.83* (0.51)	0.63 (0.52)
Improved health	−0.19 (0.42)	−0.42(0.42)	−0.44 (0.44)
Reduction in overflowing cesspits	1.02** (0.47)	1.27*** (0.48)	1.23*** (0.48)
Avoid groundwater contamination	−1.03** (0.49)	−1.08** (0.49)	−1.05** (0.51)
Reduced river pollution	0.202 (0.42)	0.16 (0.43)	−0.04 (0.44)
Lower degradation of river ecosystem	0.55(0.49)	0.51 (0.49)	0.83(0.52)

Source: Wastewater treatment household survey (2016)
*** indicates significance at 1% level ** indicates significance at 5% level, and * indicates significance at 10% level

regression results are reported in Table 9.2. The ordered Logit and Tobit models were estimated incorporating the awareness of the existence of STP. No variation in the estimated models was observed. People are WTP more for positive levels of income. Age is positive but not significant. More is the level of water quality, the more the people are willing to pay. Moreover, the more is the number of children in the family; the less the people are willing to pay for wastewater treatment. The level of education is only significant in one scenario change from D to B.

The households are also willing to pay for a reduction in overflowing cesspits. Overflowing cesspits have more weightage to the households than groundwater contamination. Therefore, the households feel sewage outfalls might not be the leading factor to groundwater contamination.

The marginal effects are reported in Table 9.3. As the household WTP increases, more he can avoid overflowing cesspits from D to C to D to B (See Table 9.3).

9.4.1 Estimating Average WTP Bids

Figure 9.1 that for the lowest level of income or the first category, i.e., up to Rupees 5000 per month, the WTP bids for all scenario changes are zero, i.e., people are not willing to pay. The justified reason is that they cannot afford to pay higher sewerage taxes since they barely afford the necessities follow from Fig. 9.1, that environmental problems are nowhere seen as a priority for people in the low-income category. This could be true for low-educated respondents, too—the middle-income category, i.e., Rs. 5000–30,000, the average bids are lowest in scenario change D to C, whereas it is highest for the scenario change D to A.

Table 9.3 Marginal effects for wastewater treatment

Variables	Scenario change D to C	Scenario change D to B	Scenario change D to A
Family size	0.054	0.053	0.033
Number of children	−0.165	−0.150	−0.088
Avoid overflowing cesspits	0.163	0.198	0.183
Avoid groundwater contamination	−0.164	−0.170	−0.155

Source: Wastewater treatment household survey (2016)

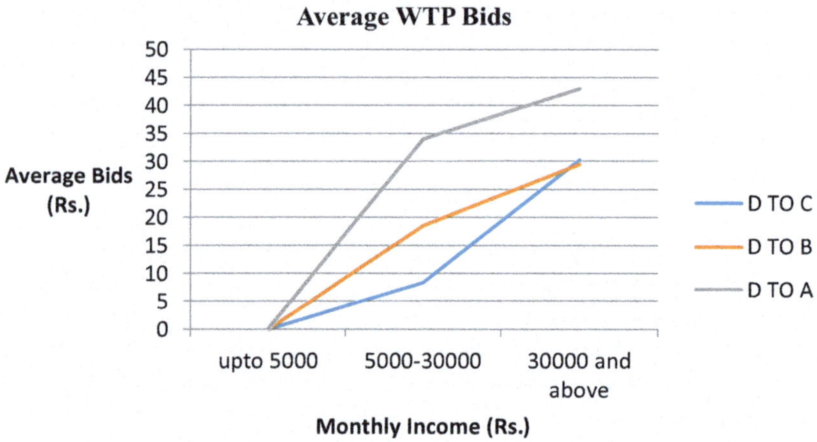

Fig. 9.1 Average Willingness to pay Bids. Source: Wastewater treatment household survey (2016)

The high-income category, i.e., category 3 in the graph below, represents people who earn between Rs. 30,000 and above, the average WTP bids almost converge for the first two scenario changes, and for the last scenario change, i.e., D to A, it is the highest of all bids.

The average WTP bids, including all three scenarios, ranging from none, i.e., zero valuation, to a maximum of Rs. 43 per month.

Based on education, the picture is a little different. The numbers look fine for scenario change D to C, since the average WTP is relatively lower for those in the lower education category. However, in the other two scenarios, the numbers for the lowest education category are the highest. However, it is possible that even though the respondent is not educated, some other family members could have a higher education which gives this high number in poorly educated cases too (See Table 9.4).

Table 9.4 Average WTP bids based on the education level

Education category	Scenario D to C	Scenario D to B	Scenario D to A
Up to the primary level of education	10.5	25.2	43.5
Up to higher secondary	8.1	18.4	29.8
Graduation and above	20.8	22.2	39.1

Source: Wastewater treatment household survey (2016)

The benefit estimates reported in this study reveal that an average household in the sample would be willing to pay Rs. 43, and when aggregated over the entire population, it amounts to Rs. 199,809,476 per month (the population of Chennai as per the 2011 Census is 4,646,732) for an improvement in Scenario from A to D. Yearly in amounts to around 24 crores.

Based on this detailed design and technical details, the Chennai City River Conservation Project initiated an estimated cost of Rs. 720.00 crore in the first phase for 2011 involving upgradation of the STPs and sewage pumping stations.

This "back-of-the-envelope" cost-benefit analysis (CBA) would suggest that even though the residents' welfare would increase as a result of an improvement of the current STPs, the water and sewerage tax revenues may not be sufficient to meet the costs and hence need additional financial sources for the financing of this endeavor.

9.5 Conclusion

The current study is an attempt to perceive the willingness of the communities for wastewater treatment and options for public investment at a regional level. In addition, the CV scenario was articulated based on the quality of the Adyar River, which directly impacts the coastal water quality of the Bay of Bengal.

An appreciable number of respondents (around 42.5%) were willing to pay for an improvement in the water quality of the Adyar River. The study is primarily focused on addressing the benefits associated with installing such a project. However, given the high installation cost, the co-financing option can be explored. A detailed cost-benefit analysis can be further explored where the policymakers can prioritize the urgency and plan for future management and investment options in the region.

Acknowledgments The paper is a part of the EU-India Consortium Project titled "*Supporting consolidation, replication and up-scaling of sustainable wastewater treatment and reuse technologies for India (SARASWATI)*" funded by the Department of Science and Technology for the Indian partners.

Appendix 1: Ordered Logit Model on Determinants of WTP for Wastewater Treatment Programs

Explanatory variable	Scenario D to C	Scenario D to B	Scenario D to A
Age	0.007768 (0.019153)	0.006795 (0.019389)	0.017881 (0.020343)
No. of household members	0.339382* (0.186677)	0.342594* (0.185522)	0.224029 (0.189154)
No. of children	−1.015029** (0.411722)	−0.951437** (0.397765)	−0.608522 (0.386321)
Household income	0.000084*** (0.000024)	0.000080*** (0.000024)	0.000067*** (0.000024)
Female	−0.512865 (0.556819)	−0.600106 (0.560946)	−0.944297 (0.611072)
Water quality	1.192592** (0.611566)	1.222202** (0.620934)	1.636683** (0.744243)
Education	0.834090* (0.509582)	0.844973* (0.512973)	0.597033 (0.525195)
Odor	−0.167019 (0.407415)	−0.0306102 (0.410465)	−0.054152 (0.415473)
Improved health	−0.183116 (0.416515)	−0.418212 (0.425942)	−0.449193 (0.441634)
Reduction in overflowing cesspits	1.027839** (0.466910)	1.26793*** (0.474778)	1.235831*** (0.484402)
Avoid groundwater contamination	−1.046993** (0.485291)	−1.088584** (0.491142)	−1.03783** (0.514766)
Reduce river pollution	0.178370 (0.425340)	0.157291 (0.429693)	−0.024605 (0.441486)
Reducing the degradation of river ecosystem	0.528780 (0.488331)	0.503245 (0.493264)	0.842794 (0.521612)
Awareness of STP	−0.256046 (0.460997)	−0.063570 (0.459784)	0.160494 (0.473710)

*10% level of significance; **5% level of significance; ***1% level of significance

Appendix 2: Tobit Model on Determinants of WTP for Wastewater Treatment Programs

Explanatory variable	Scenario D to C	Scenario D to B	Scenario D to A
Age	−2.746687*** (0.795237)	−2.676953*** (0.786195)	−3.64808*** (0.990085)
No. of household members	0.004199 (0.010162)	0.003464 (0.010111)	0.009537 (0.012174)
No. of children	0.208681** (0.099790)	0.205280** (0.098475)	0.161675 (0.115113)

Explanatory variable	Scenario D to C	Scenario D to B	Scenario D to A
Household income	−0.581288*** (0.219170)	−0.549227*** (0.212986)	−0.426808* (0.240011)
Female	0.000046*** (0.000013)	0.000044*** (0.000012)	0.000044*** (0.000015)
Water quality	−0.356112 (0.296871)	−0.384468 (0.293501)	−0.577830 (0.363237)
Education	0.643016** (0.330665)	0.672737** (0.332346)	1.07107** (0.456926)
Odor	0.476623* (0.268591)	0.474103 (0.266655)	0.373805 (0.313487)
Improved health	−0.133047 (0.22041)	−0.065596 (0.217057)	−0.063649 (0.250772)
Reduction in overflowing cesspits	−0.117498 (0.225846)	−0.234496 (0.226613)	−0.287848 (0.267921)
Avoid groundwater contamination	0.547486** (0.247302)	0.644432*** (0.244583)	0.728718*** (0.291402)
Reduce river pollution	−0.554640** (0.259832)	−0.561602** (0.257588)	−0.599591** (0.304593)
Reducing the degradation of river ecosystem	0.106355 (0.227640)	0.095807 (0.225995)	0.014492 (0.264459)
Awareness of STP	0.261098 (0.261473)	0.260513 (0.260238)	0.521644* (0.314952)
	−0.150652 (0.250661)	−0.057580 (0.247477)	0.0787705 (0.288920)

*10% level of significance; **5% level of significance; ***1% level of significance

References

Arrow, K., Solow, R., Portney, P. R., Leamer, E. E., Radner, R., & Schuman, H. (1993). Report of the NOAA panel on contingent valuation. *Federal register, 58*(10), 4601–4614.

Bouzit, M., Das, S., & Cary, L. (2018). Valuing treated wastewater and reuse preliminary implications from a meta-analysis. *Water Economics and Policy*, *4*(02), 1650044.

Breaux, A., Farber, S., & Day, J. (1995). Using natural coastal wetlands systems for wastewater treatment: an economic benefit analysis. *Journal of environmental management*, *44*(3), 285-291.

Can, Ö., & Alp, E. (2012). Valuation of environmental improvements in a specially protected marine area: A choice experiment approach in Göcek Bay, Turkey. *Science of the Total Environment*, *439*, 291-298.

Central Pollution Control Board (2015) River stretches for restoration of water quality. A Ministry of Environment, Forest, and Climate change report

Giridhar V (2001) Coastal ocean pollution monitoring program. In: Proceedings of the UGC course on Marine sciences and environment conducted by the Department of Applied Geology, University of Madras, Chennai, India

Halkos, G., & Matsiori, S. (2018). Environmental attitudes and preferences for coastal zone improvements. *Economic Analysis and Policy*, *58*, 153-166.

Indab, A. L., Guzman, A. I., & Bagarinao, R. T. (2003). *Effluent charge for Sarangani Bay, Philippines: an ex-ante assessment*. EEPSEA, IDRC Regional Office for Southeast and East Asia, Singapore, SG.

Jakobsson, K. M., & Dragun, A. K. (1996). *Contingent valuation and endangered species: methodological issues and applications*. Edward Elgar Publishing.

Kontogianni, A., Skourtos, M.S., Langford, I.H., Bateman, I.J., and Georgiou, S. (2001). Integrating stakeholder analysis in non-market valuation of environmental assets. Ecological Economics, 37.1, 123-138.

MacDonald, D. H., Ardeshiri, A., Rose, J. M., Russell, B. D., & Connell, S. D. (2015). Valuing coastal water quality: Adelaide, South Australia metropolitan area. *Marine Policy*, *52*, 116-124.

Perni, A., Martínez-Carrasco, F., & Martínez-Paz, J. M. (2011). Economic valuation of coastal lagoon environmental restoration: Mar Menor (SE Spain). *Ciencias marinas*, *37*(2), 175-190.

Shanmugam, R. T., & Babu, S. C. (2017). Incorporating social and sustainability aspects into economics of agroforestry: a case study from South India. *Journal of Sustainable Development*, *1*(2), 20–31.

Wattage, P., & Mardle, S. (2008). Total economic value of wetland conservation in Sri Lanka identifying use and non-use values. *Wetlands ecology and management*, *16*(5), 359-369.

Whittington, D., & Swarna, V. (1994). The economic benefits of potable water supply projects to households in developing countries.

Chapter 10
Blue Carbon Potential of India: The Present State of the Art

Abhra Chanda and Tuhin Ghosh

10.1 Introduction

The fire was perhaps the most breakthrough invention, which marked the beginning of the modernization of the human race. This invention also led to the indiscriminate exploitation of Earth's resources. To meet the growing needs and demands of humankind, we mined and unearthed several minerals and fossil fuels and changed the landscape by rampant deforestation for the sake of agriculture and human settlements. In the year 1760, the first industrial revolution acted as a catalyst for this ongoing environmental degradation. Since the industrial revolution, the concentration of several greenhouse gases is continuously increasing in the atmosphere, out of which carbon dioxide (CO_2) happens to be the single largest contributor (Yoro and Daramola 2020). The atmospheric CO_2 concentration has increased from 270 ppm from the pre-industrial revolution era to ~412.55 ppm in August 2020 (Lan et al. 2020; NOAA 2020). Even if we implement stringent initiatives to reduce carbon emissions, the Intergovernmental Panel on Climate Change (IPCC 2007) estimated that the atmospheric CO_2 concentration by the end of the year 2050 would most likely rise to 480 ppm. Despite recognizing the potential harm that this ongoing carbon emissions can bring upon us in the form of global warming, climate change, glacial melting, relative sea-level rise, and more, we cannot put an end right now to all the anthropogenic activities that lead to such emission. However, the occurrence of such global phenomena made us realize that it is high time we looked and cared for the environment and mother nature and find out a solution for such a problematic scenario.

Scientists all over the world unanimously gathered pieces of evidence that show the global forests, the oceans, and the lithosphere (especially the top layers close to the atmosphere) act as a natural sink for carbon (Ciais et al. 2013). Global

A. Chanda (✉) · T. Ghosh
School of Oceanographic Studies, Jadavpur University, Kolkata, West Bengal, India

S. Hazra, A. Bhukta (eds.), *The Blue Economy*,
https://doi.org/10.1007/978-3-030-96519-8_10

international organizations like United Nations Framework Convention on Climate Change (UNFCCC) took initiatives for climate change mitigation like Reducing Emissions from Deforestation and Forest Degradation (REDD+) to safeguard, conserve, and proliferate the carbon sink potential of these natural carbon reservoirs like the terrestrial forests (Lund et al. 2017). However, it is only since the last decade; when Nellemann et al. (2009) officially coined the term "blue carbon," research and characterization of the carbon dynamics in the marine ecosystems have increased manifold. Blue carbon refers to the CO_2 utilized from the atmosphere and stored by the coastal ocean ecosystems, mostly mangroves, salt marshes, seagrasses, and potentially macro-algae, through plant growth and the accumulation and burial of organic matter in the soil. The scientific community recognized the potential of marine ecosystems in sequestering substantial quantities of carbon long ago. However, it was the report furnished by Nellemann et al. (2009) that mentioned more than 55% of the global green carbon, that is, the CO_2 fixed by photosynthesis and transformed to organic carbon, is present in the blue carbon ecosystems like mangroves, seagrass, tidal flats, and salt marshes. Today, blue carbon conservation and restoration have come out as one of the strongest contenders to combat the evil of climate change, with increased research and emphasis on this concept (Lovelock and Duarte 2019).

10.1.1 Importance of Blue Carbon

The coastal vegetated ecosystems all around the world store substantial organic carbon of varying strength. Mangroves thrive in the tropics and subtropics, and seagrass meadows and salt marshes flourish from the equator to the poles, with seagrass mainly confined to the temperature regions (Pendleton et al. 2012). Together, these ecosystems cover an area of about 490,000 km^2. Apart from acting as a repository for carbon, these ecosystems provide a wide range of ecosystem services like the nursing ground for fishes, shoreline protection from natural disasters, and mitigation of aquatic pollution (Barbier et al. 2011). Despite being of such global significance, these ecosystems have witnessed a decline of almost 50% of their aerial cover in the last 100 years (McLeod et al. 2011). These ecosystems experienced habitat conversion in the past century due to multifarious anthropogenic and climate-induced activities like conversion to agriculture, aquaculture, and industrial pollution, clearing for urbanization, sea-level rise, dredging, and eutrophication of the overlying water column (Pendleton et al. 2012). There are mainly two reasons for which the blue carbon ecosystems need special attention. First, these ecosystems are capable of storing disproportionately high amounts of carbon compared to the area covered (Fourqurean et al. 2012; Jennerjahn 2020), which essentially warrants that degradation of a small area of such ecosystems can lead to substantially high carbon emission. Pendleton et al. (2012) estimated that the deterioration and degradation of blue carbon ecosystems lead to a release of about 0.15–1.02 Pg of CO_2, equivalent to an economic loss of around 6–42 billion US dollars (according to 2007 US dollars valuation). Second, these ecosystems owing to their substantially high carbon sequestration potential can mitigate climate change by absorbing the anthropogenic CO_2, provided we increase the area of these ecosystems, by a proper restoration, afforestation, conservation, and management of these ecosystems, wherever and however possible. Both of these aspects make blue carbon an essential weapon to fight out the ongoing climate change. According to Ullman et al. (2013), the inclusion of blue carbon in market-based climate policy mechanisms is essential to raise funds needed to restore and manage these ecosystems.

Ullman et al. (2013) further advocated the need for a national blue carbon policy to focus research on the elementary functioning of these ecosystems and to seek avenues to enhance their carbon sequestration potential and estimate the country-wise emission and sink strength of these blue carbon ecosystems, followed by promoting the concept of the blue carbon market.

10.2 Blue Carbon Ecosystems of India

India encompasses a coastline of ~7500 km comprising the mainland and the islands of Lakshadweep and Andaman and Nicobar (Jena et al. 2019), which shelters all the three conventional blue carbon ecosystems: mangroves, seagrasses, and salt marshes. According to the latest estimates, India has a mangrove forest area of 4795 km^2, mainly concentrated in the deltaic regions adjoining the perennial rivers of the country along with the Andaman and Nicobar Islands (Anneboina and Kumar 2017; ISFR 2019). The seagrass cover in India is approximately 500 km^2, mainly situated at the Lakshadweep Islands, Andaman and Nicobar Islands, Gulf of Mannar, Gulf of Kutch, and Palk Bay (Thangaradjou and Bhatt 2018). Salt marshes in India cover approximately 1600 km^2, and the spatial distribution is concentrated along the coastlines of Gujarat, followed by Tamil Nadu, Andaman and Nicobar Islands, Andhra Pradesh, Maharashtra, Puducherry, and Daman and Diu (Patro et al. 2017). Since the onset of the twenty-first century, the mangroves of India have received substantial attention from the perspective of blue carbon stock assessment (Akhand et al. 2017; Chowdhury et al. 2018; Anand et al. 2020; ShyleshChandran et al. 2020), followed by seagrass (Banerjee et al. 2018; Ganguly et al. 2017, 2018; Thangaradjou and Bhatt 2018). However, studies on the Indian salt marshes are scarce (Banerjee et al. 2017; Kaviarasan et al. 2019) (Fig. 10.1).

10.3 The Mangroves of India

Mangroves are a specialized group of tidal halophytic plants. These plants are abundant in the coastlines throughout the tropical and subtropical regions. These plants mainly thrive along the sheltered shorelines, deltaic-flat terrains, micro- to

Spatial extent (km^2)	~ 5000	~ 500	~ 1600
Species Count	50	16	15
Scholarly articles (since 1950)	> 900	< 100	< 10
Carbon content estimation in number of sites	> 10	2	5
Proper economic valuation of carbon	Yes	Yes	Not yet

Fig. 10.1 A snapshot of the present overview on the Indian blue carbon ecosystems

macro-tidal estuaries, and coastal stretches (Tomlinson 1986). The floristic diversity of mangroves comprises 9 orders, 20 families, and about 70 species in the whole world (Ellison et al. 1999). The vegetated structure of mangrove forests includes full-grown trees, shrubs, and ferns of diverse taxonomic origin. India has a long coastline of almost 7500 km facing the Bay of Bengal in the east and the Arabian Sea in the west. There are nine states and four union territories (UTs) within India, which has a coastline. According to the India State of Forest Report (ISFR 2019), all the nine states and three union territories comprise mangroves. Nasser et al. (1999) reported the existence of mangrove patches covering only a few hectare areas in the Minicoy Island of Lakshadweep; however, the ISFR did not consider these patches. The Indian mangroves are rich in biodiversity comprising almost 50 species (belonging to 20 genera) out of the 70 species found worldwide (Jagtap et al. 1993) (Table 10.1). The mangroves of India are not only rich in the floristic diversity of mangrove plants but also shelter floral species starting from bacteria to terrestrial and manglicolous fungi, bryophytes, algae, seagrasses, and mangrove associates (Jagtap et al. 1993). The mangrove forests of India also act as an abode to a wide range of faunas like crustaceans, mollusks, wood borers, fishes, birds, reptiles, and mammals (Naskar and Mandal 1999). India also proudly shares a substantial part of the world's largest mangrove forest of Sundarbans with Bangladesh, which happens to be a UNESCO World Heritage Site.

Table 10.1 The state-wise distribution of mangrove forests and patches along with the number of species, area cover, and carbon stock in India

State	Place	No. of species	Area	Carbon in live biomass (Mg C ha^{-1})	References
West Bengal	Sundarbans	22	2112	49.54	Ray et al. (2011)
Odisha	Bhitarkanika, Mahanadi, Subarnarekha, Devi-Kauda, Dhamra, Chilika	36	251	131.06	Anand et al. (2020)
Andhra Pradesh	Coringa East, Godavari, Krishna	26	404		
Tamil Nadu	Pichavaram, Muthupet, Ramnad, Pulicat, Kazhuveli	18	45	62.81	Kathiresan et al. (2013)
Kerala	Vembanad, Kannur (North Kerala)	7	9	117.1	Harishma et al. (2020)
Karnataka	Kundapur, Dakshina Kannada/Honnavar, Karwar, Mangalore Forest Division	15	10	50.40	Suresh et al. (2013)
Maharashtra	Achra-Ratnagiri, Devgarh-Vijay	20	320	34.14	Patil et al. (2014)
Goa	Goa	17	26		
Gujarat	Gulf of Kutch, Gulf of Khambhat, Dumas-Ubhrat	9	1177	24.57	Pandey and Pandey (2013)
Andaman and Nicobar	North Andaman, Nicobar	27	616	118.3	Mall et al. (1991)
Puducherry	Puducherry	1	2	220.15	Muthukumaran et al. (2013)
Daman and Diu	Daman and Diu		3		
Lakshadweep	Minicoy Island	1	0.01		Nasser et al. (1999)
Total		50	4975		

10.3.1 Spatial Distribution

To quantify the blue carbon stock in any ecosystem, identifying the area of these ecosystems is one of the most crucial prerequisites. The conventional technique for mapping any forested ecosystems includes ground-based monitoring of the extent of such forests. However, the mangrove forests are swampy areas with dense vegetation covers, making these places difficult to access (Nandy and Kushwaha 2011). The emergence of remote sensing tools, coupled with geographic information systems, has enabled us to overcome this difficulty (Giri et al. 2014).

Though these tools and applications cannot entirely substitute the ground-based protocols, these methods facilitate mapping and monitoring of the mangroves in a synoptic scale, which is practically a cumbersome endeavor and near impossible (Dwivedi et al. 1999; Nayak et al. 2001). Due to various natural and anthropogenic factors, the aerial cover of mangroves changes incessantly over time. Remotely sensed data serves as a cheap and cost-effective tool to monitor the short-term (seasonal to inter-annual) as well as long-term (multi-decadal) temporal changes in the mangrove cover (Green et al. 1998a, 1998b; Verhayden et al. 2002), which is essential from the viewpoint of characterizing the blue carbon stock of these ecosystems. There are many studies conducted so far in India, which successfully mapped and monitored the mangrove patches from Sundarbans, West Bengal (Giri et al. 2014), Mumbai, Maharashtra (Vijay et al. 2005), Bhitarkanika mangrove forest, Odisha (Reddy et al. 2007), mangroves of Tamil Nadu and Andaman and Nicobar Islands (Ramachandran et al. 1998), and overall India (Nayak and Bahuguna 2001). As mangroves thrive in the coastal margins, pixels of soil, water, and vegetated portions remain intertwined with each other (Giri et al. 2014). Mangrove and non-mangrove vegetation often coexist together in the landward margin of these forests, and it becomes difficult to differentiate these two vegetation types. In the present day, several advanced techniques have evolved to overcome these difficulties and map the mangroves accurately (Gupta et al. 2018). In India, mangroves received ample attention in this regard, and environmental conservationists continuously monitor the temporal and spatial changes in mangrove cover.

10.3.2 Measurement of Biomass and Carbon Stock

The architecture of the mangrove forest structure is less complicated than many of the terrestrial forests. Even then, these forests can store substantially high amounts of carbon (Alongi 2012). The mangrove forests seem to have much less vegetated biomass, but these forests can lock carbon equivalent to that of tropical rainforests (one of the most productive terrestrial forests of the world). The specialty of mangrove forests is their capability to store carbon in the belowground root systems and the ambient soil substratum (Atwood et al. 2017). The aboveground compartments of the mangrove forests can store a substantial amount of carbon; however, their belowground carbon stock far exceeds any other terrestrial forests. Mangroves have developed some morphological adaptations to acclimatize under harsh conditions like high salinity and regular water submergence (Naskar and Palit 2015). One such morphometric alteration is the extensive stretch of the root system both radially outwards and deep down the sediment profile. This vast stretch of the root system enables the mangrove to cling together with the sediments and favors the sedimentation process (Ellison 1999). The mangroves store substantial quantities of carbon in these roots. In addition to the root system, mangroves have a remarkable capability to store carbon in the soils where these grow. Most of the mangrove species are evergreen trees and shed leaves throughout the year. These leaves continuously undergo microbial degradation as they encounter the incoming tidal water, which in turn provides a favorable environment for such degradation (Chanda et al. 2016). However, a part of the organic matter stored in these leaf litters shows significant resistance towards decomposition (Keuskamp et al. 2015). These recalcitrant components, mostly composed of complex organic compounds like tannin, lignin, and humin, end up in the soils lying adjacent to the mangrove plants. These compounds penetrate the soil surface and move deep down the sediment profile up to great depths (Lunstrum and Chen 2014). Mangroves usually thrive under anoxic conditions. Such anoxic conditions disfavor the oxygenation and hence remineralization of this organic carbon, making the soils of the mangrove forests a giant repository of organic carbon. If the mangrove forests do not experience any disturbance due to any form of natural or anthropogenic factors, these ecosystems can lock substantial quantities of carbon for centuries (Lovelock et al. 2017).

Quantitative estimation of blue carbon stock in these forests is one of the most integral and perhaps the most crucial aspect to reap the benefits of these ecosystems in combating the evil of climate change. We can measure the carbon stock in the different compartments of these forests following several ways, like the mean-tree method, the harvest method, and the allometric method. The weight of average mangrove trees often sums up to more than 1000 kg, and hence the harvest method, which requires complete uprooting of a tree, is not feasible to measure biomass (Komiyama et al. 2005). Moreover, this approach leads to the destruction of a tree, which is not at all desirable. The mean-tree method is best applicable for homogeneous forest stands with simple structure; however, mangroves often display a heterogeneous mixed species composition. The most common practice is to implement the allometric techniques for quantifying the biomass that is visible above the ground surface (Komiyama et al. 2008). In this technique, we usually measure various morphometric parameters of a tree like the diameter at breast height, the number of primary and secondary shoots from the trunk, the number of tertiary branches, and so on. Using tree coring, we can scoop out a part of the tree without doing much harm to the tree (Mantgem and Stephenson 2004). Using the scooped out tree parts, we measure the density (Chave 2005) as well as carbon content (Kauffman and Donato 2012) in the trees. The coupled measurement of density and carbon concentration enables us to estimate the total carbon stock of a particular tree. Unlike the aboveground compartments, carrying out such estimation for the belowground roots is cumbersome. Assessment of belowground root biomass essentially involves the excavation of soils along with the root system (Santos et al. 2017), which leads to a significant destruction of some of the representative trees chosen for sampling. Measurement of soil carbon stock involves the coring technique, whereby we manually take out a one-meter core through the surface soil. When we retrieve that core, it brings up the sediment column along with it. We can slice open it, take samples from different depths, and measure the carbon concentration in the soils. Upon extrapolation of the soil organic carbon data from various sediment cores randomly collected across the spatial extent of a mangrove forest, we can derive an estimate of the total soil carbon pool of forested land. The total of the aboveground biomass carbon stock, the belowground biomass carbon stock, and the soil carbon stock can represent the total forest carbon stock. In the present date, the advent of several remote sensing techniques has also enabled us to measure as well as monitor the temporal changes of mangrove forest biomass and carbon stock having a synoptic coverage (Bindu et al. 2020). Several scholars have applied almost all of the possible ground-based as well as remote sensing techniques to measure the biomass and carbon stock in many of the Indian mangrove patches.

10.3.3 Carbon Locked in Live Biomass

Several scholars have measured and reported the carbon stock from various mangrove patches of India (Table 10.1). West Bengal shelters around 40% of the world's largest mangrove forest of Sundarbans and encompasses the highest mangrove cover compared to any of the other states of India. This forest has perhaps received the most attention concerning the quantification of standing carbon stock. Mitra et al. (2011) evaluated the carbon stock in the aboveground compartments of three of the dominant mangrove species of Sundarbans, namely, *Excoecaria agallocha*, *Avicennia alba*, and *Sonneratia apetala*. They observed significant variation in biomass between the different species and observed spatial variation in biomass between the central, eastern, and western zone of this vast stretch of mangrove forests. Mitra et al. (2011) also pointed out that the varying environmental conditions like sediment salinity played a deciding role in regulating the biomass and carbon stock of the mangrove stands. Mitra et al. (2011) could only throw light on the aboveground carbon stock, but Ray et al. (2011) carried out a comprehensive analysis of the entire Sundarbans forest's carbon stock. They reported that Sundarbans, unlike many other mangrove forests of the world or even India, have comparatively lesser carbon stock in the belowground biomass. They observed an average carbon stock in the aboveground and belowground biomass of 39.93 ± 14.05 Mg C ha^{-1} and 9.61 ± 3.37 Mg C ha^{-1}, respectively. They also observed a varying rate of carbon sequestration in the live biomass (1.69 Mg C ha^{-1} year^{-1}) and the soil pool (0.012 Mg C ha^{-1} year^{-1}). Upscaling their results for the entire Indian part of Sundarbans, Ray et al. (2011) observed that this mangrove forest holds as much as 26.5 Tg C (1 Tg C = 10^{12} g C) (Fig. 10.2).

Besides the Sundarbans in West Bengal, Bhitarkanika mangrove forest in Odisha, Pichavaram mangrove forest in Tamil Nadu, and the mangroves of Andaman are some of the most prominent mangrove patches of India. Though the total area coverage of the Bhitarkanika and Mahanadi mangroves of Odisha and the mangroves of Andaman is much less, compared to that of Sundarbans, the carbon content per unit area is substantially higher in the former forests compared to the later (Table 10.1). Banerjee et al. (2020) recently measured species-specific aboveground carbon stock and soil organic carbon stock in the Bhitarkanika and the Mahanadi mangrove system. They observed significant variability in the carbon stock among

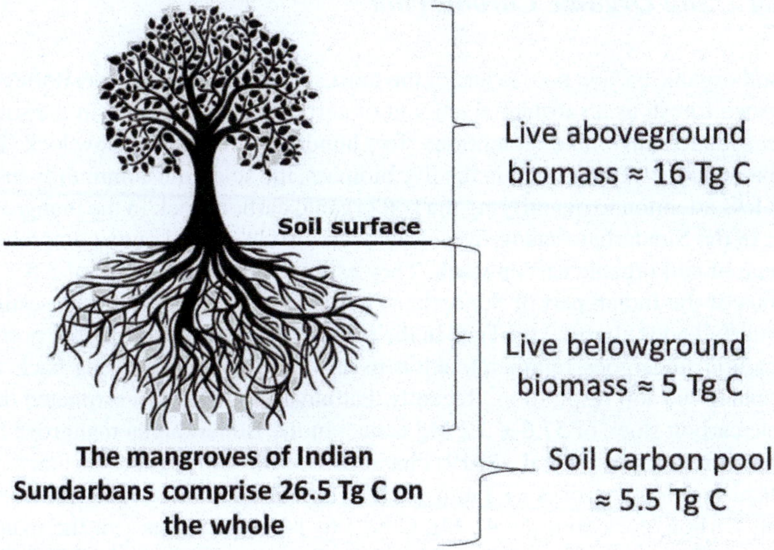

Fig. 10.2 The carbon stock in the different compartments of the Sundarbans mangrove forest

the five dominant species, with a mean carbon content per unit area in this forest of 124.11 ± 30.14 Mg C ha^{-1}. Anand et al. (2020) also observed a similar magnitude while estimating the carbon content in the same region with the aid of remote sensing tools. Sahu et al. (2016), while working in the mangroves of the entire Mahanadi wetlands, observed that besides the natural mangrove stands, the planted mangroves also had substantial carbon content per unit area. They also observed a statistically significant positive correlation between the biomass of the mangrove stands and the soil organic carbon in the topsoil. They estimated a total carbon content of 0.98 Tg C for the entire Mahanadi mangrove wetlands. Kathiresan et al. (2013), while working in the Pichavaram mangroves and the mangroves of Vellar-Coleroon estuarine complex, Tamil Nadu, observed a positive relationship between carbon sequestration potential of these mangroves and various allometric parameters like tree age, height, and diameter at breast height. They also recorded a positive relationship between the rate of carbon accrual in the trees and the total biomass, canopy photosynthetic rate, growth rate, and leaf area index. Gujarat has the second-highest mangrove area after West Bengal. However, the carbon content in these mangrove patches per unit area was lowest compared to that of all the other states of India. Despite that, the mangroves of the Gujarat store an estimated 8.11 Tg C of carbon (Pandey and Pandey 2013). The available literature shows that there are no estimates of carbon content per unit area for the mangroves of Goa, Daman and Diu, Lakshadweep, and Andhra Pradesh. Several other states, like Karnataka and Kerala, have received much less attention.

10.3.4 Soil Organic Carbon Pool

The soil organic carbon pool is one of the most significant carbon repositories of the mangrove forest, as a substantial amount of carbon remains locked in the soil substratum under favorable conditions for hundreds of years (Lovelock 2008). Compared to the carbon stock in the live biomass, the scientific community has paid much less attention to quantifying the soil organic carbon stock in the mangroves of India. In the Sundarbans mangroves, Ray et al. (2011) carried out a comprehensive estimate of soil organic carbon stock. They estimated a carbon stock of 5.5 Tg C in the soils of the Indian part of Sundarbans in the top 30 cm. They further estimated that litterfall adds almost 2.07 Tg C in the topsoil. However, a substantial portion of this carbon undergoes remineralization to CO_2 and eventually goes back to the atmosphere via soil respiration. Recently, Pattnayak et al. (2019) estimated the soil organic carbon stock of 57.6 ± 3.2 Mg C ha^{-1} in the Bhitarkanika mangrove forest, Odisha. Gnanamoorthy et al. (2019) characterized the soil organic carbon stock in the Pichavaram mangroves of Tamil Nadu. They observed a varying range of soil organic carbon stock from 57.41 Mg C ha^{-1} to 146.1 Mg C ha^{-1} in the mangrove stands of varying age from 12 to 21 years. Gnanamoorthy et al. (2019) further observed a higher burial rate of organic carbon in the soils of natural mangrove stands compared to that observed in the restored mangrove stands of Pichavaram. Harishma et al. (2020) observed a mean soil organic carbon stock of 81.3 Mg C ha^{-1} in the mangroves of Kerala. Similarly, Patil et al. (2014) measured the organic carbon in the mangrove soils around the Thane creek of Mumbai city and observed a varying range from 0.07 Mg C ha^{-1} to 2.07 Mg C ha^{-1}. Except for the few studies mentioned in this paragraph, no other notable endeavors of characterizing the soil organic carbon exist in the form of scholarly articles, especially for the states like Andhra Pradesh, Karnataka, Goa, and Andaman and Nicobar Islands.

10.4 The Seagrasses of India

Seagrasses are marine angiosperms that typically thrive in shallow coastal waters, and this floral community completes their entire life cycle while remaining submerged under the water. The seagrass meadows are available all along the coastal margins of the world across varying latitudes, and their habitat includes bay, estuary, lagoons, shallow impoundments, open sea, continental shelves, and backwaters (Geevarghese et al. 2018). This marine ecosystem acts as an exclusive abode to some of the unique marine fauna like dugong and green turtles, which belongs to the endangered category of the IUCN Red List (Broderick et al. 2006; Cullen-Unsworth et al. 2018). Conservation efforts for the seagrass ecosystems mainly originated to safeguard the only habitat of this critically endangered marine fauna. However, by the end of the twentieth century, we realized the carbon sink potential of the seagrass beds (Duarte and Chiscano 1999). Ever since Nellemann et al. (2009) coined the term blue carbon, the seagrass beds have become a point of prime interest. Short et al. (2011) reported the existence of 72 species of seagrass all over the world. In India, pieces of evidence confirmed that 16 species thrive in the Indian seagrass meadows (Thangaradjou and Bhatt 2018). In India, the scientific community focused more on exploring the biotechnological applications of seagrass, which shadowed the restoration and conservation-related studies compared to that observed in the case of mangroves.

10.4.1 *Spatial Distribution*

Identifying and mapping any submerged vegetation is far more challenging than that of terrestrial vegetation due to difficulty in physically accessing the habitats. Seagrass thrives at varying depths from the near water surface to more than 20 m (Erftemeijer and Shuail 2012). The advent of several remote sensing tools and features has enabled us to overcome this difficulty, but seagrass zone mapping continues to be a cumbersome endeavor. Several factors like the degree of transparency of the water column, the density of the seagrass vegetation, the type of substratum where the seagrass grew, and the variability in the bathymetry need serious consideration to reduce the uncertainty of seagrass bed mapping using remotely sensed images (Roelfsema et al. 2013). The past three decades have witnessed considerable efforts to map the seagrass beds in the significant locations surrounding the Indian coastlines. The seagrass beds of the Lakshadweep were perhaps the first to receive the attention of remote sensing technology. Jagtap and Inamdar (1991) elaborately mapped an area of 112 ha distributed in six islands around the Lakshadweep. They also studied the species diversity and estimated the biomass of the standing crop. In the years to follow, IRS 1D LISS III and IRS P6 LISS III images explored almost 3300 ha of seagrass cover in the Gulf of Mannar (Thangaradjou et al. 2008; Umamaheswari et al. 2009). These studies also tried to quantify the loss of seagrass meadows that underwent in recent times due to both natural and anthropogenic disturbances, which is essential to understand the degree of threats these ecosystems are facing. Paulose et al. (2013) estimated a seagrass cover of almost 29 km^2 in and around the Andaman and Nicobar Islands. They also estimated the Indian Ocean tsunami that occurred in the year 2004 wiped out an estimated area of 16.19 km^2 of seagrass bed. Overall, the locations in India with extensive seagrass beds have received a larger share of attention concerning mapping and temporal monitoring. The smaller patches remain understudied even today.

10.4.2 *Carbon Stock in the Live Biomass and Sediments of Seagrass Meadows*

Environmentalists and marine scientists have unanimously acknowledged the capability of seagrass meadows to remove atmospheric CO_2 and lock it in organic form in their biomass and the underlying beds (Fourqurean et al. 2012). The seagrass meadows, covering only <0.1% of the ocean floor, account for almost 10–18% of the total organic carbon burial (48–112 Tg C $year^{-1}$) in the oceans (Duarte et al. 2005; Kennedy et al. 2010; McLeod et al. 2011). The seagrass species usually exhibit very high primary productivity, and these plants have the remarkable ability to slow down the water current in the nearshore regions and subsequently trap carbon of allochthonous origin (Hansen and Reidenbach 2012; Serrano et al. 2018). India being one of the countries that ratified the Paris Agreement aims to fulfill its nationally determined contribution (MoEFCC 2015), and several authors believe that seagrass could be an excellent option for India to create some additional carbon sinks (Koshy et al. 2018; Ramesh et al. 2019). However, the assessment of carbon stock in Indian seagrass meadows has not geared up to that extent. In the last 2–3 years, Ganguly et al. (2017, 2018) measured the carbon stock and primary productivity rates in two sites, namely, Palk Bay and Chilika lagoon, which have substantially high seagrass vegetated areas. Ganguly et al. (2017) reported that the seagrass of Palk Bay could sequester 99.31 ± 45.13 mM C m^{-2} d^{-1}. They also estimated that the entire seagrass-dominated area of the Palk Bay (~330 km^2) stores 0.03 Tg C in the aboveground biomass and almost three times of that in the belowground biomass (0.09 Tg C). However, the carbon locked in the sediments was nearly 38 times that of the carbon locked in the live biomass. Considering the carbon concentrations up to a depth of 1 m, Ganguly et al. (2017) inferred that the seagrass beds of Palk Bay store as much as 4.6 Tg C.

While working in both Palk Bay and Chilika lagoon, Ganguly et al. (2018) observed a carbon capture rate ranging from 8.44 Mg CO_2 ha^{-1} $year^{-1}$ to 15.9 Mg CO_2 ha^{-1} $year^{-1}$. They also confirmed that the seagrass sediments played a substantially large role in sequestering carbon compared to the live biomass.

10.4.3 Atmosphere-Hydrosphere CO_2 and CH_4 Exchange from the Seagrass Water Column

Apart from the seagrass biomass and underlying sediments, the water column associated with the seagrass meadows is also an integral part of the seagrass ecosystem. The estuaries throughout the world mostly emit CO_2 towards the atmosphere. However, most of the continental shelf waters act as a net sink of CO_2 (Laruelle et al. 2010; Chen et al. 2012). The presence of submerged vegetation can significantly alter the behavior of the water column adjoining these ecosystems using regulating the partial pressure of CO_2 (pCO_2(water)) in these water columns. Research conducted in many of the seagrass meadows indicated that the seagrass water column acts as both sources as well as a sink for CO_2 (Macklin et al. 2019; Akhand et al. 2021). In India, only Banerjee et al. (2018) observed the CO_2 and CH_4 exchange between the seagrass adjacent hydrosphere and the atmosphere. They observed that the seagrass water column in the seagrass-dominated regions of the Chilika lagoon acted as a net source of both CO_2 and CH_4 (Fig. 10.3). However, they

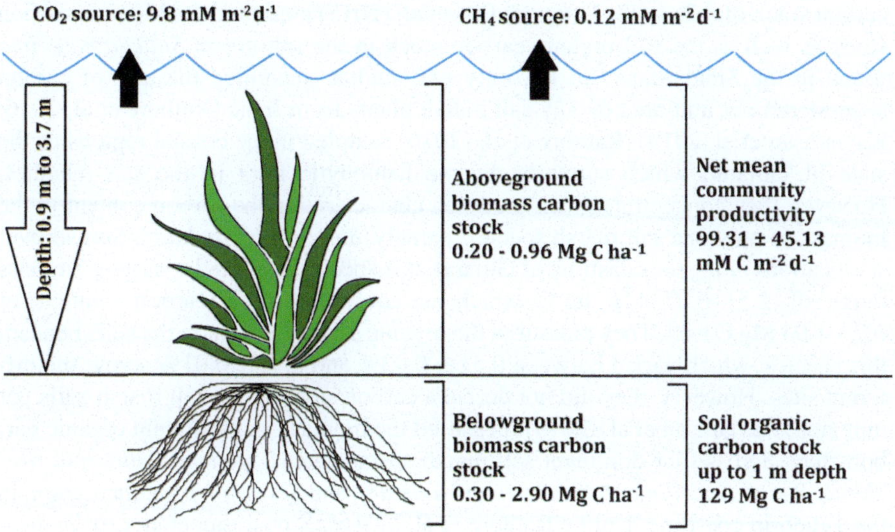

Fig. 10.3 The estimates of carbon stock and ecosystem productivity from the Indian seagrass meadows

also observed that the presence of seagrass meadows reduced the net air-water CO_2 and CH_4 exchange towards the atmosphere by almost 14 times compared to the regions where there was no seagrass. Such observation shows that the seagrass meadows not only store a substantial amount of carbon in their biomass and sediments below but also help reduce the CO_2 source character of the above lying water column.

10.5 Salt Marshes and Tidal Flats

Salt marshes are an ecological group of halophytic rooted plants comprising almost 500 species (Silliman 2014) that prefer to thrive in the low-energy environment surrounding the coastal periphery in partially submerged to emerged flat terrains with soft substratum (Mcowen et al. 2017). India has a much higher spatial coverage of salt marshes than that of seagrasses, yet this ecosystem has received the least attention in India among the three conventional blue carbon ecosystems. Studies on the carbon dynamics of this crucial ecosystem from India were absent until the last decade. Jana et al. (2013) were perhaps the first to point out the carbon sequestration potential of the salt marsh species *Porteresia coarctata*, which mostly remains associated with the mangroves of Sundarbans as a mangrove associate. In the following years, Sivakumar et al. (2014) studied the species-specific carbon stock (*Sesuvium portulacastrum*, *Salicornia brachiata*, and *Suaeda monoica*) in the salt marshes associated with the Muthupet mangroves. They observed a very high soil organic carbon stock of 4.2–4.8%. Later on, Das et al. (2015) measured the biomass carbon stock as well as the soil organic carbon stock in the salt marsh *Suaeda maritima* found in the Sundarbans. So far, only two studies quantified the unit of carbon sequestered per unit area by any salt marsh plants from India (Rathore et al. 2016; Kaviarasan et al. 2019). Rathore et al. (2016) sampled in the coastal regions of the state of Gujarat, which supports diverse halophytic flora in the salt marshes. However, they concentrated their study on characterizing the carbon content in the biomass of only one salt marsh species, namely, *Salicornia brachiata*. Spread over six locations along the coastline of Gujarat, this species exhibited a varying biomass range of 2.51–6.07 Mg ha^{-1}, which is equivalent to a carbon content of 0.77–1.93 Mg C ha^{-1}. They measured the organic carbon content in the soils beneath this species, which varied between 0.51 ± 0.03% and 0.91 ± 0.01% across the different sites. However, they did not derive a carbon stock in the salt marsh soils per unit area. Kaviarasan et al. (2019) measured the biomass and sediment organic carbon stock across four dominant salt marsh species areas, namely, *Salicornia brachiata*, *Suaeda maritima*, *Arthrocnemum indicum*, and *Sesuvium portulacastrum*, in the Tuticorin coastline. Unlike Rathore et al. (2016), Kaviarasan et al. (2019) measured the sediment organic carbon on a per unit area basis, which varied from 8.42 ± 0.64 Mg C ha^{-1} to 54.46 ± 1.46 Mg C ha^{-1}. Though they measured the biomass per unit area for all the four species, they did not report their findings in terms of carbon content in the live biomass. They observed that the aboveground biomass and the belowground biomass varied from 6.32 g cm^{-2} to 14.87 g cm^{-2} and 1.86 g cm^{-2} to 4.49 g cm^{-2}, respectively. Apart from the observations discussed above, there are no more measurements in any other salt marsh patches of this country.

10.6 Economic Evaluation of the Blue Carbon Stock

Given the importance of these blue carbon repositories, we have realized long ago that these ecosystems need preservation. However, converting the theoretical considerations to practical implementation needs the involvement of money, wherein comes the scope of finance and market and hence economics. Blue carbon economics is an emerging discipline, which originated with the principal aim of devising a proper finance mechanism to ensure the conservation, restoration, and management endeavors for these ecosystems (Locatelli et al. 2014). Any physical attention towards these ecosystems inevitably involves manual labor or compromise from a section of society directly dependent on these ecosystems. We cannot also deny an expectation of profit from the stakeholders who are in the business of safeguarding these ecosystems. To meet the necessities of the stakeholders as well as ensure the sustainability of the blue carbon ecosystems, financial mechanisms like the payments for ecosystem services (PES) and the markets for ecosystem services (MES) have emerged. PES denotes a suite of endowment by which the direct beneficiaries of environmental services prize the stakeholders through subsidies or incentives.

MES refers to all the trade and commerce along with the transactions between the beneficiaries and the stakeholders (Thomas 2014). Both climate finances and markets are equally important, as the finance comes in the form of investments required to execute research and development projects on these ecosystems, which in turn can show us avenues to seek their sustainability shortly. The market bridges the interested buyers of environmental services without whom there will be no source of money (Ullman et al. 2013). The first and foremost challenge to come up with any plausible approach to amalgamate economics with the blue carbon ecosystems is to derive a monetary cost of the ecosystem services that they furnish towards human beings. This chapter has exclusively focused on only one of the regulating services of these ecosystems, carbon sequestration. Usually, while quantifying the net worth of any such ecosystem per unit area, we should consider all the possible ecosystem services provided by these ecosystems (Vo et al. 2012). However, the global scientific community has already concocted plans to estimate the monetary worth of carbon locked by these ecosystems.

We refer to the monetary worth of carbon locked in any natural ecosystems that needs conservation as the social cost of carbon (SCC). Tol (2011), using an exhaustive review, derived at a mean SCC of US$80, taking into account a wide range of peer-reviewed works. The estimate of Tol (2011) was much lower than the mean value reported by a conglomeration of non-peer-reviewed articles. Pendleton et al. (2012) estimated the net worth of blue carbon loss throughout the world considering an SCC of US$41 per ton of CO_2 (according to the value of US$ in the year 2007). This monetary figure refers to a central estimate about the expenses that we need to bear to recover from the economic damages if we emit an additional ton of CO_2 into the atmosphere by the year 2020. Lately, Nordhaus (2017) re-estimated the SCC by implementing a revised dynamic integrated model of climate and the economy (DICE), and it came out to be US$31 per ton of CO_2 (according to the value of US$ in the year 2015).

Several authors have estimated the monetary worth of blue carbon locked in any ecosystem using different protocols. To the best of the author's knowledge, there is still no uniform protocol to measure the monetary values of blue carbon. The two main ways that are in practice are (i) estimation of net new carbon addition to the system and (ii) estimation based on the total carbon locked in the system. Pendleton et al. (2012) underlined that the carbon already locked in the blue carbon ecosystem, if degraded, would return to the atmosphere as CO_2 eventually. Mineralization of 12 g of carbon (atomic weight of carbon) releases 44 g of CO_2 (molecular weight of CO_2) towards the atmosphere. Thus, 1 ton of carbon is equivalent to 3.67 tons of carbon dioxide. Hence, if we can estimate the total weight of organic carbon locked in any blue carbon ecosystem, we can multiply the carbon content with 3.67 to calculate the potential CO_2 emission, most likely to occur if we lose this carbon stock forever. We can compute the net worth of the carbon by multiplying the total CO_2 emission with the social cost of carbon. Very few studies, in this context, exist from India. Akhand et al. (2017) estimated the area of mangrove cover lost during the years 1975 to 2013 in the Indian part of Sundarbans. Using the carbon content per unit area of Ray et al. (2011) for this region, Akhand et al. (2017) estimated a blue carbon loss of US$64.29 million during the abovementioned span of 38 years. Another way to measure the cost of blue carbon is to take into account the rate of net primary productivity of an ecosystem. The multiplicative product of the net ecosystem productivity and the social cost of carbon per unit CO_2 can fetch us the monetary worth of new carbon that an ecosystem incorporates at present. Ganguly et al. (2017) measured the net ecosystem productivity in two seagrass patches of India. Based on the observations carried out in these study sites, Ganguly et al. (2017) derived a conservative estimate of 109–146 million US dollars' worth of carbon in the seagrass meadows of India. However, besides these two regions, the carbon estimation in the other seagrass meadows of India continues to remain constrained, which can lead to considerable uncertainties in the monetary valuation of these seagrass carbon contents. No such attempt is available in the context of the Indian salt marshes.

10.7 Summary and Conclusion

From the present collation of data and observations, we can infer that India, as a country, has a promising potential of blue carbon strength. However, several aspects related to this topic need more attention. We have enumerated the key points that need more endeavors to capitalize on the benefits of the marvelous ecosystems that India is blessed to have in its coastal periphery:

1. The environmental conservationists and scientists have mapped out the spatial extent for all the three conventional blue carbon ecosystems of this country, i.e., the mangroves, the seagrasses, and the salt marshes. However, monitoring the changes of the spatial extents, especially in the smaller patches, needs more emphasis. Studies indicate that the aerial coverage of salt marshes could be underestimated, given the potential of the country's coastline to shelter such ecosystems.
2. Studies reporting the carbon stock measurement in many of the blue carbon sites of this country are presently available in the form of scientific literature. However, again the smaller patches remain neglected in this regard. So far, the Indian government maintains a report on the spatial extent of these ecosystems. The government keeping in mind the importance of the Paris Agreement and the promised nationally determined contributions should strive hard to include an estimate of carbon repository within these ecosystems.
3. Measurement of carbon stock in all the possible compartments of these ecosystems still has many gaps. The sediment carbon stock in many of the mangrove forests and salt marsh patches is yet to receive any attention.
4. Due to multifarious anthropogenic activities and climate change-driven disturbances, the blue carbon ecosystems of this country are incessantly undergoing degradation and are under severe threat. Conservation of these ecosystems by safeguarding the present aerial covers needs a priority to ensure that degradation

of these does not contribute to CO_2 emission. Once we can restore the state of these ecosystems, we should desperately look for viable options to increase the aerial cover of these ecosystems to enable the absorption of more CO_2 from the atmosphere.

5. Coupling of financial mechanisms with the science of blue carbon conservation engaging the local communities is perhaps the only future that can lead to effective management strategies. However, the entire concept from the perspective of India's present scenario is still in its infancy. Academicians and policy managers should join hands to carry out small-scale pilot projects to test the efficacy of such amalgamation for the betterment of these ecosystems.

References

Akhand, A., Mukhopadhyay, A., Chanda, A., Mukherjee, S., Das, A., Das, S., ... & Rao, K. H. (2017). Potential CO 2 emission due to loss of above ground biomass from the Indian Sundarban mangroves during the last four decades. Journal of the Indian Society of Remote Sensing, 45(1), 147-154.

Akhand, A., Watanabe, K., Chanda, A., Tokoro, T., Chakraborty, K., Moki, H., ... & Kuwae, T. (2021). Lateral carbon fluxes and CO2 evasion from a subtropical mangrove-seagrass-coral continuum. Science of The Total Environment, 752, 142190.

Alongi, D. M. (2012). Carbon sequestration in mangrove forests. Carbon management, 3(3), 313-322.

Anand, A., Malhi, R. K. M., Pandey, P. C., Petropoulos, G. P., Pavlides, A., Sharma, J. K., & Srivastava, P. K. (2020). Use of Hyperion for Mangrove Forest Carbon Stock Assessment in Bhitarkanika Forest Reserve: A Contribution Towards Blue Carbon Initiative. Remote Sensing, 12(4), 597.

Anneboina, L. R., & Kumar, K. K. (2017). Economic analysis of mangrove and marine fishery linkages in India. Ecosystem Services, 24, 114-123.

Atwood, T. B., Connolly, R. M., Almahasheer, H., Carnell, P. E., Duarte, C. M., Lewis, C. J. E., ... & Serrano, O. (2017). Global patterns in mangrove soil carbon stocks and losses. Nature Climate Change, 7(7), 523-528.

Banerjee, K., Paneerselvam, A., Ramachandran, P., Ganguly, D., Singh, G., & Ramesh, R. (2018). Seagrass and macrophyte mediated CO2 and CH4 dynamics in shallow coastal waters. PloS One, 13(10), e0203922.

Banerjee, K., Sahoo, C. K., Bal, G., Mallik, K., Paul, R., & Mitra, A. (2020). High blue carbon stock in mangrove forests of Eastern India. Tropical Ecology, 1-18.

Banerjee, K., Sappal, S. M., Ramachandran, P., & Ramesh, R. (2017). Salt Marsh: Ecologically Important, Yet Least Studied Blue Carbon Ecosystems in India. Journal of Climate Change, 3(2), 59-72.

Barbier EB, Hacker SD, Kennedy C, Koch EW, Stier AC et al. (2011) The value of estuarine and coastal ecosystem services. Ecological Monographs 81: 169–193.

Bindu, G., Rajan, P., Jishnu, E. S., & Joseph, K. A. (2020). Carbon stock assessment of mangroves using remote sensing and geographic information system. The Egyptian Journal of Remote Sensing and Space Science, 23(1), 1-9.

Broderick, A. C., Frauenstein, R., Glen, F., Hays, G. C., Jackson, A. L., Pelembe, T., ... & Godley, B. J. (2006). Are green turtles globally endangered?. Global Ecology and Biogeography, 15(1), 21-26.

Chanda, A., Akhand, A., Manna, S., Das, S., Mukhopadhyay, A., Das, I., ... & Dadhwal, V. K. (2016). Mangrove associates versus true mangroves: a comparative analysis of leaf litter decomposition in Sundarban. Wetlands Ecology and Management, 24(3), 293-315.

Chave, J. (2005). Measuring wood density for tropical forest trees. A field manual for the CTFS sites. Toulouse, France, 1-7.

Chen, C. T. A., Huang, T. H., Fu, Y. H., Bai, Y., & He, X. (2012). Strong sources of CO2 in upper estuaries become sinks of CO2 in large river plumes. Current Opinion in Environmental Sustainability, 4(2), 179-185.

Chowdhury, A., Naz, A., Bhattacharyya, S., & Sanyal, P. (2018). Cost–benefit analysis of 'Blue Carbon' sequestration by plantation of few key mangrove species at Sundarban Biosphere Reserve, India. Carbon Management, 9(6), 575-586.

Ciais, P., Sabine, C., Bala, G., Bopp, L., Brovkin, V., Canadell, J., … Thornton, P. (2013). Carbon and other biogeochemical cycles. In T. F. Stocker et al. (Eds.), Climate Change 2013: The Physical Science Basis. Contribution of Working Group I to the Fifth Assessment Report of the Intergovernmental Panel on Climate Change (pp. 465–570). Cambridge, UK and New York: Cambridge University Press. https://doi.org/10.1017/CBO9781107415324.015

Cullen-Unsworth, L. C., Jones, B. L., Seary, R., Newman, R., & Unsworth, R. K. (2018). Reasons for seagrass optimism: local ecological knowledge confirms presence of dugongs. Marine Pollution Bulletin, 134, 118-122.

Das, S., Zaman, S., Pramanick, P., Pal, N., & Mitra, A. (2015). Suaeda maritima: a potential carbon reservoir of coastal zone. International advanced research journal in science. Engineering and Technology, 2(5), 61-65.

Duarte CM, Middelburg JJ, Caraco N (2005) Major role of marine vegetation on the oceanic carbon cycle. Biogeosciences 2: 1–8. https://doi.org/10.5194/bgd-2-1-2005.

Duarte, C. M., & Chiscano, C. L. (1999). Seagrass biomass and production: a reassessment. Aquatic botany, 65(1-4), 159-174.

Dwivedi RS, Rao BRM, Bhattacharya S (1999) Mapping wetlands of the Sundarban delta and its environs using ERS-1 SAR data. Int J Remote Sens 20:2235–2247

Ellison, A. M., Farnsworth, E. J., & Merkt, R. E. (1999). Origins of mangrove ecosystems and the mangrove biodiversity anomaly. Global Ecology and Biogeography, 8(2), 95-115.

Ellison, J. C. (1999). Impacts of sediment burial on mangroves. Marine Pollution Bulletin, 37(8-12), 420-426.

Erftemeijer, P. L., & Shuail, D. A. (2012). Seagrass habitats in the Arabian Gulf: distribution, tolerance thresholds and threats. Aquatic Ecosystem Health & Management, 15(sup1), 73-83.

Fourqurean, J. W., Duarte, C. M., Kennedy, H., Marbà, N., Holmer, M., Mateo, M. A., ... & Serrano, O. (2012). Seagrass ecosystems as a globally significant carbon stock. Nature geoscience, 5(7), 505-509.

Ganguly, D., Singh, G., Purvaja, R., Bhatta, R., Selvam, A. P., Banerjee, K., & Ramesh, R. (2018). Valuing the carbon sequestration regulation service by seagrass ecosystems of Palk Bay and Chilika, India. Ocean & Coastal Management, 159, 26-33.

Ganguly, D., Singh, G., Ramachandran, P., Selvam, A. P., Banerjee, K., & Ramachandran, R. (2017). Seagrass metabolism and carbon dynamics in a tropical coastal embayment. Ambio, 46(6), 667-679.

Geevarghese, G. A., Akhil, B., Magesh, G., Krishnan, P., Purvaja, R., & Ramesh, R. (2018). A comprehensive geospatial assessment of seagrass distribution in India. Ocean & Coastal Management, 159, 16-25.

Giri, S., Mukhopadhyay, A., Hazra, S., Mukherjee, S., Roy, D., Ghosh, S., ... & Mitra, D. (2014). A study on abundance and distribution of mangrove species in Indian Sundarban using remote sensing technique. Journal of Coastal Conservation, 18(4), 359-367.

Gnanamoorthy, P., Selvam, V., Ramasubramanian, R., Chakraborty, S., Pramit, D., & Karipot, A. (2019). Soil organic carbon stock in natural and restored mangrove forests in Pichavaram south-east coast of India. Indian Journal of Geo Marine Sciences Vol. 48 (05), pp. 801-808

Green EP, Clark CD, Mumby PJ, Edwards AJ, Ellis AC (1998a) Remote sensing techniques for mangrove mapping. Int J Remote Sens 19:935–956

Green EP, Mumby PJ, Ellis AC, Edwards AJ, Clark CD (1998b) The assessment of mangrove areas using high resolution multispectral air borne imagery. J Coast Res 14:433–443

Gupta, K., Mukhopadhyay, A., Giri, S., Chanda, A., Majumdar, S. D., Samanta, S., ... & Hazra, S. (2018). An index for discrimination of mangroves from non-mangroves using LANDSAT 8 OLI imagery. MethodsX, 5, 1129-1139.

Hansen JCR, Reidenbach MA (2012) Wave and tidally driven flows within Zostera marina seagrass beds and their impact on sediment suspension. Mar Ecol Prog Ser 448: 271–287. https://doi.org/10.3354/meps09225.

Harishma, K. M., Sandeep, S., & Sreekumar, V. B. (2020). Biomass and carbon stocks in mangrove ecosystems of Kerala, southwest coast of India. Ecological Processes, 9(1), 1-9.

IPCC (2007) Contribution of Working Group III to the Fourth Assessment Report of the Intergovernmental Panel on Climate Change; Metz, B., Davidson, O.R., Bosch, P.R., Dave, R., Meyer, L.A., Eds.; Cambridge University Press: New York, NY, USA, p. 851.

ISFR (2019) India State of Forest Report. 2019. Chapter 3 Mangrove cover. Forest Survey of India. Ministry of Environment Forest and Climate Change.

Jagtap, T. G., & Inamdar, S. N. (1991). Mapping of seagrass meadows from the Lakshadweep Islands (India), using aerial photographs. Journal of the Indian Society of Remote Sensing, 19(2), 77-82.

Jagtap, T. G., Chavan, V. S., & Untawale, A. G. (1993). Mangrove ecosystems of India: a need for protection. Ambio, 252-254.

Jana, H., Zaman, S., Chakraborty, S., Pramanick, P., Mondal, K. C., & Mitra, A. (2013). Spatiotemporal variation of stored carbon in Porteresia coarctata along the East and West Coast of India. International Journal of Engineering and Management Sciences, 4(3), 377-381.

Jena, B. K., Arunraj, K. S., Suseentharan, V., Tushar, K., & Karthikeyan, T. (2019). Indian coastal ocean radar network. Current Science, 116(3), 372.

Jennerjahn, T. C. (2020). Relevance and magnitude of 'Blue Carbon' storage in mangrove sediments: Carbon accumulation rates vs. stocks, sources vs. sinks. Estuarine, Coastal and Shelf Science, 107027.

Kathiresan K, Anburaj R, Gomathi V, Saravankumar K (2013) Carbon sequestration potential of *Rhizophora mucronata* and *Avicennia marina* as influenced by age, season, growth and sediment characteristics in southeast coast of India. J Coast Conserv 17: 397-408.

Kauffman, J. B., & Donato, D. C. (2012). Protocols for the measurement, monitoring and reporting of structure, biomass, and carbon stocks in mangrove forests (pp. 50). Bogor, Indonesia: CIFOR.

Kaviarasan, T., Dahms, H. U., Gokul, M. S., Henciya, S., Muthukumar, K., Shankar, S., & James, R. A. (2019). Seasonal Species Variation of Sediment Organic Carbon Stocks in Salt Marshes of Tuticorin Area, Southern India. Wetlands, 39(3), 483-494.

Kennedy H, Beggins J, Duarte CM, Fourqurean JW, Holmer M et al. (2010) Seagrass sediments as a global carbon sink: isotopic constraints. Glob Biogeochem Cycles 24: 1–8.

Keuskamp, J. A., Hefting, M. M., Dingemans, B. J., Verhoeven, J. T., & Feller, I. C. (2015). Effects of nutrient enrichment on mangrove leaf litter decomposition. Science of the Total Environment, 508, 402-410.

Komiyama, A., Ong, J. E., & Poungparn, S. (2008). Allometry, biomass, and productivity of mangrove forests: A review. Aquatic Botany, 89(2), 128-137.

Komiyama, A., Poungparn, S., & Kato, S. (2005). Common allometric equations for estimating the tree weight of mangroves. Journal of Tropical Ecology, 471-477.

Koshy, N. E., Bhatt, J. R., & Vakily, J. M. (2018). Synthesis of the Conference on Management and Conservation of Seagrass Ecosystems in India. Ocean & Coastal Management, 159, 3-6.

Lan X, Hall BD, Dutton G, Mühle J, Elkins JW (2020). Atmospheric composition [in State of the Climate in 2018, Chapter 2: Global Climate]. Special Online Supplement to the Bulletin of the American Meteorological Society, Vol. 101, No. 8, August, 2020.

Laruelle, G. G., Dürr, H. H., Slomp, C. P., & Borges, A. V. (2010). Evaluation of sinks and sources of CO2 in the global coastal ocean using a spatially-explicit typology of estuaries and continental shelves. Geophysical Research Letters, 37(15).

Locatelli, T., Binet, T., Kairo, J. G., King, L., Madden, S., Patenaude, G., ... & Huxham, M. (2014). Turning the tide: how blue carbon and payments for ecosystem services (PES) might help save mangrove forests. Ambio, 43(8), 981-995.

Lovelock, C. E. (2008). Soil respiration and belowground carbon allocation in mangrove forests. Ecosystems, 11(2), 342-354.

Lovelock, C. E., & Duarte, C. M. (2019). Dimensions of blue carbon and emerging perspectives. Biology letters, 15(3), 20180781.

Lovelock, C. E., Fourqurean, J. W., & Morris, J. T. (2017). Modeled CO2 emissions from coastal wetland transitions to other land uses: tidal marshes, mangrove forests, and seagrass beds. Frontiers in Marine Science, 4, 143.

Lund, J. F., Sungusia, E., Mabele, M. B., & Scheba, A. (2017). Promising change, delivering continuity: REDD+ as conservation fad. World Development, 89, 124-139.

Lunstrum, A., & Chen, L. (2014). Soil carbon stocks and accumulation in young mangrove forests. Soil Biology and Biochemistry, 75, 223-232.

Macklin, P. A., Suryaputra, I. G. N. A., Maher, D. T., Murdiyarso, D., & Santos, I. R. (2019). Drivers of CO2 along a mangrove-seagrass transect in a tropical bay: Delayed groundwater seepage and seagrass uptake. Continental Shelf Research, 172, 57-67.

Mall LP, Singh VP, Garge A (1991) Study of biomass, litter fall, litter decomposition and soil respiration in monogeneric mangrove and mixed mangrove forests of Andaman Islands. Trop Ecol 32: 144-152.

Mantgem, P. J. V., & Stephenson, N. L. (2004). Does coring contribute to tree mortality?. Canadian Journal of Forest Research, 34(11), 2394-2398.

McLeod E, Chmura GL, Bouillon S, Salm R, Bjork M et al. (2011) A blueprint for blue carbon: toward an improved understanding of the role of vegetated coastal habitats in sequestering CO2. Frontiers in Ecology and the Environment 9: 552–560.

Mcowen, C. J., Weatherdon, L. V., Van Bochove, J. W., Sullivan, E., Blyth, S., Zockler, C., ... & Fletcher, S. (2017). A global map of saltmarshes. Biodiversity data journal, (5).

Mitra A, Sengupta K, Banerjee K (2011) Standing biomass and carbon storage of above-ground structures in dominant mangrove trees in the Sundarbans. Forest Ecology and Management 261: 1325-1335.

MoEFCC, 2015. India's Intended Nationally Determined Contribution: Working towards Climate Justice. India's INDC submission to UNFCCC. http://www4.unfccc.int/Submissions/INDC/Published%20Documents/India/1/INDIA%20INDC%20TO%20UNFCCC.pdf, Accessed date: 14 March 2017.

Muthukumaran, M., Rao, A. V. B., & Nagalakshmama, K (2013). Loss of above Ground Biomass (AGB) on Avicennia Marina (Forsk.) and CO2 Emission Due to Thane Cyclone in Puducherry Coast, South India.

Nandy S, Kushwaha SPS (2011) Study on the utility of IRS 1D LISS-III data and the classification techniques for mapping of Sundarban mangroves. J Coast Conserv 15:123–137

Naskar, K., & Mandal, R. (1999). Ecology and biodiversity of Indian mangroves (Vol. 1). Daya Books.

Naskar, S., & Palit, P. K. (2015). Anatomical and physiological adaptations of mangroves. Wetlands ecology and management, 23(3), 357-370.

Nasser, A. K. V., Kunhikoya, V. A., & Aboobaker, P. M. (1999). Mangrove ecosystems of Minicoy Island, Lakshadweep. Marine Fisheries Information Service, Technical and Extension Series, 159, 8-10.

Nayak S, Sarangi RK, Rajawat AS (2001) Application of IRS P4 OCM data to study the impact of cyclone on coastal environment of Orissa. Curr Sci 80:1208–1213

Nayak, S., & Bahuguna, A. (2001). Application of remote sensing data to monitor mangroves and other coastal vegetation of India.

Nellemann C, Corcoran E, Duarte CM et al. (2009) Blue carbon: a rapid response assessment. United Nations Environmental Programme, GRID-Arendal, Birkeland Trykkeri AS, Birkeland

NOAA (2020) Global Monitoring Laboratory, Earth System Research Laboratories, https://www.esrl.noaa.gov/gmd/ccgg/trends/mlo.html (accessed on 06 October 2020)

Nordhaus, W.D., 2017. Revisiting the social cost of carbon. PNAS 114 (7), 1518–1523.

Pandey, C. N., & Pandey, R. (2013). Carbon sequestration in mangroves of Gujarat, India. International Journal of Botany and Research, 3(2), 57-70.

Patil, V., Singh, A., Naik, N., & Unnikrishnan, S. (2014). Estimation of carbon stocks in Avicennia marina stand using allometry, CHN analysis, and GIS methods. Wetlands, 34(2), 379-391.

Patro, S., Krishnan, P., Samuel, V.D., Purvaja, R. and Ramesh, R., 2017. Seagrass and Salt Marsh Ecosystems in South Asia: An Overview of Diversity, Distribution, Threats and Conservation Status. In: Wetland Science. Springer India.

Pattnayak, S., Kumar, M., Sahu, S. C., Dhal, N. K., & Behera, R. K. (2019). Comparison of soil characteristics and carbon content of contrastingly different moist-mixed deciduous and evergreen mangrove forest in Odisha, India. Geology, Ecology, and Landscapes, 3(4), 239-246.

Paulose, N. E., Dilipan, E., & Thangaradjou, T. (2013). Integrating Indian remote sensing multispectral satellite and field data to estimate seagrass cover change in the Andaman and Nicobar Islands, India. Ocean Science Journal, 48(2), 173-181.

Pendleton, L., Donato, D. C., Murray, B. C., Crooks, S., Jenkins, W. A., Sifleet, S., ... & Megonigal, P. (2012). Estimating global "blue carbon" emissions from conversion and degradation of vegetated coastal ecosystems. PloS One, 7(9), e43542.

Ramachandran, S., Sundaramoorthy, S., Krishnamoorthy, R., Devasenapathy, J., & Thanikachalam, M. (1998). Application of remote sensing and GIS to coastal wetland ecology of Tamil Nadu and Andaman and Nicobar group of islands with special reference to mangroves. Current Science, 236-244.

Ramesh, R., Banerjee, K., Paneerselvam, A., Raghuraman, R., Purvaja, R., & Lakshmi, A. (2019). Importance of seagrass management for effective mitigation of climate change. In Coastal Management (pp. 283-299). Academic Press.

Rathore AP, Chaudhary DR, Jha B (2016) Biomass production, nutrient cycling, and carbon fixation by Salicornia brachiata Roxb.: a promising halophyte for coastal saline soil rehabilitation. International Journal of Phytoremediation 18(8):801–811

Ray, R., Ganguly, D., Chowdhury, C., Dey, M., Das, S., Dutta, M. K., ... & Jana, T. K. (2011). Carbon sequestration and annual increase of carbon stock in a mangrove forest. Atmospheric Environment, 45(28), 5016-5024.

Reddy, C. S., Pattanaik, C., & Murthy, M. S. R. (2007). Assessment and monitoring of mangroves of Bhitarkanika Wildlife Sanctuary, Orissa, India using remote sensing and GIS. Current Science, 1409-1415.

Roelfsema, C., Kovacs, E. M., Saunders, M. I., Phinn, S., Lyons, M., & Maxwell, P. (2013). Challenges of remote sensing for quantifying changes in large complex seagrass environments. Estuarine, Coastal and Shelf Science, 133, 161-171.

Sahu, S. C., Kumar, M., & Ravindranath, N. H. (2016). Carbon stocks in natural and planted mangrove forests of Mahanadi Mangrove Wetland, East Coast of India. Current Science, 2253-2260.

Santos, D., Estrada, G. C., Fernandez, V., Estevam, M. R., Souza, B. T., & Soares, M. L. (2017). First assessment of carbon stock in the belowground biomass of Brazilian mangroves. Anais da Academia Brasileira de Ciências, 89(3), 1579-1589.

Serrano, O., Almahasheer, H., Duarte, C. M., & Irigoien, X. (2018). Carbon stocks and accumulation rates in Red Sea seagrass meadows. Scientific reports, 8(1), 1-13.

Short, F. T., Polidoro, B., Livingstone, S. R., Carpenter, K. E., Bandeira, S., Bujang, J. S., ... & Erftemeijer, P. L. (2011). Extinction risk assessment of the world's seagrass species. Biological Conservation, 144(7), 1961-1971.

ShyleshChandran, M. S., Ravi, A., John, S. M., Sivan, S., Asha, M. S., Mammen, P. C., ... & Sruthi, S. N. (2020). Ecosystem Carbon Stock of Selected Mangrove Forests of Vypin–Cochin Region, Southwest Coast of India. Wetlands, 1-11.

Silliman, B. R. (2014). Salt marshes. Current Biology, 24(9), R348-R350.

Sivakumar S, Krishnakumar P, Lakshumanan C (2014) Estimation of carbon stocks in above ground biomass in Muthupet mangrove, Southeast coast of India. International Journal of Intellectual Advancements and Research in Engineering Computations 5(2): 139–150

Suresh HS, Bhatt DM, Ravindranath NH, Sukumar R (2013) Species diversity, above ground biomass and standing carbon stocks in different mangrove forest patches of coastal Karnataka: In Mangroves in India: their biology and uses: 191-198.

Thangaradjou T, Sridhar R, Senthilkumar, S., & Kannan, S. (2008). Seagrass resource assessment in the Mandapam coast of the Gulf of Mannar Biosphere Reserve, India. Applied ecology and environmental research, 6(1), 139-146.

Thangaradjou, T., & Bhatt, J. R. (2018). Status of seagrass ecosystems in India. Ocean & Coastal Management, 159, 7-15.

Thomas, S. (2014). Blue carbon: Knowledge gaps, critical issues, and novel approaches. Ecological Economics, 107, 22-38.

Tomlinson, P.B. (1986) The botany of mangroves. Cambridge University Press, Cambridge

Ullman, R., Bilbao-Bastida, V., & Grimsditch, G. (2013). Including blue carbon in climate market mechanisms. Ocean & Coastal Management, 83, 15-18.

Umamaheswari, R., Ramach, S., & Nobi, E. P. (2009). Mapping the extend of seagrass meadows of Gulf of Mannar Biosphere Reserve, India using IRS ID satellite imagery. International Journal of Biodiversity and Conservation, 1(5), 187-193.

Verhayden A, Dahdouh-Guebas F, Thomaes K, De Genst W, Hettiarachchi S, Koedam N (2002) High-resolution vegetation data for mangrove research as obtained from aerial photography. Environ Dev Sustain 4:113–133

Vijay, V., Biradar, R. S., Inamdar, A. B., Deshmukhe, G., Baji, S., & Pikle, M. (2005). Mangrove mapping and change detection around Mumbai (Bombay) using remotely sensed data.

Vo, Q. T., Kuenzer, C., Vo, Q. M., Moder, F., & Oppelt, N. (2012). Review of valuation methods for mangrove ecosystem services. Ecological indicators, 23, 431-446.

Yoro, K. O., & Daramola, M. O. (2020). CO_2 emission sources, greenhouse gases, and the global warming effect. In Advances in Carbon Capture (pp. 3-28). Woodhead Publishing.

Chapter 11
The Conservation of Marine Biodiversity in South Asia and the Blue Economy

Clement A. Tisdell and Somnath Hazra

11.1 Introduction

Economic studies demonstrate that natural marine ecosystems (and the biodiversity which they protect) are globally very valuable. Costanza et al. (2014), for example, estimated that marine ecosystems provided 39% (US$49.7 trillion) of the global economic value of the flow of ecosystem services in 2011. If the economic value of the flow of ecosystem services generated by tidal marshes and mangroves is added to this figure, marine ecosystems account for 59.7% of the value of the global flow of ecosystem services. Unfortunately, no similar valuations are available for the economic value of marine ecosystems in South Asia. However, it is known that the value of ecosystem services provided by marine ecosystems in South Asia is considerable. These services include provisioning, regulating, cultural and supporting services of the type identified in the Millennium Ecosystem Assessment (2005). They are especially important for many coastal and small island communities. For example, they are also particularly important for the economy of the Maldives (which relies heavily on marine-based tourism and fishing) (Sathiendrakumar and Tisdell 1987, 1989) as well as for the welfare of the Indian island communities of Lakshadweep.

Although the estimates provided by Costanza et al. (2014) have their limitations (Tisdell 2017, Chap. 10), there is little doubt that marine biodiversity is very valuable from an economics' point of view. Consequently, it is important that steps be taken to conserve marine ecosystems and thereby sustain their economic value. At present, anthropocentric activities are diminishing the stock of this form of natural

C. A. Tisdell (✉)
School of Economics, The University of Queensland, Brisbane, Australia
e-mail: c.tisdell@uq.edu.au

S. Hazra
School of Oceanographic Studies, Jadavpur University, Kolkata, India

S. Hazra, A. Bhukta (eds.), *The Blue Economy*,
https://doi.org/10.1007/978-3-030-96519-8_11

capital. This stock is being reduced by pollution (much of which originates from land-based activities), over-exploitation of its resources (e.g. overfishing) and the degradation or destruction of marine habitats (e.g. the replacement of tidal marshes with prawn (shrimp) farms). In addition, climate change (which is largely a result of economic activity) looms as a threat to the maintenance of marine ecosystems.

These developments give rise to several questions. What economic, political and administrative failures in South Asia have contributed to a decline in the value of its marine diversity and a reduction in the economic value of its marine resources? Why do they occur? To what extent has the establishment of marine protected areas in South Asia been adopted as a means to conserve marine biodiversity? Is their establishment very effective in conserving biodiversity? To what extent is marine pollution in South Asia a threat to the maintenance of marine ecosystems and biodiversity? How are threatened marine turtle species faring in South Asia? Each of these aspects will be considered in turn, and a discussion will follow. Plastics will be given particular attention in relation to the last two questions.

11.2 Market, Political and Administrative Failures in Managing Marine Resources in the Context of South Asia: Relevant Theory

In South Asia, as in many developing countries, market failures resulting in socially unfavourable environmental effects are compounded by political and administrative failures. Arif and Karim (2013) review the legal framework of the five coastal nations of South Asia (Bangladesh, India, the Maldives, Pakistan and Sri Lanka) and find that marine pollution is not being seriously addressed and that existing laws are being leniently implemented. In addition, as discussed later, the proportion of the marine area of South Asia which has been designated as marine protected areas is relatively small, and furthermore, the available evidence indicates that most of these areas are not well protected (Rajasuriya et al. 2004; Singh 2003, p. 232). These outcomes can be traced primarily to the economic situation faced by low-income communities in South Asia.

These communities face a low-income trap from which it is difficult to escape by implementing environmental reforms. In turn, this situation appears to make it politically and administratively very difficult to change the environmental status quo. In particular, it is difficult to implement policies to reduce externalities from human activities which can have an adverse effect on the conservation of marine biodiversity and to regulate open access to marine resources as well as the way in which common pool resources (e.g. migratory fish stocks) are utilized. The problem can be illustrated by a couple of theoretical examples.

The first example shows how it can be difficult in developing countries to regulate an economic activity which has a negative external impact on the productivity of marine resources when the incomes of those benefiting from this externality are near subsistence level or quite low. It is, for instance, usually politically impractical to reduce the economic activity giving rise to such an adverse spillover by taxing it or by mandating by law that it be reduced because these measures depress (for some time) the already low incomes of the beneficiaries of the negative externality. Furthermore, those adversely affected by the externality may have already adjusted to it, or each may only be affected by a small amount. Therefore, they are politically not responsive to its removal. Furthermore, they are likely to be unwilling to lobby for its removal because of the tyranny of large numbers (Olson 1965). This situation favours political inaction.

Figure 11.1 illustrates the basics of this problem. The line ABC represents the private value of the marginal product (PVMP) of an activity X, which has an adverse effect on the value of marine productivity. The line ADF represents the social value of the marginal product (SVMP) of this activity. The difference between these two lines measures the marginal loss in marine productivity caused by activity X. The line GH represents the marginal willingness of entities to engage in activity X. Assuming that those engaging in activity X have private (or communal) property rights in these, the equilibrium outcome in this model corresponds to point B. Consequently, the entities will engage in a level of economic activity equivalent to x_2. However, this is not socially optimal given the presence of the marine spill-over from activity X. The socially optimal point corresponds to point D. Hence, social optimality would be achieved by reducing the level of activity X from x_2 to x_1. If such a reduction is mandated by law and effective, the beneficiaries of the adverse externality will have a reduction in their income equivalent to the area of the

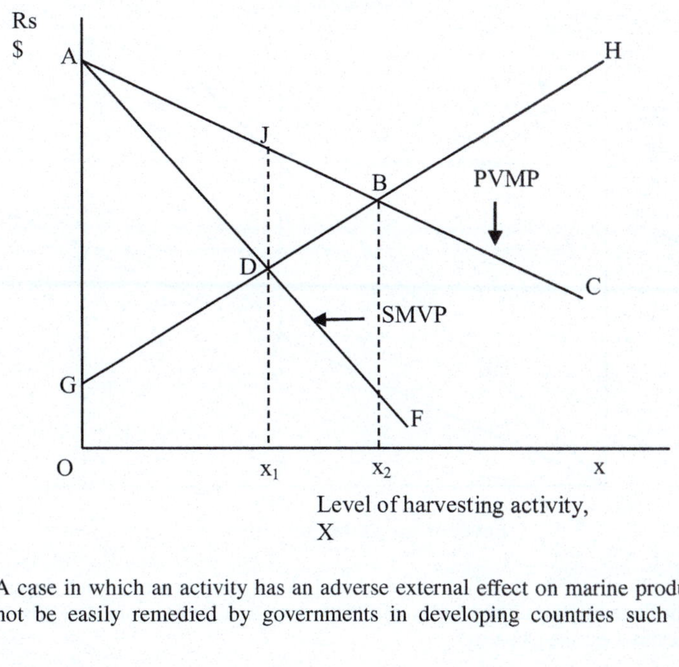

Fig. 11.1 A case in which an activity has an adverse external effect on marine productivity and which cannot be easily remedied by governments in developing countries such as those in South Asia

hatched triangle DBJ. If their incomes are already low, this may cause considerable hardship. Greater hardship would occur if the economic correction is obtained by imposing a tax of DJ on each unit of the offending activity. Consequently, it can be very difficult for governments in developing countries to alter the status quo in situations like this because of income distribution constraints.

The second example is one involving open access to marine resources by a subsistence community. This is illustrated in Fig. 11.2. Suppose that a subsistence community depends on marine resources for its existence and that its population increases up to the means of subsistence as postulated by the Malthusian law of population growth. Furthermore, suppose that the community's effort in harvesting its marine resources increases in proportion to its population and that they all share equally in its total level of harvest of these resources. In Fig. 11.2, let the relationship OABCD represent the total harvest of the community as a function of its total population, P. Then if $x = \lambda P$ is assumed to be its corresponding effort in harvesting its marine resources, this community reaches an equilibrium corresponding to point C. This is supposing that the slope of ray OC corresponds to the subsistence level of income. Per capita income is then equal to y_2/P_1.

Note that the type of production function displayed in Fig. 11.2 is similar to that used in Schaefer's fishing effort model (Schaefer 1957). It is particularly relevant to situations where marine resources are sedentary or relatively so, e.g. mangroves, many species of molluscs and non-migratory fish stocks.

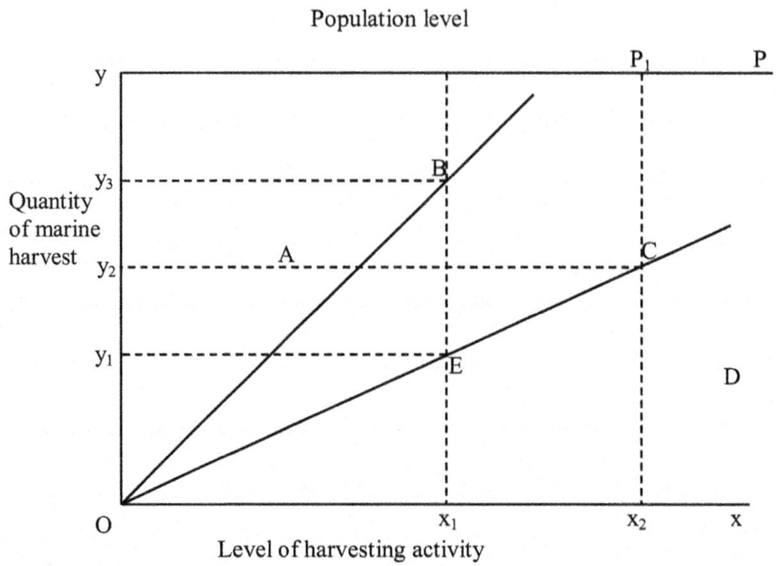

Fig. 11.2 As illustrated by this model, some South Asian communities which depend on marine resources for their livelihood may be caught in an open-access environmental poverty trap from which they cannot escape (see text)

Now if the level of harvesting activity could be reduced to x_1, income per capita would be increased if the level of the population remained at P_1 because y_3 is the maximum sustainable yield of the marine resource. The increase in per capita income would be equal to $(y_3-y_2)/P_1$. There would then be a greater stock of marine resources. However, dynamics make this situation difficult or impossible to achieve. This is because reducing effort will in the short run reduce income per capita to below subsistence level since when marine resource stocks are over-exploited, they take time to recover and to become more productive. Consequently, in the very short run, the harvest may decline in proportion to the decrease in the amount of effort. For example, if effort is rapidly reduced from x_2 to x_1, total output could fall to y_1 resulting in a per capita income of y_1/P_1, and this community would not survive.

Even if the effort put into harvesting the marine resources is reduced to x_1 and the community survives, its level of population would still increase (if the Malthusian law of population growth applied) at the new higher level of output and therefore eventually reduce its income per capita to subsistence level. Hence, the average standard of living of this community would not increase. On the other hand, the stocks of its marine resources would be larger than they otherwise would have been. Furthermore, on average, members of this community would need to make less effort to achieve their subsistence level of income compared to the open-access situation corresponding to point C in Fig. 11.1. This is because the higher level of the marine harvest (y_3) corresponding to point B can support a higher level of population than the level of harvest (y_2) corresponding to point C. Consequently, this community would be better off than at point C because each would have more leisure. However, there could be social pressures to increase the rate of exploitation of the larger stock of marine resources for short-term gain once point B is reached. In the longer term, this would result in reduced output and a population crisis; that is, the level of population of the community would no longer be sustainable in the long run.

The main point is that given the dynamics of adjustment of marine ecosystems (i.e. the time lags involved in increasing their productivity once the extent of their exploitation is reduced), subsistence communities dependent on these resources cannot bear the economic burden of the adjustment involved. This explains why in many developing countries, including those in South Asia, it is difficult or impossible to escape from the poverty trap created by open access to natural resources. There is further discussion of this problem in Tisdell (2006, 2009a, 2009b).

Note that the above models do not capture all the reasons why over-exploitation of marine resources occurs in South Asia and is so difficult to remedy. In some cases, corruption, for instance, facilitates the over-exploitation of these resources. The open-access model, however, probably helps to explain why the marine protected areas of South Asia constitute a comparatively small proportion of its total marine area and why use of their resource is poorly regulated.

11.3 Marine Protected Areas in South Asia

11.3.1 General Observations on Marine Protected Areas

Marine protected areas rely on spatial zoning to manage the conservation of marine resources. Their effectiveness as conservation measures depends on several factors, one of which is their size and another is the extent of the biodiversity contained within them. Apart from ecological factors of this type, their effectiveness depends on how well they are managed. The determination of their economic value is complex because it requires both ecological and economic considerations to be taken into account. Furthermore, the nature of marine protected areas can be very diverse.

11.3.2 Notes on the Extent and the Status of Marine Protected Areas in South Asia

Figure 11.3 specifies for each South Asian country (having a sea border) the total space occupied by the marine protected areas of each. In comparison to their land areas, the Maldives and Bangladesh have set aside the largest aggregate areas for marine protection. Pakistan is the most deficient in this respect and Sri Lanka has also been very reluctant to adopt this form of marine protection. There are, of course, other policy measures which can be adopted to protect and conserve marine resources. Other methods can include restrictions on the nature of fishing gear, seasonal closures of marine areas, harvesting quotas, controls on harvesting methods

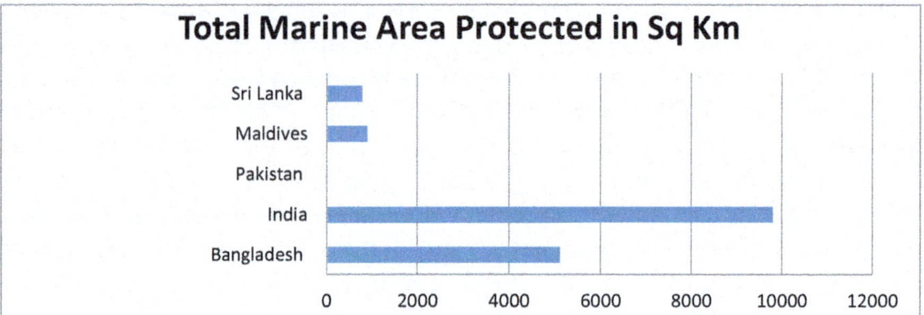

	Bangladesh	India	Pakistan	Maldives	Sri Lanka
Total Marine Area Protected in Sq Km	5098.02	9795.82	6.7	894.48	774.52

Fig. 11.3 Total marine areas protected in South Asia as reported in 2017, by country (in sq. km). *Source*: IUCN (2017): The Newsletter of IUCN World Commission on Protected Areas, South Asia, Regional Vice-Chair, IUCN-WCPA South Asia, Wildlife Institute of India, Chandrabani, Dehradun – 248,001 (Uttarakhand) India

and taxes on catch. It is, however, not possible to include coverage of all of these methods in this chapter due to space limitations.

11.3.2.1 Present Status of MPAs in India

The total area of India's marine protected area is the largest in South Asia. This partly reflects the fact that India has the largest maritime area in South Asia. However, a large proportion of its overall marine protected area is located in the Andaman and Nicobar Islands. These are located in Southeast Asia rather than in South Asia. Therefore, including these areas in the South Asian total inflates the figures for South Asia. Some of India's marine protected areas seem to be too small to have a significant positive impact on biological conservation, for example, the one in the Lakshadweep.

11.3.2.2 Present Status of MPAs in Pakistan

On the other hand, Pakistan only has one marine protected area and in relation to its total marine area has the smallest proportion of marine area protected in this way in South Asia.

11.3.2.3 Present Status of MPAs in Bangladesh

At present, Bangladesh has 13 wildlife sanctuaries and 15 national parks, out of which 7 are associated with marine ecosystems, especially mangrove ecosystems (IUCN 2010). Before 2014, except for a few protected areas used only for hilsa fish management (BOBLME 2015), Bangladesh had no MPAs. Based on the Environment Conservation Act of 1995, the government has now established some other protected areas named 'Ecologically Critical Areas' (ECA). Within Bangladesh's marine zone, four significant ECAs have been established. These are the coastal area of Cox's Bazar, including the Teknaf Peninsula and St. Martin's Island (IUCN 2012). In October 2014, the country's first MPA was declared in the 'Swatch of no ground' area, which is very near the Sundarbans. This MPA aims to preserve the breeding grounds and nursing habitats of a few species of marketable marine fish by maintaining their marine ecosystems. The aim of this is to sustain the livelihood of communities dependent on the harvesting of these species. Bangladesh is now actively trying to conserve its marine biological diversity in order to achieve improved economic outcomes. In June 2019, the Government of Bangladesh declared a large new MPA, namely the Nijhum Dweep Marine Protected Area (NDMPA). This 3188 sq. km NDMPA area covers the coastal and offshore areas of Bhola, Noakhali and Patuakhali districts. It includes important areas for hilsa migration and spawning and is of significance for dolphins, as well as some critically endangered migratory bird species.

11.3.2.4 Present Status of MPAs in the Maldives

The Maldives is a country in the Indian Ocean and consists of a double chain of 26 atolls oriented north to south off India's Lakshadweep islands, between Minicoy Island and the Chagos Archipelago. The Maldives consists entirely of coral reefs, and it has one of the most diverse marine ecosystems in the world. The Maldives is composed of approximately 1190 individual coral structures (Hoon et al. 1997). Its increasing human habitation is concentrated on only a few islands. Due to high tourist demand, its resort islands are increasing to accommodate tourists. Coral reefs and marine biodiversity are the drivers of income through fishing and tourism. As a consequence of population growth and hotel construction, the demand for sand mining, coral mining and reef fishing is also increasing (Sattar et al. 2014). Climate change is a severe threat, mainly characterized by bleaching events such as that in 1998 (Wilkinson et al. 1999), and sea-level rise is also an inundation risk.

The Maldives' government has recognized the value of the coral reefs for its tourism industry (Environment Protection and Preservation Act of Maldives 4/93). In 1995, the government established 15 marine protected areas (MPAs), and later in this year, it established another 10 MPAs. Now, the Maldives has 29 MPAs spread across 12 atolls. Nevertheless, most of the decisions to declare these MPAs were based on the popularity of dive sites with little attention being paid to other aspects of these areas. More recently, the government, United Nations Development Programme (UNDP) and Global Environment Facility (GEF) completed the Atoll Ecosystem Conservation Project (AEC) to design an effective atoll ecosystem conservation and management system and to promote sustainable development (AEC 2009).

11.3.2.5 Present Status of MPAs in Sri Lanka

Realization of the importance of MPAs has come late in Sri Lanka. Some marine and coastal habitats situated within the boundaries of terrestrial protected areas (TPAs) are protected. On the other hand, some of these habitats (tidal and intertidal areas), such as wetlands, mangroves and estuaries, are not protected as a part of the TPAs. Therefore, these areas are not officially acknowledged as MPAs. In Sri Lanka, the first MPA was declared in 1961 at Hikkaduwa as a fisheries protected area (HSAMMSCC 1996). Afterwards, in 1979, the Hikkaduwa Marine Sanctuary was created with an area of 44.5 ha (Rajasuriya 1995). In 1998, it was declared to be a nature reserve, and it was extended to 104 ha. In 2002, it was upgraded to a national park (Rajasuriya and Karunarathna 2002). This series of declarations was carried out to provide a more robust legal mandate for management. After the establishment of Hikkaduwa, several other MPAs have been established by Sri Lanka. Out of several sites identified as desirable protected areas by an Inter-Ministerial Committee on Marine Parks and Sanctuaries, many have not yet been designated as protected areas (De Silva 1985; Rajasuriya 1995). Pigeon Island, currently an MPA, was first declared a sanctuary in 1974 but did not include the surrounding coral reefs until 2003.

According to the report of the BOBLME 'Marine Protected Areas Working Group Meeting,' Sri Lanka has 6 national parks, 13 marine sanctuaries and 12 fishery management areas. Since 2010, Sri Lanka has not declared any new MPAs. Out of all the MPAs, three coastal and marine sites (Pigeon Island, Uppar Lagoon and Pottuvil-Panama sand dunes) are managed through co-management. The largest marine protected area in Sri Lanka is 306.7 sq. km (BOBLME 2014).

11.3.3 How Well Are South Asian Protected Marine Areas Protected?

The oceans are the storehouse of some exceptional and valuable habitats and support human beings through their provision of ecosystem goods and services. No ocean areas remain without any human interventions (Halpern et al. 2008). It is observed that the health of the marine ecosystem is declining due to the multiple impacts of either natural or human intervention (Lasagna et al. 2014), and this is threatening the entire ecosystem systems and their goods and services, including fisheries (Lester and Halpern 2008). As a consequence of these damaging impacts on marine ecosystems and habitats, there is a need for ecosystem-based approaches to the management of marine ecosystems and a call for the expansion of marine protected areas (Lester and Halpern 2008; Lubchenco et al. 2003). However, some marine resource-dependent populations have objected to measures to protect marine resources as marine reserves. This creates a lot of difficulties for establishing marine protection strategies (Lester and Halpern 2008).

Throughout the globe, many coastal communities are dependent on marine natural resources. This dependence is more prevalent in developing countries where the level of education is low and poverty is pervasive along with minimal options for livelihood diversification. Consequently, in low-income families, establishing sustainable harvesting is challenging because every single penny is vital for this poor group who depend on marine resources for their livelihood (Russ 2002). In response to the targets agreed under the CBD (Convention on Biological Diversity), the declaration of the number of MPAs is increasing, but it is observed that the management of these MPAs is challenging (Edgar et al. 2014) and establishing them in LDCs faces many social obstacles.

More than 50% of the world's terrestrial plant and animal species can be found in Asia (Rice 1988). All ecosystems are economically and ecologically important in this terrestrial region, and regulatory frameworks are in place. However, coastal and marine ecosystems are still under threat in South Asia. Several developmental activities have negative impacts on coastal and marine biodiversity across the region. Threats are increasing due to over-exploitation of resources, land transformation, loss of habitat, marine pollution and climate change. Sometimes the number of threatened species is higher at country level: for example, in India, 1256 higher plant species, of more than 15,000 species, are threatened (Sukumar et al. 1995); in Sri Lanka, also 1000 plant species and 380 animal species are threatened.

It is observed that the coral reefs play a crucial role in fisheries and in coastline protection from waves and erosion (Middleton 1999; Ruddle et al. 1988). The habitat richness of Asian coral reefs is declining rapidly (Cesar et al. 1997; Nie et al. 1997; Pennisi 1998), with a substantial decrease of corals and damage to entire reefs. Reasons include unsustainable and wrong fishing techniques, reef mining, siltation, sedimentation and marine pollution with contaminants, including their contamination with untreated wastewater (Glynn 1996; Middleton 1999; UNEP 1999). Extreme temperatures, climate change and sea-level rise are also a severe threat to coral reefs. Consequently, the conditions of the environment have changed, and the reef community structure in the coastal areas of the Andaman Islands of India, the Maldives and Sri Lanka has been increasingly converted from dominated fast-growing branching species to one monopolized by physically rigorous and slow-growing corals (Wilkinson 1998).

Mangrove forests in South Asia are degrading due to severe anthropogenic pressures (Farnsworth and Ellison 1997). These forests are highly vulnerable due to climate change-induced sea-level rise, which changes water salinity. It has been observed that around 7500 ha of mangroves have been inundated due to sea-level rise in the Sundarbans National Park (Bangladesh and India). These coastal mangrove forests provide a habitat for species such as the Bengal tiger, the Eurasian otter, the spotted deer, the wild boar, the estuarine crocodile, fiddler crabs, mud crabs, monitor lizards and marine turtles (Green 1990). According to a research study, a 1 m rise in the sea level will destroy the Sundarbans, resulting in the extinction of the tiger and other species in this area (Smith et al. 1998).

Increased storms are the primary cause of physical damage to seagrass meadows (Short and Neckles 1999). Coastal storms are the primary reason for the sediment movement, which adversely affects seagrass meadows. This turbidity, combined with the massive disturbance in sediments, diminishes the seagrass growth (Kennedy and Björk 2009). Increasing sea-level rise also reduces the productivity and distribution of seagrasses (Short and Neckles 1999).

South Asia's wetland biodiversity is seriously threatened due to sea-level rise (Nicholls et al. 1999). In Gujarat, the Great Rann of Kutch is a seasonal salt lake that supports large populations of the greater flamingo in Asia (Ali 1985; Bapat 1992). Due to sea-level rise, these salt marshes and mudflats may be submerged (Bandyopadhyay 1993). Consequently, it may reduce the extent of the habitat available for the breeding flamingos and lesser floricans (Sankaran et al. 1992) and reduce the amount of habitat available to wild asses.

11.3.4 Further Observations on the Protection of Marine Resources in South Asia

It has been observed that South Asian countries have long coastlines with islands and ocean areas. Many of these areas have a high diversity in marine ecosystems like mangroves, coral reefs, intertidal mudflats, seagrass meadows and others. India is endowed with over 7000 km of coastline comprising 131 MPAs. Bangladesh holds a mega-diversity region. Its extensive Sundarbans mangrove forests are shared with neighbouring country India. The Maldives is also endowed with over 4000 km^2 of coral reef area and many marine resources, giving massive economic services to the Maldives. Very recently, Pakistan declared their first marine protected area. It has rich marine resources (IUCN 2017).

The Bangladesh Forest Department has conducted a tiger census using camera technology to determine the tiger population in the Bangladesh Sundarbans region. In 2015, the tiger population was only 106 in the Bangladeshi Sundarbans, declining from 440 in 2004 (IUCN 2017).

NASA's Earth Observatory has taken Landsat images at two different times (1988 and 2017) of the Thane Creek (or Mumbai's Flamingo Sanctuary). Comparing these, it was found that urbanization is encroaching on the mangroves in the northern part of the creek. In winter, a lot of migratory birds are using this sanctuary (IUCN 2017).

The Odisha State Government of India declared a 7-month ban (November 1, 2017, to May 31, 2018) on fishing along the coast within 20 km off the coastline near turtle nesting sites in order to protect olive ridley sea turtles because in winter they nest in this area. The endangered olive ridley turtles come every year for mass nesting in the Rushikulya river mouth and Gahirmatha Marine Sanctuary (IUCN 2017).

The government of West Bengal has developed a detailed record of the animal and protozoan species of the Indian Sundarbans, and this was published for the first time by the Zoological Survey of India. The publication is called Fauna of Sundarban Biosphere Reserve with details of 2626 species in all (IUCN 2017).

A research group has been studying different locations throughout the Maldives and reported that some of the regions have a remarkable resilience to increasing water temperatures, but other areas are under stress from various causes. In the Maldives, bleaching was reported in 2015. It happened due to El Niño unusual weather patterns throughout 2015 and was aggravated in 2016. It was found that the temperatures at 10 m depth in Baa Atoll and South Malé Atoll were abnormally high and that they gradually increased between March and April (IUCN 2017).

In 2015, the 'Dugong and Seagrass Conservation Project' was launched by Sri Lanka to conserve dugongs (a highly threatened marine mammal and very rare) and seagrass habitats. The Department of Wildlife Conservation of Sri Lanka wants to establish an MPA to protect dugongs (IUCN 2017).

These snippets of information indicate conservation issues involving marine areas are subject to continual change.

11.4 Pollution and Threats to Marine Biodiversity in South Asia: Plastic Wastes and Other Pollutants

11.4.1 General Observations

The full extent to which marine biodiversity is threatened by the disposal of wastes and pollutants is imperfectly known. These costs originate both from the land and on the sea, but the major source is from the land. A report by the South Asia Co-operative Environment Programme (SACEP 2007) identified the following types of wastes and pollutants as having significant negative effects on marine environments in South Asia:

- Sewage discharges.
- Agricultural chemicals, 90% of which eventually reach marine areas.
- Sediments from soil erosion.
- Oil hydrocarbons.
- Solid waste which come from land sources, e.g. plastics, and wastes dumped at sea, e.g. from ships and fishing vessels.

The largest proportion of these wastes comes from land-based human sources, but all have significant consequences for the conservation of marine ecosystems. To the above list should be added industrial wastes and pollutants which are often discharged into waterways in South Asia and which eventually enter maritime areas. Leaking of these wastes from their land-based disposal can also be a problem.

Empirical data is lacking for the extent of economic losses from damages caused to marine ecosystems by the above-mentioned types of litter in South Asia (SACEP 2007, p. viii). Furthermore, according to SACEP (2007, p. viii), 'there are also no systematic assessments of damages to ecosystems, tourism, or public health and safety due to coastal and marine litter in the region'. In fact, having such estimates is a prerequisite for the satisfactory economic assessment of the litter problem. Funding this type of research does not seem to be a high priority in South Asia.

In addition, there appears to be a political reluctance in South Asia to address marine pollution problems, probably due in part to the issues raised in Sect. 11.2 of this chapter. In addition, no doubt, lobbying by special interest groups also is consequential.

11.4.2 South Asia as a Source of Plastic Waste in Marine Areas

It is impossible to consider all sources of marine pollution in this chapter. Therefore, we focus on marine plastic pollution in South Asia. Kapinga and Chung (2020, p. 5) state that their research finds that the main driving forces behind the severe marine plastic pollution in the South Asia region are primarily due to poor practices of waste management systems and an inefficient informal plastic recycling sector. The consequences of marine plastic pollution in South Asia has been enormous, leading to threats to wildlife, community health concerns and economic losses due to the severity of plastic litter and associated clean-up cost.

The World Bank (2020) has pointed out that South Asia is the third largest regional contributor to global plastic waste and that this is predicted to double by 2050 if preventative measures are not adopted.

In 2010, Pakistan was the major generator of plastic wastes (6.41 million tonnes) in South Asia, followed by India (4.49 million tonnes), Sri Lanka (2.62 million tonnes), Bangladesh (1.89 million tonnes) and the Maldives (43,134 tonnes) (Kapinga and Chung 2020, p. 21). On a per capita basis, Sri Lanka generated more plastic wastes than India and Bangladesh and also Pakistan. The major method of disposing of plastics in South Asia is by dumping in the open. Approximately 75% of all plastics disposed of in South Asia are dumped in the open (World Bank Group 2020). Plastics which are dumped in the open tend to be transported relatively quickly to marine areas.

11.4.3 Economic Issues

Although several South Asian countries have national government regulations banning single-use plastic carry bags, these regulations lack enforcement, for example, in Bangladesh, India and Pakistan (Kapinga and Chung 2020, p. 31). Based on data in Beaumont et al. (2019) and Jambeck et al. (2015), Kapinga and Chung (Kapinga and Chung 2020) estimated that in 2010 Bangladesh, India, Pakistan and Sri Lanka as a whole emitted 3.46 million metric tonnes of plastic into their neighbouring oceans resulting in an annual capital loss of nature in marine areas of US$11.4 billion. Beaumont et al. and Jambeck et al. estimated that each tonne of plastic litter entering maritime areas results in an annual loss to maritime nature of US$3300–33,000.

Kapinga and Chung transferred the lower of their estimates to the South Asian situation. However, this may not be an accurate measure of the actual cost of plastic marine pollution in South Asia because this figure is not specific to South Asia. Nevertheless, it does highlight the potential economic seriousness of the problem of plastic litter in marine areas and the magnitude of its threat to nature conservation in these areas. On the other hand, this cost figure does not provide sufficient information to determine appropriate economic policies to address the problem of plastic rubbish in South Asian marine areas.

This is because it does not provide information about the economic benefits of alternative policies to deal with the problem. Furthermore, information is needed about the prospects of effectively implementing alternative policies in South Asia and the comparative costs of managing these policies. For example, taxing the supply of plastics has been raised as one policy possibility by Kapinga and Chung

(2020). However, this may not be politically feasible because of its likely adverse income distribution consequences for the suppliers of plastics. They are likely to lobby against it. They may prefer restrictions on the type of plastics in use because this is likely to reduce their incomes by less than a tax on supply (see Fig. 11.1 and surrounding discussion).

South Asia (as well as Indonesia) imports large amounts of plastic wastes from higher-income countries. In India, for example, many poor families are engaged in sorting this waste for their livelihood. It is sorted into items that are reusable and those that are not. This import undoubtedly adds to the amount of plastic waste reaching marine areas in South Asia. It results in higher-income countries exporting their plastic waste problem to lower-income countries, such as those in South Asia. It would be possible for South Asian nations to ban this import (as China has done), but this would adversely affect the income of the poor who depend on the sorting of this trash for their living. It would also be possible to only allow the import of plastic wastes of a particular quality, but this would require effective monitoring of imports.

There is also a range of other policies which can be considered. These include the safer disposal of used plastics, more recycling of plastics and educational campaigns about improved ways of disposing of plastics and about the plastic waste problem generally. Greater effort might also be made to reduce the use of particular types of plastic which are of greatest threat to marine life. For example, Schuyler et al. (2012a, 2012b, 2016, 2019)) have identified clear soft plastics as a form of plastic most likely to be ingested by sea turtles. This ingestion is often lethal or sublethal to marine turtles. Coloured plastics pose a lower risk to these turtles than clear plastics. Therefore, replacement of clear soft plastics with coloured plastics, where feasible, is worthy of consideration.

11.5 The Blue Economy and Threats to Sea Turtles

According to the IUCN Red List, most species of sea turtles are threatened with extinction. Among other things, their populations are being reduced by the development of the blue economy. Apart from ingesting oceanic litter, especially plastics, many become entangled in discarded, abandoned or lost fishing nets or lines (ghost gear). Lomas (2020) reports that nets of this type drift into Maldivian waters from neighbouring countries causing peak entanglements in the period December to April. Also in South Asia, foreshore areas where sea turtles nest are being lost or are becoming less suited for turtle nesting as a result of the increased economic utilization of these areas. In some areas, also, the consumption of turtle eggs and turtles themselves is having a negative impact on the populations of marine turtles in South Asia.

At present, the economic value of conserving marine turtles is not well understood. The extent to which they are a keystone species is not clear, that is species, the loss of which would result in the demise of many other species. It is known that they do have both consumptive use value in some societies (e.g. for their eggs and meat and in the case of the hawksbill turtle, its shell) and non-consumptive use value as well, e.g. for tourism (Tisdell and Wilson 2005a; Wilson and Tisdell 2001). For many people, they are also a species with considerable non-use value, e.g. their continuing existence is valued in itself. Furthermore, they have become a flagship species for marine conservation of biodiversity (Tisdell and Wilson 2005b). For example, many NGOs support their conservation or use them as part of their fund-raising efforts. A reduction in sea turtle populations due to marine litter reduces the economic value of sea turtles for all the above-mentioned purposes.

The oceanic areas surrounding South Asia have been identified as global hotspots for the ingestion of plastics by sea turtles. The highest risk is in the Bay of Bengal extending to the south of Sri Lanka. The main species at risk in this area are loggerheads, olive ridley, hawksbill and leatherback turtles. In the Arabian Sea (and to its south), green, olive ridley and hawksbill turtles have been identified as facing an intermediate threat from plastic ingestion. Carnivorous turtle species are less likely, however, than other species to ingest anthropogenic debris (Schuyler et al. 2013).

It should, however, be remembered that it is not only sea turtle populations that can be adversely affected by marine debris. Other oceanic species can, for example, become entangled in ghost nets and die as a result. Furthermore, other marine pollutants (such as oil spills), e.g. in the Arabian Sea, have adverse consequences for the survival of marine life. Threats to the conservation of marine biodiversity have and continue to increase with economic growth and with increasing populations in the South Asian region and elsewhere.

11.6 Concluding Comments

The substantial value of conserving marine resources has been pointed out. The satisfactory conservation of marine resources is essential for the sustainable development of the blue economy in South Asia. Relevant theory has been outlined which predicts failures in the satisfactory conservation of marine resources and their biodiversity in South Asia. Market, political and administrative failures occur because in South Asian countries (which have marine boundaries), incomes are low and socioeconomic conditions generally are not very favourable to the conservation of marine resources. This is borne out by the available empirical evidence involving the establishment of marine protected areas. In addition, the less developed status of South Asian nations is an obstacle to their effective control of marine pollution. This has been exemplified by the presence of marine pollution caused by plastic wastes. Marine pollution reduces the economic value of the goods and services supplied by the blue economy and is a threat to the maintenance of marine biodiversity in South Asia. If these nations could reach a higher level of income, many of these problems might disappear, but irreversible losses in marine biodiversity and resources could occur before this happens, if it does.

References

AEC. (2009). Baa Atoll Conservation Programme: 5 year Strategy. *2009-2013 Atoll Ecosystem Conservation Project,* Ministry of Environment, Energy and Water, Government of Maldives, UNDP-GEF. Available from http://broffice.gov.mv/en/files/Baa_Atoll_Conservation_Programme_web_version(1).pdf.

Ali, S. (1985). *The Fall of a Sparrow,* Oxford University Press, New Delhi, India.

Arif, A. and Karim, E. (2013). Marine pollution and the South Asian coastal status: A legal appraisal. *MqJICEL, 2*, 1-23.

Bandyopadhyay, M. (1993). Impact of rising sea levels along the Indian coastline. Pp. 153-166 in M. Lal (Ed.), *Global Warming: Concern for Tomorrow*, Tata, McGraw-Hill Publishing Company Limited, New Delhi, India.

Bapat, N.N. (1992). A visit to the 'flamingo city' in the Great Rann of Kutch, Gujarat. *Journal of the Bombay Natural History Society, 89*, 366-367.

Beaumont, N.J., Aanesen, M., Austen, M.C., Börger, T., Clark, J.R., Cole, M., Hooper, T., Lindeque, P.K., Pascoe, C. and Wyles, K.J. (2019). Global ecological, social and economic impacts of marine plastic. *Marine Pollution Bulletin, 142*, 189-195. https://doi.org/10.1016/j.marpolbul.2019.03.022

BOBLME. (2014). Report of the Marine Protected Areas Working Group Meeting, 11-12 February 2014. BOBLME-2014-Ecology-02, World Fish, Penang, Malaysia. Available from http://aquaticcommons.org/19147/1/BOBLME-2014-Ecology-02.pdf.

BOBLME. (2015). Final Report, Bay of Bengal - Large Marine Ecosystem Project, Compilation of national and regional workshop. SBOBLMEP Pub. /Rep 4, Bangladesh Fisheries Research Institute, Dhaka. Available from https://www.boblme.org/.

Cesar, H., Lundin, C.G., Bettencourt, S. and Dixon, J. (1997). Indonesia coral reefs - an economic analysis of a precious but threatened resource. *Ambio, 26*, 345-350.

Costanza, R., De Groot, R., Sutton, P., van der Ploeg, S., Anderson, S.J., Kubiszewski, I., Farber, S. and Turner, R.K. (2014). Changes in the global value of ecosystem services. *Global Environmental Change, 26*, 152-158.

De Silva, M.W.R.N. (1985). *Status of the Coral reefs of Sri Lanka.* Proceedings of the Conference on Fifth International Coral Reef Congress, held in Tahiti.

Edgar, G.J., Stuart-Smith, R.D., Willis, T.J., Kininmonth, S., Baker, S.C., Banks, S., Barrett, N.S., Becerro, M.A., Bernard, A.T.F., Berkhout, J., Buxton, C.D., Campbell, S.J., Cooper, A.T., Davey, M., Edgar, S.C., Försterra, G., Galván, D.E., Irigoyen, A.J., Kushner, D.J., Moura, R., Parnell, P.E., Shears, N.T., Soler, G., Strain, E.M.A. and Thomson, R.J. (2014). Global conservation outcomes depend on marine protected areas with five key features. *Nature, 506*(7487), 216-220. https://doi.org/10.1038/nature13022

Farnsworth, E.J. and Ellison, A.M. (1997). The global conservation status of mangroves. *Ambio, 26*(6), 328-334.

Glynn, P.W. (1996). Coral reef bleaching: facts, hypotheses and implications. *Global Change Biology, 2*(6), 495-509. https://doi.org/10.1111/j.1365-2486.1996.tb00063.x

Green, J.B. (1990). *IUCN Directory of South Asian Protected Areas*, The World Conservation Union (IUCN), Gland, Switzerland.

Halpern, B.S., Walbridge, S., Selkoe, K.A., Kappel, C.V., Micheli, F., Agrosa, C., Bruno, J.F., Casey, K.S., Ebert, C., Fox, H.E., Fujita, R., Heinemann, D., Lenihan, H.S., Madin, E.M.P., Perry, M.T., Selig, E.R., Spalding, M., Steneck, R. and Watson, R. (2008). A global map of human impact on marine ecosystems. *Science, 319*(5865), 948. https://doi.org/10.1126/science.1149345

Hoon, V., Chong, K.-C., Roy, R., Bierhuizen, B. and Kanvinde, J.R. (1997). Regional Workshop on The Conservation and Sustainable Management of Coral Reefs. M. S Swaminathan Research Foundation and FAO, Chennai, India. Available from http://www.fao.org/3/x5627e/x5627e00.htm.

HSAMMSCC (1996). Special Area Management Plan for the Hikkaduwa Marine Sanctuary and Surrounding Area, Sri Lanka. *Coastal Resources Management Project,* [Hikkaduwa Special Area Management and Marine Sanctuary Coordinating Committee] Coast Conservation Department, National Aquatic Resources Agency, Colombo, Sri Lanka.

IUCN. (2010). National Framework for Establishing and Managing Marine Protected Areas (MPAs) in Bangladesh. IUCN, International Union for Conservation of Nature, Bangladesh Country Office, Dhaka, Bangladesh. Available from https://portals.iucn.org/library/sites/library/files/documents/2015-054.pdf.

IUCN. (2012). Guidelines for applying the IUCN Protected Area Management Categories to Marine Protected Areas. IUCN, Gland, Switzerland.

IUCN. (2017). The Newsletter of IUCN World Commission on Protected Areas, IUCN-WCPA South Asia, Wildlife Institute of India, Chandrabani, Dehradun - 248001 (Uttarakhand) India.

Jambeck, J.R., Geyer, R., Wilcox, C., Siegler, T.R., Perryman, M., Andrady, A., Narayan, R. and Law, K.L. (2015). Plastic waste inputs from land into the ocean. *Science, 347*(6223), 768. https://doi.org/10.1126/science.1260352

Kapinga, C.P. and Chung, S.H. (2020). Marine plastic pollution in South Asia. *Development Paper 20-02*. UN ESCAP, New Delhi, India. Accessed 2 December, 2020, from https://www.unescap.org/resources/sswa-development-paper-20-02-marine-plastic-pollution-south-asia

Kennedy, H. and Björk, M. (2009). Seagrass meadow. Pp. 23-30 in D. Laffoley and G. D. Grimsditch (Eds.), *The Management of Natural Coastal Carbon Sinks*, IUCN, Gland, Switzerland.

Lasagna, R., Gnone, G., Taruffi, M., Morri, C., Bianchi, C.N., Parravicini, V. and Lavorano, S. (2014). A new synthetic index to evaluate reef coral condition. *Ecological Indicators, 40*, 1-9. https://doi.org/10.1016/j.ecolind.2013.12.020

Lester, S.E. and Halpern, B.S. (2008). Biological responses in marine no-take reserves versus partially protected areas. *Marine Ecology Progress Series, 367*, 49-56. https://doi.org/10.3354/meps07599. Available at https://www.int-res.com/abstracts/meps/v367/p49-56/

Lomas, C. (2020). Conservation and rehabilitation of sea turtles in the Maldives. *Testudo, 9*(2), 59-68. Available at https://oliveridleyproject.org/wp-content/uploads/2020/11/conservation-rehabilitation-sea-turtles-maldives_lomas_testudo-vol9_no2_-bcg.pdf

Lubchenco, J., Palumbi, S.R., Gaines, S.D. and Andelman, S. (2003). Plugging a hole in the ocean: The emerging science of marine reserves. *Ecological Applications, 13*(sp1), 3-7. https://doi.org/10.1890/1051-0761(2003)013[0003:PAHITO]2.0.CO;2

Middleton, N. (1999). *The Global Casino: An Introduction to Environmental Issues,* 2nd Edn., Arnold, London.

Millennium Ecosystem Assessment. (2005). *Ecosystems and Human Well-being: A Synthesis*. World Resources Institute, Washington, DC.

Nicholls, R.J., Hoozemans, F.M.J. and Marchand, M. (1999). Increasing flood risk and wetland losses due to global sea-level rise: regional and global analyses. *Global Environmental Change, 9*, S69-S87. https://doi.org/10.1016/S0959-3780(99)00019-9

Nie, B.F., Chen, T.G. and Lien, M.T. (1997). *The Relationship Between Reef Coral and Environmental Changes of Nasha Islands and Adjacent Regions*, China Science Press, Beijing, China. (In Chinese with English abstract).

Olson, M. (1965). *The Logic of Collective Action,* Harvard University Press, Cambridge, MA.

Pennisi, E. (1998). New threat seen from carbon dioxide. *Science, 279*(5353), 989. https://doi.org/10.1126/science.279.5353.989. Available at http://science.sciencemag.org/content/279/5353/989.abstract

Rajasuriya, A. (1995). *The present status of marine protected areas in Sri Lanka.* Proceedings of the Conference on Proceedings of the Review of Protected Area Systems of the Indomalayan Realm Regional Workshop, held in Cisarua, Bogor, Indonesia. Asian Bureau for Conservation.

Rajasuriya, A. and Kurunarathna, C. (2002). Postbleaching status of the coral reefs of Sri Lanka Pp. 54-63 in D. Souter, D. Obura and O. Linden (Eds.), *CORDIO status report 2000*, University of Kalmar, Stockholm, Sweden.

Rajasuriya, A., Zahir, H., Muley, E.V., Subramanian, B.R., Venkataraman, K., Wafar, M.V.M., Munjurul, S.M., Khan, H. and Whittingham, E. (2004). Status of coral reefs in South Asia: Bangladesh, India, Maldives and Sri Lanka. Pp. 95-115 in C. Wilkinson (Ed.), *Status of Coral Reefs of the World, 2000*, Australian Institute of Marine Science, Townsville, Australia.

Rice, C.G. (1988). Habitat, population dynamics, and conservation of the Nilgiri tahr, *Hemitragus hylocrius*. *Biological Conservation, 44*(3), 137-156. https://doi.org/10.1016/0006-3207(88)90099-7

Ruddle, K., Morgan, W.B. and Pfafflin, J.R. (1988). *The Coastal Zone: Man's Response to Change,* Harwood Academic Publishers, Chur, Switzerland.

Russ, G.R. (2002). Yet another review of marine reserves as reef fishery management tools. Pp. 421-443 in P. F. Sale (Ed.), *Coral Reef Fishes*, Academic Press, San Diego. https://doi.org/10.1016/B978-012615185-5/50024-4

SACEP. (2007). Marine Litter in the South Asian Seas Region. South Asia Co-Operative Environment Programme, Sri Lanka. Available from http://www.sacep.org.

Sankaran, R., Rahmani, A.R. and Ganguli-Lachungpa, U. (1992). The distribution and status of the lesser florican *Sypheotides indica* in the Indian subcontinent. *Journal of the Bombay Natural History Society, 89*, 156-179.

Sathiendrakumar, R. and Tisdell, C.A. (1987). Optimal economic fishery effort in the Maldivian tuna fishery: An appropriate model. *Marine Resource Economics, 4*(1), 15-44. https://doi.org/10.1086/mre.4.1.42628975. Available at https://www.journals.uchicago.edu/doi/abs/10.1086/mre.4.1.42628975

Sathiendrakumar, R. and Tisdell, C.A. (1989). Tourism and the economic development of the Maldives. *Annals of Tourism Research, 16*(2), 254-269. https://doi.org/10.1016/0160-7383(89)90071-6

Sattar, S.A., Wood, E., Islam, F. and Najeeb, A. (2014). Current Status of the Reef Fisheries of Maldives and Recommendations for Management. Darwin Reef Fish Project Marine Research Centre/Marine Conservation Society, UK. Available from https://www.mrc.gov.mv/assets/Uploads/February-2014-Current-Status-of-the-Reef-Fisheries-of-Maldives-and-Recommendations-for-Management.pdf.

Schaefer, M.B. (1957). Some considerations of population dynamics and economics in relation to the management of marine fisheries. *Journal of Fisheries Research Board of Canada, 14*, 669-681. https://cdnsciencepub.com/doi/pdf/10.1139/f57-025

Schuyler, Q., Hardesty, B.D., Wilcox, C. and Townsend, K. (2012a). To eat or not to eat? Debris selectivity by marine turtles. *PLOS ONE, 7*(7), e40884. https://doi.org/10.1371/journal.pone.0040884

Schuyler, Q., Hardesty, B.D., Wilcox, C. and Townsend, K. (2012b). Correction: To eat or not to eat? Debris selectivity by marine turtles. *PLOS ONE, 7*(10). https://doi.org/10.1371/annotation/0215f07d-0265-485c-966f-aee192a18313

Schuyler, Q., Hardesty, B.D., Wilcox, C. and Townsend, K. (2013). Global analysis of anthropogenic debris ingestion by sea turtles. *Conservation Biology, 28*(1), 129-139. https://doi.org/10.1111/cobi.12126

Schuyler, Q., Hardesty, B.D., Wilcox, C. and Townsend, K. (2019). Correction: To eat or not to eat? Debris selectivity by marine turtles. *PLOS ONE, 14*(8), e0221846. https://doi.org/10.1371/journal.pone.0221846

Schuyler, Q.A., Wilcox, C., Townsend, K.A., Wedemeyer-Strombel, K.R., Balazs, G., van Sebille, E. and Hardesty, B.D. (2016). Risk analysis reveals global hotspots for marine debris ingestion by sea turtles. *Global Change Biology, 22*(2), 567-576. https://doi.org/10.1111/gcb.13078

Short, F.T. and Neckles, H.A. (1999). The effects of global climate change on seagrasses. *Aquatic Botany, 63*(3), 169-196. https://doi.org/10.1016/S0304-3770(98)00117-X

Singh, H.S. (2003). Marine protected areas in India. *Indian Journal of Marine Sciences, 32*(3), 266-233. http://hdl.handle.net/123456789/4271

Smith, J.B., Rahman, A., Haq, S. and Mirza, M.Q. (1998). *Considering Adaptation to Climate Change in the Sustainable Development of Bangladesh,* World Bank, Washington, D.C.

Sukumar, R., Suresh, H.S. and Ramesh, R. (1995). Climate change and its impact on tropical montane ecosystems in South India. *Journal of Biogeography, 22*, 533-536. Available at https://www.jstor.org/stable/pdf/2845951.pdf

Tisdell, C.A. (2006). Poverty, political failure and the use of open-access resources in developing countries. *Indian Development Review, 4*(2), 441-450.

Tisdell, C.A. (2009a). *Resource and Environmental Economics: Modern Issues and Applications,* World Scientific, Singapore, Hackensack (New Jersey), London, Beijing.

Tisdell, C.A. (2009b). Poverty, policy reforms for resource-use and economic efficiency: neglected issues. *Singapore Economic Review, 54*, 155-166. https://doi.org/10.1142/S0217590809003318

Tisdell, C.A. (2017). *Economics and Environmental Change: The Challenges We Face,* Edward Elgar, Cheltenham, UK, Northampton, MA, USA.

Tisdell, C.A. and Wilson, C. (2005a). Does tourism contribute to sea turtle conservation? *MAST/Maritime Studies - Special Issue Marine Turtles as Flagships, 3*(2), 145-167.

Tisdell, C.A. and Wilson, C. (2005b). Do open-cycle hatcheries relying on tourism conserve sea turtles? Sri Lankan developments and economic–ecological considerations. *Environmental Management, 35*(4), 441-452. Doi: https://doi.org/10.1007/s00267-004-0049-2

UNEP (1999). *Global Environment Outlook 2000,* Earthscan, London.

Wilkinson, C., Linden, O., Cesar, H., Hodgson, G.M., Rubens, J. and Strong, A.E. (1999). Ecological and socioeconomic impacts of 1998 coral mortality in the Indian Ocean: An ENSO impact and a warning of future change? *Ambio, 28*(2), 188-196.

Wilkinson, C.R. (Ed.). (1998). *Status of Coral Reefs of the World.* Australian Institute of Marine Science, Western Australia, Townsville, Australia.

Wilson, C. and Tisdell, C. (2001). Sea turtles as a non-consumptive tourism resource especially in Australia. *Tourism Management, 22*(3), 279-288. https://doi.org/10.1016/S0261-5177(00)00059-5.

World Bank. (2020). South Asia ramps up eco-innovation to curb marine plastic pollution. World Bank, Accessed 2 December, 2020, from https://www.worldbank.org/en/news/press-release/2020/06/08/south-asia-ramps-up-eco-innovation-to-curb-marine-plastic-pollution

World Bank Group. (2020). World Bank Open Data. Accessed 2 December 2020 from https://data.worldbank.org/

Chapter 12
Valuation of Mangrove Ecosystems in South Asian Countries: A Review

Anindya Bhukta and Rikhia Bhukta

12.1 Introduction

Every being lives in one or another ecosystem. Ecosystems are the natural habitat of all the floras and faunas all over the world. These ecosystems provide them with almost everything, from food to shelter, they need for their survival. But everywhere human beings have started over-exploiting these ecosystems over the last 50 years just not for need but from greed. The Millennium Ecosystem Assessment (MEA) reported that approximately 60% of the ecosystems examined during their assessment have seen to be degraded or used unsustainably (Millennium Ecosystem Assessment 2005). These services include freshwater, air and water purification, capture of fishes, regulation of regional and local climate, etc. One, among the ecosystems in which these impacts can be noticed remarkably, is the mangrove ecosystem. This Millennium Ecosystem Assessment report raised concern a lot, and eventually evaluating ecosystem services all over the world gathered momentum as a result.

12.2 Concept of Ecosystem Services

The concept of ecosystem services is relatively recent, appeared first in the 1970s as 'environmental services' (Wilson and Matthews 1970; Lele et al. 2013), and then it was renamed (Ehrlich and Ehrlich 1981) and started systematically used (Ehrlich

A. Bhukta (✉)
Department of Economics, Netaji Mahavidyalaya, Arambagh, West Bengal, India

R. Bhukta
Department of Economic Sciences, Indian Institute of Technology, Kanpur, UP, India

S. Hazra, A. Bhukta (eds.), *The Blue Economy*,
https://doi.org/10.1007/978-3-030-96519-8_12

and Mooney 1983) as 'ecosystem services' only in the 1980s. However, researches on ecosystem services have gained momentum from 1997 onwards after the publication of an edited volume by Gretchen Daily (Daily 1997) and a paper by Costanza and others (Costanza et al. 1997) on the value of economic services.

The term ecosystem service is used to mean all direct and indirect contributions of any ecosystem, natural or modified, in the form of goods and services for the sustenance of life on this Earth (Costanza and Folke 1997; Vo, Tuan et al. 2012). However, several other researchers define ecosystem services in different ways also. In 2005, the MEA report described ecosystem services as 'the benefits people obtain from ecosystems.' But this definition of the MEA is not accepted univocally. Where support in favour of this definition came from the economists like Polasky and Segerson (2009), the economists like Boyd and Banzhaf (2007) opposed it. Those who oppose the MEA definition argue that if we consider 'economic benefit provided by an environmental good or service' as 'the sum of what all members of society would be willing to pay for it' (Mendelsohn and Olmstead 2009), then it would be misleading to characterize all ecosystem services as benefits (Barbier et al. 2011).

There are four types of ecosystem services as identified by Millennium Ecosystem Assessment, namely provisioning services, regulating services, supporting services and cultural services. This typology separates services based on functional lines (Pagiola et al. 2004). Provisioning services refer to the benefits that people receive as consumers of various goods and services (LUC 2015). Regulating services are the biophysicochemical processes that help to sustain life-support systems (Thrush et al. 2013). Cultural services, an intangible contribution of ecosystems, are defined by the MEA as 'the nonmaterial benefits people obtain from ecosystems through spiritual enrichment, cognitive development, reflection, recreation, and aesthetic experiences.' Finally, supporting services, as defined by FAO, provide living spaces for plants or animals of all ecosystems and also maintain a diversity of complex processes that underpin the other ecosystem services.

12.3 Valuation of Ecosystem Services

Although ecosystem services are the lifeline of the residents of various ecosystems, valuation of these services was never thought to be essential at least before the last half of the past century. But due to the rapid and excessive depletion of natural resources, especially during the second half of the twentieth century, and their negative externalities on human health and wealth, academicians started feeling that it is actually the failure of proper valuation of ecosystem services that led to this excessive loss of environmental resources. This realization made ecosystem valuation popular since the 1990s. In a paper published in *Nature* (Costanza et al. 1997), the value of the entire biosphere of this world was estimated by a group of experts. This estimated annual value of global ecosystem services was US$16–54 trillion, and the estimated average was US$33 trillion. The estimate, which was claimed by the authors as a minimum estimate, was greater than the then global GDP (Costanza et al. 2017). This estimated value has however again been revised to $125 trillion per annum assuming updated unit values and revised biome areas (Costanza et al. 2014).

Subsequent to the study by Costanza and his group, a few studies also tried to estimate the TEV of mangrove forests globally. For example, a study by Alongi (2002) estimated this value as US$181 billion. In another study, Rönnbäck (Rönnbäck 1999) estimated that the annual market value of mangrove-based fisheries (that include crustaceans, fish and molluscs) ranges from US$750 to US$16,750 per hectare per year.

Ecosystem services were valued even before the 1990s. But in conventional economic accounting, contributions of ecosystems were valued only when they were harvested and marketed (Costanza et al. 2017). But the question like 'what will happen to the estimation of services offered by a forest, for example, in regulating climate, preventing flood and soil erosion?' led the concept of ecosystem valuation to have appeared as an approach of assigning quantitative values to different goods and services provided by natural resources whether the market value is available or not (Barbier 1997). In fact, proper valuation of ecosystem services is somewhat difficult because of this non-marketed character of most of the ecosystem services (Daily 1997). Most of the ecosystem services are not marketed. The marketed portion is only a negligible fraction of the total economic value of any ecosystem (MEA, Vol. I, p. 136). One more difficulty faced in valuing ecosystem services is the quantification of different ecosystem services. For example, the aesthetic value of coastal beaches varies from person to person. It cannot be quantified and hence its monetary valuation is impossible. Once we are able to derive appropriate values of different ecosystem services, they can be incorporated into the decision-making process of the policymakers in order to correct the market signals (Costanza et al. 1989).

For economic valuation of ecosystem services generally, the total economic value (TEV) framework, based on the presumption that individuals can hold multiple values for ecosystems, is used (National Research Council 2005). The advantage of the economic valuation method is that value concepts in this method incorporate the relationship between ecosystem products and humankind, but the disadvantage is that it often neglects the ecological interdependencies of different ecosystem entities by inadequately addressing the internal structure of different ecosystems (Winkler 2006).

The TEV framework classifies values of ecosystem services primarily into two—use value and non-use value. Use values arise out of the present (direct or indirect use value) or future (option value) anthropogenic use of environmental resources. Direct use values are the outcome of direct use of ecosystem services by the people, which include consumptive use such as harvesting of crops or timbers and non-consumptive use such as enjoying ecotourism. In brief, direct use values are best exemplified by provisioning and cultural services. Indirect use values, on the other hand, are derived from ecosystem services like storm protection offered by mangrove ecosystems which provide benefits outside the ecosystem itself (MEA, Vol. I).

Regulating and supporting services are examples of indirect use values. Finally, option values refer to the option of using the service by oneself in the future (Carson and Bergstrom, 2003). Provisioning, regulating and cultural services, which are not consumed presently but kept for future consumption, are all examples of option value.

Non-use values, on the other hand, arise either out of current direct or of current indirect non-use of ecosystem services (Aylward and Barbier 1992). For example, it may arise either out of the satisfaction an individual derives from its mere existence (existence value) or from his desire to preserve it for future (Spaninks and Van Beukering 1997) generations (bequest value). A hierarchical position of different types of values of ecosystem services is shown in Fig. 12.1.

Valuation of ecosystem services is done in two ways, namely the revealed preference approach and the stated preference approach. In the first approach, consumer behaviour is observed in the market for private goods to deduce the value of environmental goods. The revealed preference approach includes market price method, productivity method, hedonic pricing method, travel cost method, damage cost avoidance method, etc. In the stated preference, on the other hand, the survey method is applied for the valuation of environmental goods (National Research Council 2005). The stated preference approach includes contingent valuation, conjoint analysis, choice modelling, etc.

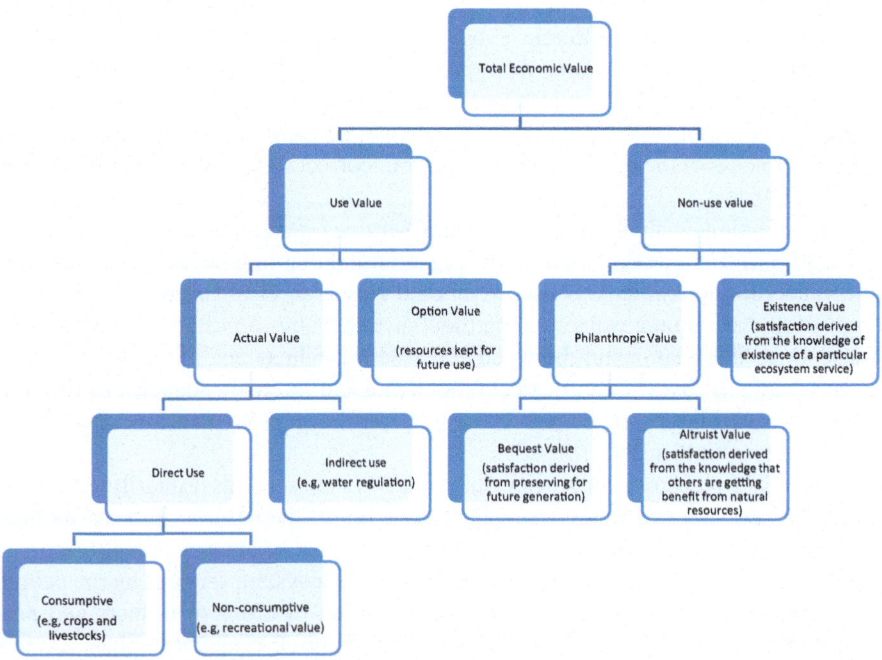

Source: Adopted from Pascual, U et.al (2010) and modified

Fig. 12.1 Total economic value of an ecosystem

Most of the methods stated above are used to capture provisioning and cultural services. But since regulatory and supporting services do not have any direct use value, their valuation is more difficult and complicated. In fact, valuation is simple so long as direct use values are concerned. Thereafter, it becomes increasingly difficult as we move on to indirect use value, option value and non-use value (MEA, Vol. I). Values of ecosystem services, like regulatory and supporting services, which have indirect values only, can exclusively be valued by the stated preference approach. In the stated preference approach, a questionnaire is designed to survey people regarding their willingness to pay for a hypothetical use of an environmental good or service. As a result, this method is most appropriate for valuing indirect uses of ecosystem services.

Many economists opine that simultaneous application of both revealed preference and stated preference approaches should be an effective approach to valuation of environmental goods and services (Keske 2011). However, it should also be noted in this context that there still remain many ecosystem services for which the valuation methods are yet to be explored.

12.4 Mangrove Ecosystem: An Overview

Mangrove, one of the most productive ecosystems of this Earth, consists of halophytic, woody trees and shrubs that grow mainly in tropical and subtropical estuaries where environmental conditions are extremely harsh characterized by scorching temperatures, extreme tides, highly saline and sometimes acidic soil, sediment-laden waters, etc. Mangroves grow in estuaries only because their growth is mainly driven by a steady supply of sediment flowing down from rivers.

Mangrove forests are seen to be distributed over 5 continents, namely Asia (39.0%), Africa (21.0%), North and Central America (15.0%), South America (12.6%) and Oceania (12.4%), and 123 countries (Nanjo 2020). Out of nearly 150,000 km^2 mangrove area distributed all over the world, more than 20% can be found (Fig. 12.2) only in Indonesia (ITTO 2012).

However, what is most frustrating is that mangrove ecosystems, over the entire world, are shrinking at a vulnerable rate due to various reasons (Hutchings and Saenger 1987; Valiela et al. 2001). The rate of decline is so alarming that it is apprehended from some corner that mangroves may be lost within the next hundred years or so (Duke et al. 2007). If immediate steps cannot be taken to reverse this declining trend, then this loss will severely impede the capacities of the mangrove ecosystems to support human needs. We can have a glimpse of the rate of decline of mangrove forests in different regions of the world over the period 1980–2000 and that of in different countries of Asia, respectively from Table 12.1 and Table 12.2.

Although nature has a definite role in this rapid disappearance of mangrove ecosystems, the role of anthropogenic activities is much more crucial. Aquaculture expansion along the coastal line, especially the establishment of shrimp culture farms, plays a major role (Barbier and Cox 2003). Aquaculture is estimated to

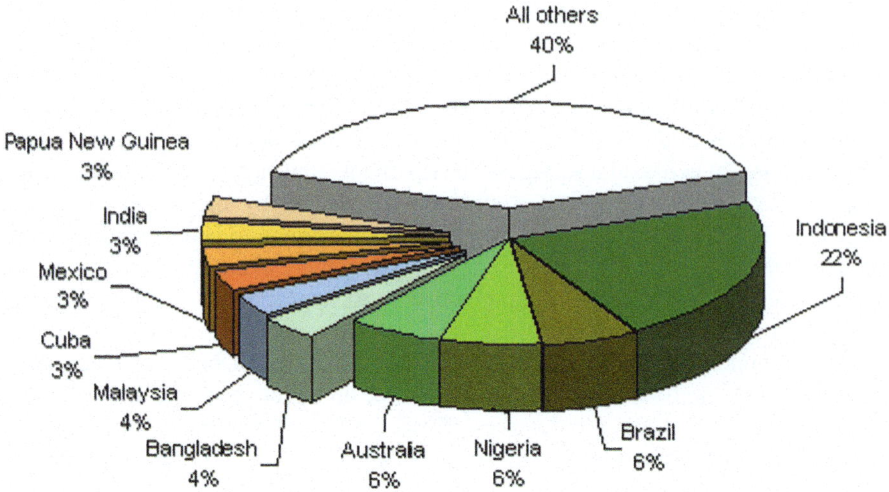

Fig. 12.2 Mangrove extent per country (percentage). (Source: http://www.fao.org/3/j1533e/J1533E02.htm)

Table 12.1 World mangrove coverage

Region	1980 (in '000 ha)	1990 (in '000 ha)	2000 (in '000 ha)	Annual change 1980–1990 (in %)	Annual change 1990–2000 (in %)
Africa	3659	3470	3351	−0.5	−0.3
Asia	7857	6689	5833	−1.6	−1.4
Oceania	1850	1704	1527	−0.8	−1.1
N and C America	2641	2296	1968	−1.4	−1.5
South America	3802	2202	1974	−5.3	−1.1
World total	19,809	16,361	14,653	−1.9	−1.1

Source: http://www.fao.org/3/j1533e/J1533E02.htm

account for 52% of global mangrove loss, where shrimp farming alone responsible for 38% loss. The other important nonsense human activities responsible for this loss include the use of forest products for industrial and woodchip uses (26%), freshwater diversion (11%) and reclamation of land for other uses (5%), and the rest consists of impact of herbicide, agriculture, salt ponds and other coastal developments (Valiela et al. 2001). In India, however, the degradation of mangrove forests has happened mostly due to their conversion into agricultural lands, mainly for rice cultivation (Paul et al. 2018). It is estimated by a study, using satellite remote sensing data, that conversion of mangrove forests into agricultural fields accounts for 81% of the destruction of mangrove ecosystems (Giri et al. 2008). In a nutshell,

Table 12.2 Area under mangrove forests in Asian countries

Country/area	Ha	Reference year
Bahrain	100	1992
Bangladesh	622,482	1996
Brunei Darussalam	17,100	1992
Cambodia	72,835	1997
China	36,882	1994
India	487,100	1997
Indonesia	3,493,110	1988
Islam. Rep. of Iran	20,700	1994
Japan	400	1980
Kuwait	2	2000
Malaysia	587,269	1995
Maldives	n.a.	n.a.
Myanmar	452,492	1996
Oman	2000	1992
Pakistan	207,000	1990
Philippines	127,610	1990
Qatar	500	1992
Saudi Arabia	20,400	1985
Singapore	500	1990
Sri Lanka	8688	1992
Thailand	244,085	2000
Timor-Leste	3035	2000
United Arab Emirates	4000	1999
Vietnam	252,500	1983
Yemen	927	1993
Total Asia	**6,661,717**	**1991**

Source: http://www.fao.org/3/j1533e/J1533E02.htm

there may be differences in percentile data over the conversion of mangrove forests into either aquaculture or agricultural land, but what is undeniable is that such conversion is happening all over the world. This unfortunate event happens in spite of scientists' argument that long-run values of mangrove ecosystems are far greater than its value for any alternative use (Rizal 2018).

Among the most important natural factors responsible for deforestation of mangroves, we must mention the falling supply of sediments. The growth of mangroves, as we have already argued, depends heavily on the supply of sediments. However, over the last three decades or so, this delivery has been declining remarkably in most of the rivers (https://www.downtoearth.org.in/news/wildlife-biodiversity/mangrove-coverage-is-declining-but-there-is-hope-flags-study-74211).

12.5 Mangrove Ecosystem as Service Provider

Mangrove ecosystems provide coastal communities with various kinds of goods and services essential for their livelihood. The list includes nutrient cycling, soil formation fisheries, biodiversity conservation, ecotourism, carbon storing, protection from storm surges, regulation of water quality of estuarine and coastal areas through sedimentation and nutrient uptake, etc. (https://www.itto.int/sustainable_forest_management/mangroves). These services help in maintaining ecological sustainability (by organic matter cycling), enhancing economic prosperity (by providing food and various kinds of forest products that can be marketed for earning livelihood), ensuring environmental security (by protecting from storm surges like a tsunami), etc. Examples of different ecosystem services in the mangrove ecosystem are listed in the following table (Table 12.3).

Mangrove ecosystems play important economic and environmental functions. One of the most important contributions of mangrove forests is their provisioning services in the form of food and fuel, raw materials for different industries, etc. Among the food articles provided by mangroves, the most important are fish and shrimps, which include many commercially important species, and thus the linkage

Table 12.3 Ecosystem functions and ecosystem services of mangrove ecosystem

Ecosystem functions	Goods or services provided	Nature of service
1. Production of:		
(a) Food	Fishes, shrimps, shellfish, crabs, honey	Provisioning service
(b) Fuel	Fuelwood such as goran, hantal, wax, resin	
(c) Other biotic resources	Medicinal ingredients	
(d) Raw materials	Timbers such as sundari, keora; thatching materials like golpata, gewa for paper making	
2. Regulation of environmental quality		
(a) Climate control	Carbon sequestration	Regulatory service
(b) Prevention of soil erosion	Stabilization of sediments and slowing down water flows by the above-ground root system of mangrove trees	
(c) Flood mitigation	Protecting coastal shore from coastal floods, typhoons and tsunamis	
(d) Storm protection	Works as a buffer against storm, especially tropical storm	
(e) Nutrient cycle	Provides nutrients like N, P, K through recycling	
3. Biodiversity conservation	Provides natural habitat to existing flora and fauna; acting as the breeding ground for fish, shrimp, crabs and other shellfish	Supporting service
4. Recreation	Promotion of ecotourism	Cultural service

Source: Prepared and designed by the authors

between mangroves and nearshore and offshore fisheries has been focused on in different studies. More than 50% of global fish and shellfish harvests have been linked directly or indirectly to estuarine nurseries (Manson et al. 2005). In Southeast Asia, mangrove-associated species contribute 30% to fish and 100% to shrimp catches.

Other important goods and services that mangrove ecosystems provide us with include fuel, raw materials for various industries like the drug industry and construction industry, etc. Many species of mangrove trees are used as building materials due to their insect- and rot-resistant properties (http://www.fao.org/forestry/mangrove/3643/en/). Nipa palms (known locally as golpata), used mainly for roof thatching, have alternative uses also, like wood for fuel and ingredients of medicines.

Indirect use values that mangrove ecosystems create by carbon sequestration (Eong 1993; Alongi 2012), prevention of coastal erosion, preventing salt intrusion, purification of coastal water from being polluted, protecting coral reefs from suspended solids, stabilization of sediments and working as a buffer against storms, particularly tropical storms, and ocean waves are no less important.

Mangrove forests play a vital role in climate regulation, especially in controlling global warming and thereby help in reducing its likely impacts. Mangrove ecosystems stabilize atmospheric carbon by fixing carbon in excess of their requirement. This carbon is fixed mainly in the sediments on which mangroves grow. A study estimates that approximately 26 million tonnes of carbon sequestered per year (Yessoufou and Stoffberg 2015). In another study (Chmura et al. 2003) which compile data sets for 154 mangrove sites from both eastern and western Atlantic and Pacific coasts, the Gulf of Mexico, the Mediterranean Ocean and the Indian Ocean, found that the accumulation rate of carbon for mangrove forests is approximately 210 g per square metre per year. At this rate of accumulation, the global estimate of carbon sequestration in mangroves stands at 38 teragrams (1 teragram = 10^{12} or 1 trillion gram) carbon per year. This implies that mangrove forests sequester carbon at a much faster rate than terrestrial forests.

The mangroves stabilize sediment and retain soil in their root structure and thereby help reduce shoreline erosion (Barbier et al. 2011). On every coastline when sea waves strike the land, it erodes soil. Mangrove forests prevent this erosion by resisting the tidal water flow over the soil surface. Sea waves bring sediments with them. By slowing down the speed of the incoming water flow, mangroves subdue the capacity of this water flow to dislodge sediments and carry them out of the mangrove area (Spalding et al. 2014). Suspended sediments then settle down, which results in increased deposition of sediment.

Mangrove forests also work very efficiently as a buffer against storms and sea waves to protect coastal lives and livelihoods. It is seen that, depending on the healthy physical and ecological characteristics, mangrove ecosystems are able to absorb at least 70–90% of the energy of wind-generated waves (UNEP-WCMC 2006).

Mangrove ecosystems are natural habitat for thousands of species, both marine and terrestrial ecosystems (https://www.amnh.org/explore/videos/biodiversity/mangroves-the-roots-of-the-sea/why-mangroves-matter). The loss of habitats will definitely lead to loss of biodiversity and thereby malfunctioning of the ecosystem's productivity. Since the functioning of marine ecosystems and biodiversity is most often positively correlated, biodiversity loss could result in a malfunctioning of the ecosystems' capacity in providing goods and services (Carugati et al. 2018).

12.6 Valuation of Mangrove Ecosystem in Asian Countries: An Overview

Mangrove ecosystem is believed to provide a huge amount of ecosystem services, much of which does not come to the market and hence remains unaccounted for in the policymaking process. It is believed that the enormous benefits that a mangrove ecosystem can provide are yet to be fully realized and hence their potential contributions are generally overlooked in conventional decision making (Brander et al. 2012). Moreover, valuation of some of the ecological services like pollution control by mangroves is yet to be done (Barbier et al. 2011), which again stands on the way in incorporating values of ecosystem services in policymaking framework.

Valuation of mangrove ecosystems has been attempted by several researchers (Gilbert and Janssen 1998; Nickerson 1999; Bennett and Reynolds 1993; Atkinson et al. 2016). A brief overview of some of these studies is summarized in Table 12.4. But doubts were expressed by some that the world's mangrove habitats are always underestimated. One of the reasons that mangrove ecosystem services are often underestimated is that most of them are public goods in nature, and as a result, they do not have precise market value. In fact, this is the case for almost all the ecosystem services, and the same was explicitly acknowledged by Costanza and his co-authors in their pioneering work on the regulation of global ecosystem services. According to them, since ecosystem services are inadequately quantified in terms of comparability with economic services, too little weight is given to them in policy decisions (Costanza et al. 1997).

In valuing the mangrove ecosystem, a very few researchers try to capture total economic value. Most of the others took into account services that either have market value or have available data (Sarhan and Tawfik 2018). However, it is also pertinent to mention in this context that a few researchers took initiatives to value non-market goods and services (Vo et al. 2012).

In valuing mangrove ecosystems, the method which is mostly used is the method of benefit transfer (Himes-Cornell et al. 2018). By this method, the value of an ecosystem is estimated by assigning an already estimated value of a prototype ecosystem situated in almost a similar type of geographic location (Brander et al. 2012, 2013). This method is being used for ecosystem services where market prices are absent. The application of this method is less expensive, less time-consuming, but more prone to either overestimation or underestimation.

Table 12.4 Summary of studies on mangrove ecosystem

Author	Study area	Ecosystem services covered	Method used	Summary of the result
Badola and Hussain (2005)	Bhitarkanika Mangrove Forest, Odisha, India	Regulatory services	Damage cost avoidance method	Locals were found to be aware of the mangrove's function of protecting them from cyclones and were ready to cooperate with the forest department for mangrove restoration
Kibria et al. (2017)	Cambodia	Provisioning services, regulatory services, supporting services, cultural services	Direct valuation method, rainfall storage method, benefit transfer method	This study estimated the total annual monetary and non-monetary values of ecosystem services (ESS) of Veun Sai-Siem Pang National Park (VSSPNP) in Cambodia to be US$129.84 million. The main contribution in this amount comes from air purification, followed by water storage, soil erosion protection, soil fertility improvement, carbon sequestration, provisioning services and recreation
Hema and Devi (2015)	Mangrove areas of Ernakulam and Kannur districts, Kerala, India	Provisioning services, regulatory services, supporting services, cultural services	Contingent valuation method	The residents of the areas expressed their willingness to pay (WTP) in conserving mangroves either in cash or by voluntarily delivering labour or a combination of both. The average WTP estimated is `2308/annum, the range being `50–28,870. Accordingly, the TEV of the mangrove regions becomes `117,947 million
Das and Vincent (2009)	Orissa, India	Regulatory services	Regression analysis	The villages with wider mangroves between them experienced significantly lower death tolls than the villages with narrower/no mangroves during the 1999 Orissa super cyclone. There would have been 1.72 additional deaths per village within 10 km of the coast if mangrove width had been zero. The average price of agricultural land near mangroves is used to calculate the average opportunity cost of saving one life by retaining mangroves to be Rs 11.7 million

(continued)

Table 12.4 (continued)

Author	Study area	Ecosystem services covered	Method used	Summary of the result
Zulkarnaini and Mariana (2016)	Indragiri estuary, Indonesia	Provisioning services, regulatory services, supporting services, cultural services	Replacement cost method, benefit transfer method, contingent valuation method	The total economic value of mangrove forest in Indragiri estuary is Rp. 6,432,296,302/ha/year, which includes direct, indirect, option and existence value
Sannigrahi et al. (2020)	Six eco-regions of the Sundarbans Biosphere Reserve, India	Provisioning services, regulatory services, supporting services, cultural services	Benefit transfer method	The total ecosystem service value (ESV) of six eco-regions of the Sundarbans Biosphere Reserve measured in million US$ is estimated to be 16629.5, 20175.7, 19733.4 and 16761.3, respectively, for four reference years, viz. 1973, 1988, 2002 and 2013. The maximum (ESV) of six eco-regions of the Sundarbans Biosphere Reserve is provided by the mangrove forest, followed by water surface, cropland and sparse vegetation cover
Sathirathai and Barbier (2001)	Southern Thailand	Provisioning services, regulatory services	Economic valuation approach	The paper estimated that the economic value of the mangrove forest lies within the range $27,264–$35,921 per hectare, which includes direct use values of the mangrove forest to local people and indirect use values derived from coastline protection from shrimp farms and offshore fishery linkage. The results also pointed out that although enormous private benefits are obtainable from shrimp farming, it may not be economically viable once we consider negative externalities generated by water pollution and mangrove destruction

(continued)

Table 12.4 (continued)

Author	Study area	Ecosystem services covered	Method used	Summary of the result
Menéndez et al. (2018)	Philippines	Regulatory service	A newly proposed multidisciplinary, multi-step methodology	By piloting a rigorous, engineering-based methodology, this study evaluates the effectiveness of mangrove forests as natural defences. A comparison of flood damages for areas with and without mangroves, it is found that without mangroves damages to property, people and infrastructure would increase annually around 25% in the Philippines
Rizal (2018)	Indonesia	Direct use value, indirect use value, option value and existence value	Contingent valuation method	The paper gives the following estimates of the mangrove ecosystem of Indonesia: • Direct use value ranging from US$19.42 to US$1687.24 • Indirect use value ranging from US$637.93 to US$24,000.53 • Option value (biodiversity value) is US$15.00 • Existence value ranging from US$560.00 to US$2516.40
Uddin (2013)	Sundarbans Mangrove Ecosystem, Bangladesh	Provisioning services, regulatory services, cultural services	Direct valuation method, benefit transfer method	The overall economic value of three ecosystem services of Sundarbans, Bangladesh, is estimated to be approximately US$43 million per annum. Among the three services, the highest value is generated from regulatory services, followed by provisioning services and cultural services

The benefit transfer method can be used only when a particular ecosystem has already been estimated through primary survey. The success of benefit transfers depends on the accuracy of this initial value estimate. But the problem is that an ecosystem is a dynamic concept, and as a result, these unit value estimates may soon become outdated. Therefore, while using this method, one should keep it in mind that the unit value should not be very old in age. In brief, the source, context and age of valuation are very important for the use of the benefit transfer method. But unfortunately researchers often ignored these points (Ibid., p. 9). Therefore, the justification of the argument that most of the studies underestimated mangrove forests can easily be found from this observation.

12.7 Conclusion

Economic valuation methods are generally applied in valuing ecosystem services. Economic value emphasizes on consumer's preferences, the value which a consumer assigns to a particular good or service, and which reflects on his willingness to pay for the concerned good or service. Therefore, by economic valuation method, we can neither address the social values nor the ecological values of ecosystem services in an appropriate manner. Economic valuation methods often overlook the internal structure of ecosystems and thereby ignore ecological interdependencies among the different ecosystem entities (Winkler 2006).

Another drawback of the economic approach is that it does not value people's sentiment regarding their mental attachment with the natural environment they live in. But, for example, in supporting services like biodiversity preservation, the price of this service is simply invaluable. Hence, valuing ecosystem services only by their objective values may be misleading.

Nevertheless, in spite of these limitations, it must be acknowledged in the final note that the method of economic valuation of ecosystem services made the valuation of ecosystem services possible and thereby opened up a new avenue to bring the value of economic services under the purview of national income accounting. Since many of the ecosystem services are enjoyed outside of any market structure, their value was usually underestimated and ignored in decision making (Pascual et al. 2010). The estimation of the value of ecosystem services in monetary terms undoubtedly helped in concluding a resolution to the problem of non-inclusion of ecosystem services in mainstream economic analyses.

References

Alongi, Daniel M. (2002). Present state and future of the world's mangrove forests, *Environmental Conservation 29*(3): 331–349.

Alongi, Daniel (2012). Carbon Sequestration in Mangrove Forests. *Carbon Management, 3*, 313–322. https://doi.org/10.4155/cmt.12.20.

Atkinson, S. C., Jupiter, S. D., Adams, V. M., Ingram, J. C., Narayan, S., Klein, C. J., & Possingham, H. P. (2016). Prioritising mangrove ecosystem services results in spatially variable management priorities. *PloS One*, *11*(3), e0151992.

Aylward, B., & Barbier, E. B. (1992). Valuing environmental functions in developing countries. *Biodiversity & Conservation*, *1*(1), 34-50.

Badola, R., & Hussain, S. A. (2005). Valuing ecosystem functions: an empirical study on the storm protection function of Bhitarkanika mangrove ecosystem, India. *Environmental Conservation*, 85-92.

Barbier, E. B. (1997). The economic determinants of land degradation in developing countries. *Philosophical Transactions of the Royal Society of London. Series B: Biological Sciences*, *352*(1356), 891-899.

Barbier, E. B., & Cox, M. (2003). Does economic development lead to mangrove loss? A cross-country analysis. *Contemporary economic policy*, *21*(4), 418-432.

Barbier E. B. et al. (2011). The value of estuarine and coastal ecosystem services; Ecological Monographs, Vol. 81, No. 2, pp. 169–193; The Ecological Society of America.

Bennett, E. L., & Reynolds, C. J. (1993). The value of a mangrove area in Sarawak. *Biodiversity & Conservation*, *2*(4), 359-375.

Boyd, J., and S. Banzhaf (2007). What are ecosystem services? The need for standardized environmental accounting units; Ecological Economics 63:616–626.

Brander, Luke M. and Alfred J. Wagtendonk, Salman S. Hussain, Alistair McVittie, Peter H. Verburg, Rudolf S. deGroot, Sandervander Ploe (2012); Ecosystem service values for mangroves in Southeast Asia: A meta-analysis and value transfer application Ecosystem Services, 1(2012) 62–69.

Brander, Luke and Roy Brouwera, Alfred Wagtendonkc (2013). Economic valuation of regulating services provided by wetlands in agricultural landscapes: A meta-analysis. Ecological Engineering, 56(2013), 89–96.

Carson Rebecca M. and John C. Bergstrom (2003). A Review of Ecosystem Valuation Techniques; Faculty Series, The University of Georgia, December 2003.

Carugati, L., Gatto, B., Rastelli, E., Martire, M. L., Coral, C., Greco, S., & Danovaro, R. (2018). Impact of mangrove forests degradation on biodiversity and ecosystem functioning. *Scientific reports*, *8*(1), 1-11.

Chmura, G. L., Anisfeld, S. C., Cahoon, D. R., & Lynch, J. C. (2003). Global carbon sequestration in tidal, saline wetland soils. *Global biogeochemical cycles*, *17*(4).

Costanza, R., Farber, S. C., & Maxwell, J. (1989). Valuation and management of wetland ecosystems. *Ecological economics*, *1*(4), 335-361.

Costanza, R. and Folke, C (1997). Valuing Ecosystem Services with Efficiency, Fairness and Sustainability as Goals; Island Press, Washington, DC, pp. 49–70.

Costanza R et al. (1997). The value of the world's ecosystem services and natural capital; Nature, Vol. 387, Page 253-260; May 1997.

Costanza R, De Groot R, Sutton P, Van Der Ploeg S, Anderson SJ, Kubiszewski I, Farber S, Turner RK. 2014. Changes in the global value of ecosystem services. Global Environ Change. 26:152–158. doi:https://doi.org/10.1016/j.gloenvcha.2014.04.002

Costanza, Robert, Rudolf S de Groot, Leon Braat, Ida Kubiszewski, Lorenzo Fioramonti, Paul Sutton, Steve Farber and Monica Grasso (2017). Twenty years of ecosystem services: How far have we come and how far do we still need to go? Ecosystem Services 28 (2017) 1–16.

Daily, G.C. (1997) Introduction: What Are Ecosystem Services? In: Daily, G.C., Ed., Nature's Services: Societal Dependence on Natural Ecosystems, Island Press, Washington DC, 1-10.

Das, S., & Vincent, J. R. (2009). Mangroves protected villages and reduced death toll during Indian super cyclone. *Proceedings of the National Academy of Sciences*, *106*(18), 7357-7360.

Duke, N. C., Meynecke, J. O., Dittmann, S., Ellison, A. M., Anger, K., Berger, U., & Dahdouh-Guebas, F. (2007). A world without mangroves? *Science*, *317*(5834), 41-42.

Ehrlich, P., & Ehrlich, A. (1981). Extinction: the causes and consequences of the disappearance of species.

Ehrlich, P. and H. Mooney (1983). Extinction, substitution, and ecosystem services; *Bioscience* 33(4): 248–254.

Eong, Ong Jin (1993). Mangroves — a carbon source and sink. Chemosphere 27(6), 1097–1107.

Gilbert, A. J., & Janssen, R. (1998). Use of environmental functions to communicate the values of a mangrove ecosystem under different management regimes. *Ecological economics*, *25*(3), 323-346.

Giri, C., Zhu, Z., Tieszen, L. L., Singh, A., Gillette, S., & Kelmelis, J. A. (2008). Mangrove forest distributions and dynamics (1975–2005) of the tsunami-affected region of Asia. *Journal of Biogeography*, *35*(3), 519-528.

Hema, M and P. Indira Devi (2015). Economic Valuation of Mangrove Ecosystems of Kerala, India; Journal of Environmental Professionals Sri Lanka: 2015 – Vol. 4 – No. 1 – 1-16

Himes-Cornell, A., Grose, S. O., & Pendleton, L. (2018). Mangrove ecosystem service values and methodological approaches to valuation: where do we stand? *Frontiers in Marine Science*, *5*, 376.

Hutchings, P. and Saenger, P.(1987). Ecology of Mangroves. Brisbane, Australia: University of Queensland Press.

ITTO [International Tropical Timber Organization]. 2012. Tropical Forest Update 21(2).

Keske Catherine M (2011). How to Value Environmental and Non-Market Goods: A Guide for Legal Professionals, 39 Denv. J. Int'l L. & Pol'y 423

Kibria, A. S., Behie, A., Costanza, R., Groves, C., & Farrell, T. (2017). The value of ecosystem services obtained from the protected forest of Cambodia: The case of Veun Sai-Siem Pang National Park. *Ecosystem Services*, *26*, 27-36.

Lele, Sharachchandra et al. (2013). Ecosystem Services: Origins, Contributions, Pitfalls, and Alternatives; Conservation and Society 11(4): 343-358.

LUC (2015): Calculating the economic values of the ecosystem services provided by Natura 2000 features and sites in Wales, LIFE N2K Wales: LIFE11 NAT/UK/385, (https://cdn.cyfoethnaturiol.cymru/media/679964/action-a09-ecosystem-services-feasbility-study.pdf?mode=pad&rnd=131547122450000000)

Manson, F.J, N.R. Loneragan, G.A. Skilleter and S.R. Phinn (2005). An Evaluation of the Evidence for Linkages between Mangroves and Fisheries: A Synthesis of the Literature and Identification of Research Directions; Oceanography and Marine Biology: an Annual Review, 2005, 43, 485-515 © R. N. Gibson, R. J. A. Atkinson, and J. D. M. Gordon, Editors Taylor & Francis.

Mendelsohn, R., and S. Olmstead. 2009. The economic valuation of environmental amenities and disamenities: methods and applications. Annual Review of Environment and Resources 34:325–347.

Menéndeza, Pelayo and Iñigo J. Losadaa, Michael W. Beckb, Saul Torres-Ortegaa, Antonio Espejoa, Siddharth Narayanc, Pedro Díaz-Simala, Glenn-Marie Langed (2018). Valuing the protection services of mangroves at national scale: The Philippines. Ecosystem Services 34(2018), 24–36.

Millennium Ecosystem Assessment (2005). *Ecosystems and Human Well-being: Synthesis*; Island Press, Washington, DC.

Nanjo, Kusuto (2020). The importance of mangrove and mangrove fishes; The Innovation platform, issue 3, page 304-05, www.innovationnewsnetwork.com; accessed on November 03, 2020.

National Research Council (2005). Valuing Ecosystem Services: Toward Better Environmental Decision-Making. Washington, DC: The National Academies Press.

Nickerson, D. J. (1999). Trade-offs of mangrove area development in the Philippines. *Ecological Economics*, *28*(2), 279-298.

Pagiola, Stefano et al. (2004). Assessing the Economic Value of Ecosystem Conservation; Environment Department Paper No. 10, The World Bank Environment Department.

Pascual, U et al. (2010). The Economics of Ecosystems and Biodiversity; The Ecological and Economic Foundations.

Paul, A.K, Amrit Kanila and Ratnadip Roy (2018). Natural threats and impacts to mangroves within the coastal fringing forests of India; Threats to mangrove forests, coastal research library, Springer.

Polasky, S. and K. Segerson. 2009. Integrating ecology and economics in the study of ecosystem services: some lessons learned. Annual Review of Resource Economics 1:409–434.

Rizal, A. (2018). Science and policy in the coastal zone management. *World News of Natural Sciences*, *21*, 1-8.

Rönnbäck, P. (1999). The ecological basis for economic value of seafood production supported by mangrove ecosystems. *Ecological economics*, *29*(2), 235-252.

Sannigrahi, Srikanta and Suman Chakraborti , Anushna Banerjee, Shahid Rahmat, Sandeep Bhatt, Shouvik Jha, L.K. Singh, Saikat Kumar Paul, Somnath Sen (2020). Ecosystem service valuation of a natural reserve region for sustainable management of natural resources. *Environmental and Sustainability Indicators, 5*(2020).

Sathirathai S. and E. B. Barbier (2001). Valuing Mangrove Conservation in Southern Thailand, Contemporary Economic Policy, Volume 19, Issue 2.

Sarhan, M., & Tawfik, R. (2018, January). The Economic Valuation of Mangrove Forest Ecosystem Services: Implications for Protected Area Conservation. In *The George Wright Forum* (Vol. 35, No. 3, pp. 341-349). George Wright Society.

Spalding M, McIvor A, Tonneijck FH, Tol S and van Eijk P (2014) Mangroves for coastal defence. Guidelines for coastal managers & policy makers. Published by Wetlands International and The Nature Conservancy. 42 p

Spaninks, F., & Van Beukering, P. (1997). Economic valuation of mangrove ecosystems: potential and limitations.

Thrush SF, Townsend M, Hewitt JE, Davies K, Lohrer AM, Lundquist C, Cartner K 2013. The many uses and values of estuarine ecosystems. *In* Dymond JR ed Ecosystem services in New Zealand – conditions and trends. Manaaki Whenua Press, Lincoln, New Zealand. Pp 226–237.

Vo, Quoc Tuan and C. Kuenzer, Quang Minh Vo, F. Moder, N. Oppelt (2012). Review of valuation methods for mangrove ecosystem services. *Ecological Indicators, 23*(2012), 431–446.

Uddin, Md. Shams (2013). Economic valuation of Sundarbans mangrove ecosystems-Bangladesh, Lambert Academic Publishing.

UNEP-WCMC (2006). In the front line: shoreline protection and other ecosystem services from mangroves and coral reefs. UNEP-WCMC, Cambridge, UK 33.

Valiela, I., Bowen, J.L. &York, J.K. 2001. Mangrove forests: one of the world's threatened major tropical environments. BioScience 51, 807–815.

Vo, Quoc Tuan and C. Kuenzer, Quang Minh Vo, F. Moder, N. Oppelt (2012). Review of valuation methods for mangrove ecosystem services, *Ecological Indicators 23*(2012), 431–446.

Wilson, C.M. and W.H. Matthews (eds.). 1970. Man's impact on the global environment: report of the study of critical environmental problems (SCEP). Cambridge, MA: MIT Press.

Winkler, Ralph (2006). Valuation of ecosystem goods and services: An integrated dynamic approach, Part I, Ecological Economics · February 2006.

Yessoufou, K and G.H. Stoffberg (2015). Biogeography, threats and phylogenetic structure of mangrove forest globally and in South Africa: A review; South African Journal of Botany 107 (2016) 114–120.

Zulkarnaini and Mariana (2016). Economic valuation of Mangrove forest ecosystem in Indragiri Estuary. *International Journal of Oceans and Oceanography, 10*(1), 13–17.

Part IV
Cultural Services

Chapter 13
Economic Valuation of Ocean-Based and Ocean-Related Tourism and Recreation

Estibaliz Treviño, David Hoyos, and Elisa Sainz de Murieta

13.1 Introduction

Oceans are a source of energy, nourishment, commerce, transportation, recreation, medicines and freshwater. They also supply jobs and support industries, but the ocean economy's sustainability relies heavily on robust ocean health. Furthermore, oceans are directly affected by anthropogenic impacts which are likely to be intensified by climate change (Waycott et al. 2009). Valuing oceans' ecosystem services (ES) has proven to be a way forward to acknowledge their contribution to human well-being. However, this is not without problems, for example, the lack of a monetary value for ecosystem services, which often leads to an implicit assumption that their value is zero. Moreover, the benefits provided by ES and the costs of their degradation are often not correctly incorporated into the evaluation of projects and public policies (Jacobs et al. 2016). In practice, this has translated into various processes of destruction of natural capital and ecological services around the world. That is why policymakers, planners and managers are increasingly demanding information about the economic implications of biodiversity loss and require tools to incorporate the value of ES into their decisions.

Tourism, the largest economic sector dependent on marine ecosystem function (Ghermandi et al. 2019), also contributes to ocean degradation, especially in coastal areas. Coastal recreational activities, which have been increasing in volume and number over the past decades, occupy a unique place in coastal tourism. They

E. Treviño · D. Hoyos (✉)
University of the Basque Country (UPV/EHU), Leioa, Spain
e-mail: david.hoyos@ehu.es

E. Sainz de Murieta
University of the Basque Country (UPV/EHU), Leioa, Spain

BC3 Basque Centre for Climate Change (BC3), Leioa, Spain

S. Hazra, A. Bhukta (eds.), *The Blue Economy*,
https://doi.org/10.1007/978-3-030-96519-8_13

comprise two main types of recreational uses of coastal areas: (1) consumptive activities, such as fishing, shellfishing and shell collecting, and (2) non-consumptive activities, including swimming, diving, sailing, surfing, windsurfing, jet-skiing, birdwatching, snorkelling, etc. Tourism is one of the primary income sources in many countries and regions. However, the growing tourism industry, although providing significant amounts of investment and being considered an easy way to strengthen national economies, has pushed a constant and often uncontrolled growth of tourist activities in coastal areas around the world. The rapid growth of the recreation sector over the last two decades has also raised concern over the sustainability of its current recreation intensity, thus calling for improved regulation and management of coastal ecosystems (UNEP 2009).

In this context, environmental and natural resource valuation and wealth accounting approaches can contribute to a more sustainable use of resources (Ebarvia 2016). The economic valuation of ES allows to estimate a monetary value for the goods and services provided by nature and, at the same time, to estimate the economic impacts of human activities, taking as a reference the damage caused to ecosystems and their respective services. Additionally, the economic valuation of environmental goods and services enables the comparison of ES with market goods and services (TEEB 2010).

This chapter aims to contribute to the literature on ecosystem service valuation by assessing the benefits provided by ocean-based and coastal ecosystems to ocean-related tourism and recreation, thus helping policymakers designing more sustainable management policies. This chapter is structured as follows: Section 2 highlights the importance of oceans for tourism and recreation. Section 3 describes the main methodologies for the economic valuation of ecosystem services, and Sect. 4 reviews the ongoing literature on monetary estimates for cultural services around the world, with a special focus on tourism and recreation. A conclusion section ends the chapter.

13.2 Ocean-Based and Ocean-Related Tourism and Recreation

This section will emphasize not only the importance of oceans but also the relevance of tourism for economic growth, especially in coastal areas. The development of sustainable tourism is essential not to degrade marine ecosystems. The Earth is called the 'Blue Planet' due to the large extension of water on its surface; oceans play a crucial role in society. More than 70% of the planet is covered by water, 96.5% of which corresponds to oceans (Pidwirny 2006). The oceans are an abundant source of food, energy, medicines, commerce and recreation. They are also a means of transport and trade and a source of income and jobs (Ebarvia 2016).

Tourism is a large contributor to the world economy, making up for 10.3% of global gross domestic product (GDP)—approximately $8.9 trillion—as well as 330 million jobs, and 10.4% of total employment in 2019. Over the past 5 years, one in four of all net new jobs created worldwide has occurred in the travel and tourism sector. Moreover, tourism-related GDP growth outpaced the overall economic growth for the ninth consecutive year. The region with the highest increase in tourism is Central Asia, followed by Northeast Asia, the Middle East, Southeast Asia, South Asia, the Caribbean, North Africa, North America, the European Union, sub-Saharan Africa, Latin America and, finally, Oceania (WTTC 2020).

Much of the world's tourism is concentrated on the marine and coastal environment, and it is expanding. Trends in an ageing population, rising incomes and relatively low transport costs make coastal and oceanic locations increasingly attractive (UNEP 2014). In Europe, for instance, coastal tourism is a leading economic sector in the Mediterranean region in terms of revenues and occupation. While coastal areas around the globe represent 2% of the land area (McGranahan et al. 2007), half of the 300 million international arrivals recorded in 2011 in the Mediterranean region took place in coastal areas, accounting for a significant 15% of world figures. Benefits generated by tourism and recreational activities in coastal regions exceeded 250 billion euros. Estimates also indicate that the tourist sector in 2012 provided 3.3 million direct jobs and 8.5 million total jobs in coastal Mediterranean areas (UNEP 2016). These figures illustrate the economic importance of coastal tourism as well as the close relationship between tourism and economic growth (Sequeira and Maçãs Nunes 2008). It should also be noted that tourism is a sector vulnerable to external shocks, such as potential climate change impacts or the actual COVID-19 pandemic. According to the latest edition of the UNWTO World Tourism Barometer, the lockdown imposed in response to the pandemic led to a 98% fall in international tourist numbers in May 2020 as compared with the previous year.[1]

The marine and coastal environment is a crucial resource for the global tourism industry. It supports all aspects of the tourism development cycle, from infrastructure and the well-known 'sun, sand and sea' formula to the diverse and growing nature-based tourism field (UNEP 2014). Over the decades, coastal tourism has been identified as the largest tourism market segment globally, and it is gaining even more importance (UNEP 2011). According to Hall (2001, p. 602), 'coastal tourism embraces the full range of tourism, leisure, and recreationally oriented activities that take place in the coastal zone and the offshore coastal waters. These include coastal tourism development (accommodation, restaurants, food industry, and second homes) and the infrastructure supporting coastal development (e.g., retail businesses, marinas, and activity suppliers). Marine tourism includes ocean-based tourism such as deep-sea fishing and yacht cruising'. This definition of coastal and marine tourism is essential as it acknowledges the multiple elements involved in the tourist sector, from demand to offer, using the coastal and marine environment as the contextual background for tourism activities (Moreno and Amelung 2009).

[1] https://www.unwto.org/news/impact-of-covid-19-on-global-tourism-made-clear-as-unwto-counts-the-cost-of-standstill

Nonetheless, many coastal areas experience constant and uncontrolled growth of tourism activity, leading to the degradation of marine ecosystems. Some of these external effects of tourism include urban expansion, urbanization, habitat destruction and fragmentation, waste production, water pollution and the loss of social and cultural identity and values. Furthermore, many of these effects are likely to be aggravated by climate change, including flooding and coastal erosion, loss of biodiversity and ecosystems (coral reefs and mangroves), alteration of the productivity and distribution of wildlife (sport fish, bird migrations) and changes in the availability and quality of freshwater resources. In addition, tourism is a significant contributor to greenhouse gas emissions and therefore to climate change (Rosselló-Nadal 2014; Scott et al. 2012; UNEP 2012).

Many tourism forms and activities rely directly or indirectly on the use of environmental resources to supply tourists with various goods and services. The relationship between tourism and the environment is one of mutual dependence: not only does tourism depend heavily on the quality of the environment, but the quality of the environment is also very vulnerable to tourist development. Moreover, evidence shows that the demand for traditional mass tourism has reached a maturity stage, which encourages the demand for more responsible forms of tourism (UNEP 2009). Sustainable tourism requires, first, the rational and efficient use of local resources such as water and energy; second, proper waste management for pollution, wastewater, rubbish, gas emissions, etc.; third, the protection and conservation of fragile coastal and marine environments like dunes, wetlands, beaches, seagrass beds or coral reef assemblages; and fourth, the security and respect of local culture and lifestyles and social structures have to be included (UNEP 2016).

In this context, the UNWTO and UNEP (2005) stated that 'sustainable tourism development guidelines and management practices apply to all forms of tourism in all types of destinations, including mass tourism and the various niche tourism segments. Sustainability principles refer to the environmental, economic and socio-cultural aspects of tourism development, and a suitable balance must be established between these three dimensions to guarantee its long-term sustainability.' Three issues can be highlighted from the previous quote, namely:

1. The vital importance of making the best use of environmental assets that represent a crucial function in tourism development, retaining essential ecological processes and supporting to preserve natural heritage and biodiversity.
2. The appreciation of the socio-cultural authenticity of host communities maintains their cultural heritage and traditional values. It also includes the commitment to inter-cultural information and tolerance as vital points.
3. The essential contribution of sustainable tourism to poverty mitigation. Therefore, making crucial to ensure viable, long-time financial operations, providing evenly distributed socio-economic advantages to all stakeholders, along with stable employment and income-earning possibilities. Social services to host communities are also demanded.

So, sustainable tourism policies require the informed participation of all sectors involved and strong political leadership to ensure broad participation and consensus building. Achieving sustainable tourism is a continuous process and requires constant monitoring of impacts, introducing the necessary preventive and corrective measures whenever necessary. Sustainable tourism should also maintain a high level of tourist satisfaction and ensure a meaningful experience for tourists by raising awareness of sustainability issues and promoting sustainable tourism practices (UNWTO and UNEP 2005). A good example of sustainable tourism is the Republic of Costa Rica, one of the most-visited nations in the Central American region. Since the late 1980s, Costa Rica has become a popular nature-based tourism destination. A pioneer of ecotourism, the country draws many tourists to its extensive series of national parks and other protected areas (Honey 1999). So, it attracts ecological tourists due to its rich biodiversity and abundant wildlife. This growing tourist sector required planning to introduce sustainability principles in the main tourist activities. Box 13.1 summarizes some programmes implemented in Costa Rica to ensure sustainability within the sector.

Box 13.1: Initiatives for Sustainable Tourism in Costa Rica

- The Blue Flag Ecological Program

Costa Rica's Blue Flag Ecological Program helps in protecting the environment and social landscape of Costa Rica. This programme began in 1996 as an incentive for coastal communities to keep their beaches clean, and since then, it has grown to encompass a wide variety of destinations and categories. The Blue Flag programme is a driving force behind Costa Rica's healthy communities and ecosystems (Nature Air 2019).

- Certification for Sustainable Tourism (CST)

CST is a national programme aimed at balancing three fundamental factors within the tourist industry: the interaction of business with natural and cultural resources, improving the quality of life within local communities and the economic contribution to other national development programmes. CST encourages companies to adopt a sustainable orientation in every business decision. It includes the use of recycled products, proper waste disposal and treatment, the implementation of water and energy-saving devices, conservation and expansion of Costa Rican forests and better information management systems, among others. The rating system used by the CST is essentially a set of standards that classify and certify each company according to the degree to which its operations comply with a sustainability model (Molina Murillo 2019).

- Ethical Traveler destination

Ethical Traveler, an international non-profit organization, publishes an annual report on social and environmental policies in developing countries. According to this institution, Costa Rica was included in the 2019 list of The World's Ten Best Ethical Destinations. This country supports decarbonization and green energy, including plans to add five new marine protected areas. In September 2019, Costa Rica was one of a dozen countries to sign the historic Escazú Agreement, guaranteeing rights to a clean and healthy environment. It was one of ten pilot sites to implement Land Degradation Neutrality by 2030 (Lefevre et al. 2019).

In sum, greater attention should be given to proper planning and better integration of tourism into coastal development to minimize tourism-induced problems and ensure both the sustainability of the tourism industry and other sectors' coastal resources (UNWTO and UNEP 2005). The following section will describe different methodologies to incorporate the value of the ecosystem services provided by the oceans into economic modelling and decision-making processes.

13.3 Methodology for Valuing Coastal and Ocean-Based Ecosystem Services

Economic valuation is a means to describe the value that people ascribe to natural resources. Estimating a monetary value for the ecosystem services provided by nature, including marine and coastal ecosystems, begins by understanding the many different services that these ecosystems provide to people's well-being. The concept of ecosystem services provides a framework for identifying and quantifying the variety of benefits obtained from the environment (Salcone et al. 2016). Under the framework of the Millennium Ecosystem Assessment (MEA), ES are classified using four categories: provisioning, regulating, cultural and supporting services (MEA 2005).

Valuing the benefits of ES to society is of outstanding importance when formulating environmental or sectoral policies, such as tourism. However, the importance of preserving these ecosystems is often not fully met, partly because many ESs are considered free and unlimited. Furthermore, ecosystems offer intrinsic benefits that cannot be valued in economic terms, making it difficult to implement natural resource management policies. So, proper information on the economic implications of the loss of nature and biodiversity and tools to incorporate the value of ES could be very relevant for environmental decision making.

The value of natural resources is often considered within the total economic value (TEV) framework that represents the value of the sum of all ecosystem services flows that the natural capital generates both in the present and in the future, given an appropriate discount rate (AEMA 2010; TEEB 2010). TEV can be further divided into use and non-use values.

Use values are related to the current or future use of a particular good or service by individuals and can be further classified into direct use values, indirect use values and option values. Direct use values derive from the actual use of a resource, either consumptively (the use of resources extracted from the ecosystem) or non-consumptively (the use of services without removing any elements from the ecosystem). They refer to the benefits obtained from the direct use of ecosystem services such as raw materials, food products, leisure and tourism. Indirect use values, on the other hand, are the benefits derived from ecosystem functions, and they are usually associated with regulating services, such as disturbance regulation, nutrient cycle control, carbon sequestration or waste treatment. Finally, option values form a separate category, representing the value placed on having the option of using ES in the future, even if they have no present use.

Non-use values reflect the satisfaction that individuals may derive from ecosystem services that other people have or will have access to (Kolstad 2000). Non-use values have typically a good public character, so no market price is usually available, and they include two main categories: on the one hand, existence value refers to the amount people get from merely knowing that an environmental resource is conserved; and, on the other hand, bequest values refer to the value that individuals gain from passing a resource on to future generations even if they may not ever directly use or experience the help themselves. These can be altruistic values, which are values attributed by individuals given the knowledge that a resource can be available to other individuals in the current generation. In general, cultural services and non-use values involve the production of experiences that occur in the valuer's mind.

Different valuation methods can be found for estimating the economic value of ES. A first classification distinguishes whether market data is available or not (Abdullah et al. 2011). As for market approaches, monetary values are directly inferred through the interactions observed in markets. Under certain conditions, market prices are assumed to provide with valid monetary values of an individual's willingness to pay (WTP). Consumers are assumed to reveal their preferences through the choices they make in real markets, where they allocate limited resources among different alternatives. In this context, market valuation approaches can be divided into (1) market price-based approach, (2) cost-based approach and (3) production functions (Montenegro 2017).

In the absence of market data (which is the case in many of the goods and services provided by nature), different non-market valuation methods have been proposed. There are two approaches to estimate the economic value of non-market goods and services: revealed preference and stated preference approaches. Revealed preference methods infer the preferences of individuals by observing the choices that individuals reveal in a related market. Revealed preference methods include travel cost method and hedonic pricing. Travel cost method uses the travel costs that people incur when visiting a site, as a proxy for the price for accessing the site. A latent demand curve is estimated by accounting for the number of trips and costs associated with these trips (Kolstad 2000). Hedonic pricing methods rely on surrogate markets, e.g. the housing market, to estimate the economic value for ecosystem or environmental services that are part of such property prices (TEEB 2010).

Rather than observing a related market, stated preference methods simulate a market and the demand for ecosystem services using surveys on hypothetical policy-induced changes in the provision of ES. Individuals are typically provided with hypothetical scenarios, based on plausible outcomes and options, and their choices determine the value of the environmental good or service in question.[2] Stated preference methods are most commonly applied to non-marketed goods or services because markets cannot reveal individuals' preferences. In the contingent valuation method, respondents are directly asked to express their WTP to improve the quantity and quality of a specific good that is not exchanged in a traditional market (Hoyos and Mariel 2010). Instead of asking directly, discrete choice experiments (DCE) present respondents with different alternative hypothetical scenarios among which they need to choose their preferred option (Salcone et al. 2016). A typical DCE contains several sets of options, each containing a set of mutually exclusive alternatives from which respondents have to choose their preferred one. A set of attributes defines the other options, and each of these attributes takes on one or more levels. The levels indicate the range of the alternative. Individual choice involves implicit trade-offs between the levels of attributes in the different alternatives included in a choice set. When there is an attribute that incorporates the programme's cost, it is possible to transform marginal utility estimates into WTP estimates for changes in attribute levels. By combining the different attribute changes, Hicksian welfare measures are obtained (Hoyos 2010). For a more detailed description of this method, the reader may refer to Mariel et al. (2021). Although both revealed and stated preference methods can be used to estimate use values, only stated preference methods can be used when non-use values are involved.

[2] https://www.oceaneconomics.org/nonmarket

Table 13.1 Summary of valuation methods for different ecosystem services

Valuation method	Element of TEV captured	Ecosystem service(s) valued	Benefits of approach	Limitations of approach
Market prices	Direct and indirect use	Those that contribute to marketed products, e.g. timber, fish, genetic information	Market data readily available and robust	Limited to those ecosystem services for which a market exists
Cost-based approach	Direct and indirect use	Depends on the existence of relevant markets for the ecosystem service in question. Examples include human-made defences being used as a proxy for wetlands storm protection; expenditure on water filtration as a proxy for the value of water pollution damages	Market data readily available and robust	Can potentially overestimate actual value
Production function approach	Indirect use	Environmental services that serve as input to market products, e.g. effects of air or water quality on agricultural production and forestry output	Market data readily available and robust	Data intensive and data on changes in services and the impact on production often missing
Hedonic pricing	Direct and indirect use	Ecosystem services that contribute to air quality, visual amenity, landscape, quietness, etc.; i.e. attributes that can be appreciated by potential property buyers	Based on market data, so relatively robust figures	Very data intensive and limited mainly to services related to property
Travel cost	Direct and indirect use	All ecosystem services that contribute to recreational activities	Based on observed behaviour	Generally limited to recreational benefits. Difficulties arise when trips are made to multiple destinations
Contingent valuation	Use and non-use	All ecosystem services	Able to capture use and non-use values	Bias in responses, resource-intensive method, hypothetical nature of the market
Choice modelling	Use and non-use	All ecosystem services	Able to capture use and non-use values	Similar to contingent valuation above

Source: (Defra 2007)

Finally, benefit transfer consists of exporting previous benefit estimates from a study site to another, at one point in time, regarding the researcher's area of interest (Abdullah et al. 2011). That is, benefit transfer is a process by which economic values that have been generated in one context (the 'study site') are applied to another context (the 'policy site') for which values are required. Its main advantage is that it can reduce the need for primary valuation studies (Defra 2007).

The TEV and MEA frameworks can be complementary when categorizing ecosystem services (Defra 2007). In fact, Table 13.1 shows how both approaches can be combined. The TEV framework is a useful tool for exploring what types of values are trying to obtain for each ecosystem service. This framework helps in determining the valuation methods required to capture these values. Additionally, some advantages and drawbacks for each method are raised.

In sum, some valuation methods are more appropriate than others for valuing particular ecosystem services and eliciting specific value components. Moreover, the type of valuation technique chosen will depend not only on the kind of ecosystem service to be valued but also on the quantity and quality of data. The next section will present a review of valuation studies, where different methodologies are applied to get the value for ocean-based and ocean-related recreational and tourism services.

13.4 A Review of Valuation Studies on Ecosystem Services Related to Coastal Tourism and Recreation

The previous section has provided a general overview of economic valuation methods for valuing ecosystem services. In this section, the applicability of these methods will be discussed in the context of cultural services, specifically recreational opportunities and tourism-related activities.

Marine and coastal ecosystems offer a wide variety of passive and active recreational services. Recreational activities provided by these ecosystems include swimming, diving, snorkelling, charter fishing, fishing from the shore, recreational gleaning, kayaking, surfing, free-diving, beach activities and passive appreciation of coastal scenery (Salcone et al. 2016). Research in this area has focused mainly on the cultural services provided by coastal and marine ecosystems, emphasizing their recreational opportunities (Torres and Hanley 2016).

Recreation and tourism represent an opportunity and an essential link for managing the interaction between ecosystems and people (Berg et al. 2005). Recreational activities offer many people a chance to experience the benefits of ES directly through physical exercise, aesthetic experiences, intellectual stimulation, inspiration and other contributions to physical and psychological well-being (Daniel et al. 2012). Studies show that the high recreational benefits associated with coastal and marine ecosystems, and the positive correlation between those benefits and environmental quality, can provide an economic justification for implementing conservation strategies. This issue is particularly relevant in nature-based tourism destinations, where the recreational opportunities offered by these ecosystems are at the core of their tourism product. More importantly, a large number of studies show that the economic justification for protection can be more substantial if the non-use values that recreationists often attribute to cultural services are also considered (Torres and Hanley 2017).

Some recreational activities imply market services, such as diving and fishing for hire, and have observable market prices. Other activities are not usually traded on markets, such as swimming, beach activities and appreciation of coastal landscape. This distinction has implications for appropriate valuation methods and the extent to which values can be estimated without the need for primary data collection. In particular, the estimation of surplus consumption of non-market leisure activities by residents would require stated preference methods. Therefore, ideally all relevant tourism and recreation activities should be identified, qualitatively described and quantified before an assessment, when possible (Salcone et al. 2016).

A number of tourism-related ES valuation studies have been identified. We find that some authors address tourism-related ecosystem services in coastal areas, and a general analysis of this is described in Sect. 4.1. Others address the services provided by specific coastal ecosystems, such as wetlands, beaches and coral reefs. An overview of these ES is presented in Sect. 4.2.

13.4.1 Tourism and Recreation ES in Coastal Systems

Recreational services are the cultural services with highest presence in coastal areas. Generally, these services are non-consumptive direct use values estimated through stated preference methods (Torres and Hanley 2016). Some studies use the contingent valuation method to estimate the non-use value of recreational services in coastal areas. For example, Östberg et al. (2012) value hiking, bathing, fishing and boating/water quality, noise and littering, showing clear support for conservation of coastal areas. Evidence shows that both tourists and residents highly value the ecological features of coastal areas. Besides, recreational and tourism services of marine protected areas (MPA) have also received specific attention. MPA's type of recreational service most valued is scuba diving, followed by snorkelling, recreational fishing/angling and glass-bottom boating. In addition, other studies estimate recreation and leisure values which results can help evaluate the effects of policy measures in certain protected areas (Wielgus et al. 2009). Valuation can also help decision makers and stakeholders to justify the sustainable use and management of the coastal systems (Batel et al. 2014; Thur 2010). In regions where tourism is an important economic driver for the local economy, analysing tourists' options to integrate them into conservation management plans is found to be essential (Oh et al. 2009).

According to the TEEB database, the total monetary value of the potential sustainable use of recreational services of coastal systems has a mean value of about 7000 Int$/ha/year (2007 values), based on seven original value points (Van der

Table 13.2 Economic valuation of ocean-based and ocean-related tourism and recreation around the world. Values have been standardized to USD/ha/year (2019), unless otherwise specified

Valuation technique	Value (USD_{2019}/ha/year)	Outcome	Country	Reference
Benefit transfer	43	Individuals value for recreational services	Spain	Brenner (2007)
Travel cost	150.2 223.4 294.1 338.8	Swimming, boating, recreational fishing and bird and wildlife watching	USA	Johnston et al. (2002)
	6.25	Expected consumer surplus for visiting Jaizkibel	Spain	Hoyos and Riera (2013)
	2226,457.2	The total non-market use value associated with diving in the area	Indonesia, Thailand and Malaysia	Pascoe et al. (2014)
	61,454.8	The total annual recreation benefits	Various	Czajkowski et al. (2015)
Contingent valuation	0.07–0.25	Citizen and foreign visitors' WTP in addition to current park entrance fees, to support reef quality improvements	Kenya	Ransom and Mangi (2010)
	0.03–0.06	Mean WTP for annual access	Netherlands	Thur (2010)
	0.05 0.03	The mean WTP value for improved water quality: respondents from the East coast region and respondents from the West coast region	Sweden	Östberg et al. (2012)
	28.0–32.9 (USD/person)	Individuals are willing to pay between 10% and 29% more for guided dolphin watching tours, which leads to a total WTP	Croatia	Batel et al. (2014)
Discrete choice experiments	0.6 and 1.2 0.01 and 0.03	Recreational anglers' WTP for unit increases in fish size and numbers during an average fishing vacation (10 days) Scuba divers' WTP for unit increases in coral-associated fish and large fish	Mexico	Wielgus et al. (2009)

Ploeg et al. 2010). According to another study held by Ghermandi and Nunes (2013), coastal ecosystems' estimated recreational values range up to 71.112 Int$/ha/year. The lowest values are found at high absolute latitudes, such as the Arctic Circle, North of Canada, East Russia, South of Chile and Patagonia. The highest values are located in large cities like Los Angeles, Caracas, Rio de Janeiro, Abidjan, Hong Kong, Taipei, Tokyo and Sydney. Mainly it is situated in European Mediterranean cities (e.g. Rome, Naples, Marseille and Barcelona) and in Florida (e.g. Miami, Orlando and Tampa), along with several tropical islands (e.g. Canary Islands, Puerto Rico and the Andaman Islands).

Results clearly show support for the conservation of coastal areas. Notably, both tourists and residents highly value the ecological characteristics of coastal areas and their biodiversity. The body of literature also shows that the quality of the recreational experience influences the value that individuals place on the activities they undertake in coastal waters. Furthermore, as water quality contributes positively to the recreational experience, there is social support for its improvement. Finally, the economic valuation of the services provided by coastal waters can be used to assess the economic efficiency of different policies aimed at protecting coastal water ecosystems.

Table 13.2 shows the main findings of selected valuation studies of ocean-based and ocean-related tourism. The literature is classified in terms of the valuation technique, their main outcomes, the country and the year of such studies and the reference. All values are updated to 2019 US$ values.

13.4.2 Tourism-Related Services Provided by Wetlands, Beaches and Coral Reefs

A number of studies focus on ES provided by specific ecosystems, such as wetlands, beaches and coral reefs. In this subsection, we review the role played by these. Sharing the same structure as Table 13.2 in the previous subsection, Table 13.3 summarizes valuation studies focused on wetlands, beaches and coral reefs, with all values updated to 2019 US$ values.

13.4.2.1 Wetlands

Coastal wetlands are transition zones between marine and terrestrial environments considered to be one of the most productive and valuable ecosystems, which offer a wide variety of goods and services that have an important global socio-economic value (Barbier et al. 2011).

The services provided by wetlands include habitat for species, protection against floods, water purification, amenities and recreational opportunities such as scuba diving, recreational fishing and recreational birdwatching, among others. Because many of these services typically have no market price, non-market valuation techniques are generally employed to value the services provided by wetlands (Woodward and Wui 2001).

Overall, most studies estimate non-use values and non-consumptive direct use values associated with the final ES object valuation, since they are attached to cultural services. Some studies applied revealed preference methods, such as travel cost method (Gürlük and Rehber 2008; Shrestha et al. 2002). Stated preference methods have also been used to value the recreational value of wetlands (Faccioli et al. 2015; Westerberg et al. 2010). Globally, Van der Ploeg et al. (2010) estimate

Table 13.3 Economic valuation of tourism-related ecosystem services provided by wetlands, beaches and coral reefs. Values for wetlands have been standardized to USD/ha/year (2019), unless otherwise specified. For beach ecosystem services, values are shown in USD/person/year (2019)

Valuation technique	Value	Outcome	Country	Reference
I. Wetlands				
Contingent valuation	4.59	Total monetary value for the mangroves of Benut	Malaysia	Bann (1999)
Travel cost	97.14–156.27	Average consumer surplus per day of recreational anglers	Brazil	Shrestha et al. (2002)
	60.35	Total monetary value for the marsh recreation in Muthurajawela Wetland	Sri Lanka	Emerton et al. (2003)
	17.247,44	Annual value assigned by visitors to the Kuscenneti National Park	Turkey	Gürlük and Rehber (2008)
Discrete choice experiments	30 USD/ person/year	Active and passive recreation	France	Westerberg et al. (2010)
Benefit transfer	1.591,20	Total monetary value for wetlands in Shenzhen	China	Tianhong et al. (2010)
II. Beaches				
Travel cost	81,035.2–117,442.2 USD/ha/year	Gross recreational benefits (total recreational loss of the beach area of Zandvoort is closed for a year)	Netherlands	Nunes and van den Bergh (2004)
	5256–19,590.8 5256–38,225.9	The net benefits of a day at the beach in North Carolina for users making day trips and for users staying overnight at the beach	North Carolina	Bin et al. (2005)
	36.7 USD/ beach trip	The value of a day at the beach	San Diego	Lew and Larson (2005)
	1095.7	For British tourists, the weighted average of consumer surplus for enjoying the beach	Turkey	Blakemore and Williams (2008)
	14,595.4	Beach recreation value	Australia	Rolfe and Gregg (2012)
	489.7	The total annual recreational value of Queensland beaches	Queensland	Windle and Rolfe (2013)
	2724–3881.4 for residents, 4336.7–5415.5 for visitors	The estimated consumer surplus from a single beach visit trip	Australia	Zhang et al. (2015)

(continued)

Table 13.3 (continued)

Valuation technique	Value	Outcome	Country	Reference
Contingent valuation	857.3	The WTP for beach recreational activities as improved by beach nourishment at all beaches and among all visitor types	South Florida	Shivlani et al. (2003)
	123.5	The annual mean WTP	Netherlands	Nunes and van den Bergh (2004)
	603.6	The value of enjoying the beach for British tourists	Turkey	Blakemore and Williams (2008)
	2860.5	Visitors' WTP for additional beach access points and parking	South Carolina	Oh et al. (2008)
	21,630.6 41,258.4	The increased economic value for an increasing beach water clarity The value of eliminating trash on beaches	Puerto Rico	Loomis and Santiago (2013)
	23.6–29.3	WTP for an annual tax to contribute to preserve beaches from further deterioration	Greece	Kontogianni et al. (2014)
	14,469–1,075,915	The WTP for SAI beaches	San Andres Island (SAI)	Castaño-Isaza et al. (2015)
Discrete choice experiments	15.5	Visitors' WTP to acquire one more beach access point	South Carolina	Oh et al. (2009)
	2.4 3.5	The median WTP is for a marine protected area which allows fishing/fishing is not permitted The median WTP for an increased chance of contracting ear infection from swimming in polluted water	Tobago	Hess and Beharry-Borg (2012)
	20,428.9 39,255.6	The increased economic value for an increasing beach water clarity The value of eliminating trash on beaches	Puerto Rico	Loomis and Santiago (2013)
III. Coral reefs				

(continued)

Table 13.3 (continued)

Valuation technique	Value	Outcome	Country	Reference
Travel cost	1020.74 million to 2.3 billion USD/ year	The annual recreational benefits of the Great Barrier Reef	Australia	Carr and Mendelsohn (2003)
	274.32 million USD/year	Recreational value of coral reefs	Hawaii	Cesar and van Beukering (2004)
	297.3 USD/ trip, 367.5 USD/dive	Boat anglers' recreational benefits/scuba divers' recreational benefit	Taiwan	Chen et al. (2013)
Contingent valuation	338.5 USD/ person/year, 7.13 million USD/year	Consumer surplus, net revenues and individual WTP	Philippines	Ahmed et al. (2007)
	13.5 USD/ ticket	The projected ticket fare for boat fishing and for scuba diving	Taiwan	Chen et al. (2013)
Discrete choice experiment	1.5 USD/dive 1.5 USD/dive	The value of coral and fish diversity The marginal price of water visibility	Israel	Wielgus et al. (2003)
	80.28 USD/2-tank dive	Divers' WTP; strong aversions to fishing activity/gear encounters and divers with a low number of large fish, with WTP values over to avoid such trips	Caribe	Gill et al. (2015)

that the total monetary value of the potential sustainable use of recreation and tourism opportunities of coastal wetlands is 684 int$/ha/year, based on ten original value points.

In general, studies show support for the protection of wetlands and mangroves. It is important to denote that the value estimates vary greatly depending on the ecosystem service assessed and also on the valuation method used. The economic valuation of the recreational and tourism services provided by wetlands can contribute to more efficient wetland management. It can also serve to give guidance to policymakers in designing sustainable policies.

13.4.2.2 Beaches

Tourism is a key element in the economic development of many countries, and beaches play in many cases a central role in tourism activities. The essential services provided by beaches are recreational and amenity services. Moreover, tourism and leisure are an inherent part of setting up the desirability of beaches. As a resort-style destination, the beach is almost synonymous with the elements of modern tourism.

Water quality is an important aspect to consider when analysing tourism and recreation in beaches (Hess and Beharry-Borg 2012; Loomis and Santiago 2013). Considering congestion and noise issues in beaches is also of high relevance as they can affect residents' interest in developing and maintaining beach access and other management measures. Oh et al. (2009) carried out a DCE to analyse tourist preferences for management options and public beach access policies in South Carolina beaches. They show that tourist preferences are fundamental when dealing with management agencies to serve coastal tourists better. Although most papers estimate non-consumptive direct use values, cultural services can attach non-use values. For instance, Kontogianni et al. (2014) analyse European tourists' perceptions regarding beach rocks' impacts on their recreational activities and their WTP to preserve beaches from further deterioration due to this phenomenon in Lesvos islands in Greece.

The studies show social support for beach protection strategies: the estimated recreational value of the beach ecosystem ranges between 5 and 1,075,914.77 USD/person/year. Individuals show positive preferences for beach nourishment options. Besides, beach visitors show a greater preference for more beach access points. It is also noted that tourists prefer less crowding and noise on the beach and are willing to support specific management measures, such as introducing some beach use rules and regulations. The economic valuation of recreational services provided by beaches can also inform policymakers about the benefits of water quality improvements, as most studies show that people are willing to pay for these improvements. For this reason, the importance of considering non-use values for beach protection has also been highlighted in several studies (Ghermandi and Nunes 2013; Kontogianni et al. 2014). Other studies argue that understanding the values that visitors attach to coastal recreational access can contribute to new regulations and more sustainable resource management (Oh et al. 2008, 2009).

As shown above, the economic valuation of the ecosystem services provided by beaches offers useful information for policymakers that can contribute to the design of more efficient tourism strategies in those destinations that attract many tourists.

13.4.2.3 Coral Reefs

Coral reefs are one of the most valued ecosystems because of the variety of goods and services they provide to humans. In particular, recreational services such as diving, snorkelling and viewing are the most valued according to different studies (Chen et al. 2013; Gill et al. 2015; Van der Ploeg et al. 2010). Recreational activities related to coral reefs are non-consumptive direct use values, so the most common methodology to estimate their value is the stated preferences (Gill et al. 2015; Wielgus et al. 2003). Other studies combine a stated preference method with travel cost method for estimating the recreational value and services like tourism, recreational boating and scuba diving (Ahmed et al. 2007; Chen et al. 2013).

A meta-analysis of 52 studies conducted by Brander et al. (2007) found that the average recreational value of coral reefs can reach US$3726/ha/year. The economic valuation of the services provided by coral reefs can be used to show the importance of sustaining and appreciating these ecosystems. While most studies focus on tourism and recreation and estimate direct use values, there are some studies that highlight the importance of non-use values, showing that coral reef conservation benefits are also significant to individuals. These results can be useful not only from an ecosystem conservation perspective but also for implementing strategies to manage recreational access. Environmental authorities could use the results of assessments such as these to, for example, impose charges for damage to coral reefs. All in all, these results can serve as a tool to justify investing in conservation activities.

13.5 Conclusions

This chapter aimed to provide an overview of the value of coastal and marine ecosystem services for recreational and tourism opportunities, considering that economic valuation can play a key role in the better management of these resources. It is important to properly value and incorporate the ecosystem services into nature-based tourism development planning in order to promote lower impact activities.

The volume of tourism and coastal recreation has increased considerably worldwide in recent decades. Coastal tourism has become a significant contributor to many countries' GDP and the well-being of large coastal populations. When assessing the impact of coastal tourism and recreation, it is essential to consider that a substantial component of the well-being generated by many recreational activities is not reflected in market transactions and, therefore, is outside the scope of market-based analyses. Such activities include consumptive uses, such as fishing and shellfishing, as well as non-consumptive services, such as swimming, sunbathing, sailing, windsurfing, birdwatching or diving. The aggregation of these non-commercial values and their extension to administrative levels can lead to significant improvements in environmental conservation management.

An economic valuation can be an essential tool for valuing the services that coastal and marine ecosystems provide to society. The TEV framework makes a clear distinction between use and non-use values that may help to determine the valuation methods needed in each case. For certain ecosystem services, only some valuation methods may be suitable. Also, not all methods capture all elements of TEV. For instance, market prices are often used to value provisioning services; instead, stated preference methods are more suitable for capturing non-use values. In many valuation contexts, such as cultural services, more than one technique can be used. Whereas revealed preference methods such as travel cost may capture direct use values only, stated preference methods, such as contingent valuation or choice experiments, may also capture non-use values associated with cultural services.

As there is no direct market to observe individuals' preferences, non-market valuation methods need to be applied to value cultural services directly related to recreation and tourism. These studies consistently find that people are willing to pay to protect coastal and marine ecosystems. Recreational activities and tourism in the coastal ecosystem are estimated at around $7000/ha/year. Coral reefs, on the other hand, have been estimated at $3.726/ha/year. Finally, the services provided by beaches and wetlands are estimated between 5 and 1,075,914.77 USD/person/year and $684/ha/year, respectively. As mentioned before, these results show that ecosystem services have a noticeable benefit to society, as reflected in the vast literature on valuation studies. Including this information into the decision-making process may help in designing sustainable and efficient management policies.

As for future directions, it is vital to bear in mind that economic valuation is a methodology that still needs to be refined. Uncertainty is the main problem when assessing environmental valuation studies. This should be taken into account when establishing the scope of the results obtained. Furthermore, the environmental economic value obtained from using the methodologies depends on people's preferences and perceptions. This can vary between individuals, societies and over time.

All in all, the maintenance of biodiversity and ecosystems directly benefits people by contributing to economic prosperity, well-being and quality of life. It is crucial to recognize natural capital as a fundamental financial asset and a source of public benefits. Valuation and an appropriate accounting system can demonstrate that preserving ecosystems and protecting the environment are economically profitable. Moreover, stakeholders should incorporate the valuation of ecosystem services and environmental impacts, including climate change, to better manage natural resources, contribute to the sustainability of the region's economic growth and move towards a blue economy.

References

Abdullah, S., Markandya, A., & Nunes, P. A. L. D. (2011). Introduction to Economic Valuation Methods. En A. A. Batabyal & P. Nijkamp, *Research Tools in Natural Resource and Environmental Economics* (pp. 143-187). WORLD SCIENTIFIC. https://doi.org/10.1142/9789814289238_0005

Ahmed, M., Umali, G. M., Chong, C. K., Rull, M. F., & Garcia, M. C. (2007). Valuing recreational and conservation benefits of coral reefs—The case of Bolinao, Philippines. *Ocean & Coastal Management*, *50*(1-2), 103-118. https://doi.org/10.1016/j.ocecoaman.2006.08.010

Bann, C. (1999). *A Contingent Valuation of the Mangroves of Benut, Johor State, Malaysia.*

Barbier, E. B., Hacker, S. D., Kennedy, C., Koch, E. W., Stier, A. C., & Silliman, B. R. (2011). The value of estuarine and coastal ecosystem services. *Ecological Monographs*, *81*(2), 169-193.

Batel, A., Basta, J., & Mackelworth, P. (2014). Valuing visitor willingness to pay for marine conservation – The case of the proposed Cres-Lošinj Marine Protected Area, Croatia. *Ocean & Coastal Management*, *95*, 72-80. https://doi.org/10.1016/j.ocecoaman.2014.03.025

Berg, A. V. D., Thaya Kulenthran, Muller, S., Pitt, D., Wascher, D., Groot, R. D., & P.S. Ramakrishnan. (2005). *Millennium Ecosystem Assessment, Series Vol. 1 Ecosystems and Human Well Being: Current State and Trends. Chapter 17: Cultural and Amenity Services.* https://doi.org/10.13140/RG.2.2.32902.88644

Bin, O., Landry, C. E., Ellis, C. L., & Vogelsong, H. (2005). Some Consumer Surplus Estimates for North Carolina Beaches. *Marine Resource Economics*, *20*(2), 145-161. https://doi.org/10.1086/mre.20.2.42629466

Blakemore, F., & Williams, A. (2008). British Tourists' Valuation of a Turkish Beach Using Contingent Valuation and Travel Cost Methods. *Journal of Coastal Research*, *246*, 1469-1480. https://doi.org/10.2112/06-0813.1

Brander, L. M., Van Beukering, P., & Cesar, H. S. J. (2007). The recreational value of coral reefs: A meta-analysis. *Ecological Economics*, *63*(1), 209-218. https://doi.org/10.1016/j.ecolecon.2006.11.002

Brenner, J. (2007). *Valuation of ecosystem services in the Catalan coastal zone*. https://doi.org/10.13140/RG.2.2.20582.01602

Carr, L., & Mendelsohn, R. (2003). Valuing Coral Reefs: A Travel Cost Analysis of the Great Barrier Reef. *AMBIO: A Journal of the Human Environment*, *32*(5), 353-357. https://doi.org/10.1579/0044-7447-32.5.353

Castaño-Isaza, J., Newball, R., Roach, B., & Lau, W. W. Y. (2015). Valuing beaches to develop payment for ecosystem services schemes in Colombia's Seaflower marine protected area. *Ecosystem Services*, *11*, 22-31. https://doi.org/10.1016/j.ecoser.2014.10.003

Cesar, H. S. J., & van Beukering, P. (2004). Economic Valuation of the Coral Reefs of Hawai'i. *Pacific Science*, *58*(2), 231-242. https://doi.org/10.1353/psc.2004.0014

Chen, J.-L., Chuang, C.-T., Jan, R.-Q., Liu, L.-C., & Jan, M.-S. (2013). Recreational Benefits of Ecosystem Services on and around Artificial Reefs: A Case Study in Penghu, Taiwan. *Ocean & Coastal Management*, *85*, 58-64. https://doi.org/10.1016/j.ocecoaman.2013.09.005

Czajkowski, M., Ahtiainen, H., Artell, J., Budziński, W., Hasler, B., Hasselström, L., Meyerhoff, J., Nõmmann, T., Semeniene, D., Söderqvist, T., Tuhkanen, H., Lankia, T., Vanags, A., Zandersen, M., Żylicz, T., & Hanley, N. (2015). Valuing the commons: An international study on the recreational benefits of the Baltic Sea. *Journal of Environmental Management*, *156*, 209-217. https://doi.org/10.1016/j.jenvman.2015.03.038

Daniel, T. C., Muhar, A., Arnberger, A., Aznar, O., Boyd, J. W., Chan, K. M. A., Costanza, R., Elmqvist, T., Flint, C. G., Gobster, P. H., Gret-Regamey, A., Lave, R., Muhar, S., Penker, M., Ribe, R. G., Schauppenlehner, T., Sikor, T., Soloviy, I., Spierenburg, M., … von der Dunk, A. (2012). Contributions of cultural services to the ecosystem services agenda. *Proceedings of the National Academy of Sciences*, *109*(23), 8812-8819. https://doi.org/10.1073/pnas.1114773109

Defra. (2007). *An introductory guide to valuing ecosystem services*. Department for Environment Food and Rural Affairs.

Ebarvia, M. C. M. (2016). Economic Assessment of Oceans for Sustainable Blue Economy Development. *Journal of Ocean and Coastal Economics*, *2*(2). https://doi.org/10.15351/2373-8456.1051

EEA, European Environmental Agency (2010). *Scaling up ecosystem benefits: A contribution to The Economics of Ecosystems and Biodiversity (TEEB) study.* Office for Official Publ. of the Europ. Communities.

Emerton, L., Kekulandala, L. D. C. B., Regional Environmental Economics Programme (IUCN Asia), & IUCN Sri Lanka. (2003). *Assessment of the economic value of Muthurajawela wetland*. IUCN, Sri Lanka.

Faccioli, M., Riera Font, A., & Torres Figuerola, C. M. (2015). Valuing the Recreational Benefits of Wetland Adaptation to Climate Change: A Trade-off Between Species' Abundance and Diversity. *Environmental Management*, *55*(3), 550-563. https://doi.org/10.1007/s00267-014-0407-7

Ghermandi, A., & Nunes, P. A. L. D. (2013). A global map of coastal recreation values: Results from a spatially explicit meta-analysis. *Ecological Economics*, *86*, 1-15. https://doi.org/10.1016/j.ecolecon.2012.11.006

Ghermandi, A., Obura, D., Knudsen, C., & Nunes, P. A. L. D. (2019). Marine ecosystem services in the Northern Mozambique Channel: A geospatial and socio-economic analysis for policy support. *Ecosystem Services*, *35*, 1-12. https://doi.org/10.1016/j.ecoser.2018.10.009

Gill, D. A., Schuhmann, P. W., & Oxenford, H. A. (2015). Recreational diver preferences for reef fish attributes: Economic implications of future change. *Ecological Economics*, *111*, 48-57. https://doi.org/10.1016/j.ecolecon.2015.01.004

Gürlük, S., & Rehber, E. (2008). A travel cost study to estimate recreational value for a bird refuge at Lake Manyas, Turkey. *Journal of Environmental Management*, *88*(4), 1350-1360. https://doi.org/10.1016/j.jenvman.2007.07.017

Hall, C. M. (2001). Trends in ocean and coastal tourism: The end of the last frontier? *Ocean & Coastal Management*, *44*(9-10), 601-618. https://doi.org/10.1016/S0964-5691(01)00071-0

Hess, S., & Beharry-Borg, N. (2012). Accounting for Latent Attitudes in Willingness-to-Pay Studies: The Case of Coastal Water Quality Improvements in Tobago. *Environmental and Resource Economics*, *52*(1), 109-131. https://doi.org/10.1007/s10640-011-9522-6

Honey, M. (1999). *Ecotourism and sustainable development: Who owns paradise?* Island Press.

Hoyos, D. (2010). The state of the art of environmental valuation with discrete choice experiments. *Ecological Economics*, *69*(8), 1595-1603. https://doi.org/10.1016/j.ecolecon.2010.04.011

Hoyos, D., & Mariel, P. (2010). Contingent Valuation: Past, Present and Future. *Prague Economic Papers*, *19*(4), 329-343. https://doi.org/10.18267/j.pep.380

Hoyos, D., & Riera, P. (2013). Convergent validity between revealed and stated recreation demand data: Some empirical evidence from the Basque Country, Spain. *Journal of Forest Economics*, *19*(3), 234-248. https://doi.org/10.1016/j.jfe.2013.02.003

Jacobs, S., Dendoncker, N., Martín-López, B., Barton, D. N., Gomez-Baggethun, E., Boeraeve, F., McGrath, F. L., Vierikko, K., Geneletti, D., Sevecke, K. J., Pipart, N., Primmer, E., Mederly, P., Schmidt, S., Aragão, A., Baral, H., Bark, R. H., Briceno, T., Brogna, D., … Washbourne, C.-L. (2016). A new valuation school: Integrating diverse values of nature in resource and land use decisions. *Ecosystem Services*, *22*, 213-220. https://doi.org/10.1016/j.ecoser.2016.11.007

Johnston, R. J., Grigalunas, T. A., Opaluch, J. J., Mazzotta, M., & Diamantedes, J. (2002). Valuing Estuarine Resource Services Using Economic and Ecological Models: The Peconic Estuary System Study. *Coastal Management*, *30*(1), 47-65. https://doi.org/10.1080/08920750252692616

Kolstad, C. D. (2000). *Environmental economics*. Oxford Univ. Press.

Kontogianni, A., Damigos, D., Tourkolias, Ch., Vousdoukas, M., Velegrakis, A., Zanou, B., & Skourtos, M. (2014). Eliciting beach users' willingness to pay for protecting European beaches from beachrock processes. *Ocean & Coastal Management*, *98*, 167-175. https://doi.org/10.1016/j.ocecoaman.2014.06.019

Lefevre, N., Blansfield, K., Blakemore, M., & Greenwald, J. (2019). *The World's Ten Best Ethical Destinations for 2019*. Ethical Traveler. https://ethicaltraveler.org/reports/destinations/the-worlds-ten-best-ethical-destinations-for-2019/

Lew, D. K., & Larson, D. M. (2005). Valuing Recreation and Amenities at San Diego County Beaches. *Coastal Management*, *33*(1), 71-86. https://doi.org/10.1080/08920750590883079

Loomis, J., & Santiago, L. (2013). Economic Valuation of Beach Quality Improvements: Comparing Incremental Attribute Values Estimated from Two Stated Preference Valuation Methods. *Coastal Management*, *41*(1), 75-86. https://doi.org/10.1080/08920753.2012.749754

Mariel, P., Hoyos, D., Meyerhoff, J., Czajkowski, M., Dekker, T., Glenk, K., Jacobsen, J. B., Liebe, U., Olsen, S. B., Sagebiel, J., & Thiene, M. (2021). *Environmental Valuation with Discrete Choice Experiments: Guidance on Design, Implementation and Data Analysis*. Springer International Publishing. https://doi.org/10.1007/978-3-030-62669-3

McGranahan, G., Balk, D., & Anderson, B. (2007). The rising tide: Assessing the risks of climate change and human settlements in low elevation coastal zones. *Environment and Urbanization*, *19*(1), 17-37. https://doi.org/10.1177/0956247807076960

MEA (Ed.). (2005). *Ecosystems and human well-being: Synthesis; a report of the Millennium Ecosystem Assessment*. Island Press.

Molina Murillo, S. A. (2019). Certificación turística sostenible y los impactos socioeconómicos percibidos por hoteles en Costa Rica. *PASOS. Revista de Turismo y Patrimonio Cultural, 17*(2), 363-372. https://doi.org/10.25145/j.pasos.2019.17.025

Montenegro, N. (2017). *The economisation of nature conservation. An analysis of the pros and cons of economic valuation of ecosystem services by public institution.* Environmental Policy Group. Wageningen University.

Moreno, A., & Amelung, B. (2009). Climate Change and Coastal & Marine Tourism: Review and Analysis. *Journal of Coastal Research SI, 56*, 1140-1144.

Nature Air. (2019). *What is Costa Rica's Blue Flag Program?* http://blog.natureair.com/2019/02/19/what-is-costa-ricas-blue-flag-program/

Nunes, P. A. L. D., & van den Bergh, J. C. J. M. (2004). Can People Value Protection against Invasive Marine Species? Evidence from a Joint TC–CV Survey in the Netherlands. *Environmental & Resource Economics, 28*(4), 517-532. https://doi.org/10.1023/B:EARE.0000036777.83060.b6

Oh, C.-O., Dixon, A. W., Mjelde, J. W., & Draper, J. (2008). Valuing visitors' economic benefits of public beach access points. *Ocean & Coastal Management, 51*(12), 847-853. https://doi.org/10.1016/j.ocecoaman.2008.09.003

Oh, C.-O., Draper, J., & Dixon, A. W. (2009). Assessing Tourists' Multi-Attribute Preferences for Public Beach Access. *Coastal Management, 37*(2), 119-135. https://doi.org/10.1080/08920750802701128

Östberg, K., Hasselström, L., & Håkansson, C. (2012). Non-market valuation of the coastal environment – Uniting political aims, ecological and economic knowledge. *Journal of Environmental Management, 110*, 166-178. https://doi.org/10.1016/j.jenvman.2012.06.012

Pascoe, S., Doshi, A., Thébaud, O., Thomas, C. R., Schuttenberg, H. Z., Heron, S. F., Setiasih, N., Tan, J. C. H., True, J., Wallmo, K., Loper, C., & Calgaro, E. (2014). Estimating the potential impact of entry fees for marine parks on dive tourism in South East Asia. *Marine Policy, 47*, 147-152. https://doi.org/10.1016/j.marpol.2014.02.017

Pidwirny, M. (2006). *Introduction to the Oceans.* Fundamentals of Physical Geography. http://www.physicalgeography.net/fundamentals/8o.html

Ransom, K. P., & Mangi, S. C. (2010). Valuing Recreational Benefits of Coral Reefs: The Case of Mombasa Marine National Park and Reserve, Kenya. *Environmental Management, 45*(1), 145-154. https://doi.org/10.1007/s00267-009-9402-9

Rolfe, J., & Gregg, D. (2012). Valuing beach recreation across a regional area: The Great Barrier Reef in Australia. *Ocean & Coastal Management, 69*, 282-290. https://doi.org/10.1016/j.ocecoaman.2012.08.019

Rosselló-Nadal, J. (2014). How to evaluate the effects of climate change on tourism. *Tourism Management, 42*, 334-340. https://doi.org/10.1016/j.tourman.2013.11.006

Salcone, J., Brander, A., & Seidl, A. (2016). *Guidance manual on economic valuation of marine and coastal ecosystem services in the Pacific. Report to the MACBIO Project.* http://macbio--pacific.info/wp-content/uploads/2017/07/MACBIO_MESV_Guidance-Manual_Web.pdf

Scott, D., Gössling, S., & Hall, C. (2012). International Tourism and Climate Change. *Wiley interdisciplinary reviews: Climate Change, 3*, 213-232. https://doi.org/10.1002/wcc.165

Sequeira, T. N., & Maçãs Nunes, P. (2008). Does tourism influence economic growth? A dynamic panel data approach. *Applied Economics, 40*(18), 2431-2441. https://doi.org/10.1080/00036840600949520

Shivlani, M. P., Letson, D., & Theis, M. (2003). Visitor Preferences for Public Beach Amenities and Beach Restoration in South Florida. *Coastal Management, 31*(4), 367-385. https://doi.org/10.1080/08920750390232974

Shrestha, R. K., Seidl, A. F., & Moraes, A. S. (2002). Value of recreational fishing in the Brazilian Pantanal: A travel cost analysis using count data models. *Ecological Economics, 42*(1-2), 289-299. https://doi.org/10.1016/S0921-8009(02)00106-4

TEEB. (2010). *The Economics of Ecosystems and Biodiversity: Ecological and Economic Foundations.* (Earthscan).

Thur, S. M. (2010). User fees as sustainable financing mechanisms for marine protected areas: An application to the Bonaire National Marine Park. *Marine Policy, 34*(1), 63-69. https://doi.org/10.1016/j.marpol.2009.04.008

Tianhong, L., Wenkai, L., & Zhenghan, Q. (2010). Variations in ecosystem service value in response to land use changes in Shenzhen. *Ecological Economics, 69*(7), 1427-1435. https://doi.org/10.1016/j.ecolecon.2008.05.018

Torres, C., & Hanley, N. (2016). *Economic valuation of coastal and marine ecosystem services in the 21st century: An overview from management perspective*. MASTS.

Torres, C., & Hanley, N. (2017). Communicating research on the economic valuation of coastal and marine ecosystem services. *Marine Policy, 75*, 99-107. https://doi.org/10.1016/j.marpol.2016.10.017

UNEP (Ed.). (2009). *Sustainable coastal tourism: An integrated planning and management approach; [priority actions programme]*. United Nations Environment Programme.

UNEP. (2011). *Towards a Green Economy: Pathways to Sustainable Development and Poverty Eradication*. https://sustainabledevelopment.un.org/content/documents/126GER_synthesis_en.pdf

UNEP. (2012). *Green economy in a blue world.* United Nations Environment Programme.

UNEP. (2014). *Blue Economy Concept*. https://sustainabledevelopment.un.org/content/documents/2978BEconcept.pdf

UNEP. (2016). *Blue Economy project: Blue economy for a healthy Mediterranean- Measuring, Monitoring and Promoting an environmentally sustainable economy in the Mediterranean region.* https://planbleu.org/sites/default/files/upload/files/Scoping_Study_Blue_Economy.pdf

UNWTO, & UNEP. (2005). *Making Tourism More Sustainable—A Guide for Policy Makers*. https://www.unwto.org/sustainable-development

Van der Ploeg, S., De Groot, R. S., & Wang. (2010). *The TEEB Valuation Database: Overview of structure, data and results.* Foundation for Sustainable Development. https://www.es-partnership.org/wp-content/uploads/2016/06/ESVD.-TEEB_Database_Report.pdf

Waycott, M., Duarte, C. M., Carruthers, T. J. B., Orth, R. J., Dennison, W. C., Olyarnik, S., Calladine, A., Fourqurean, J. W., Heck, K. L., Hughes, A. R., Kendrick, G. A., Kenworthy, W. J., Short, F. T., & Williams, S. L. (2009). Accelerating loss of seagrasses across the globe threatens coastal ecosystems. *Proceedings of the National Academy of Sciences, 106*(30), 12377-12381. https://doi.org/10.1073/pnas.0905620106

Westerberg, V. H., Lifran, R., & Olsen, S. B. (2010). To restore or not? A valuation of social and ecological functions of the Marais des Baux wetland in Southern France. *Ecological Economics, 69*(12), 2383-2393. https://doi.org/10.1016/j.ecolecon.2010.07.005

Wielgus, J., Chadwick-Furman, N. E., Zeitouni, N., & Shechter, M. (2003). Effects of Coral Reef Attribute Damage on Recreational Welfare. *Marine Resource Economics, 18*(3), 225-237. https://doi.org/10.1086/mre.18.3.42629397

Wielgus, J., Gerber, L. R., Sala, E., & Bennett, J. (2009). Including risk in stated-preference economic valuations: Experiments on choices for marine recreation. *Journal of Environmental Management, 90*(11), 3401-3409. https://doi.org/10.1016/j.jenvman.2009.05.010

Windle, J., & Rolfe, J. (2013). Estimating nonmarket values of Brisbane (state capital) residents for state based beach recreation. *Ocean & Coastal Management, 85*, 103-111. https://doi.org/10.1016/j.ocecoaman.2013.09.011

Woodward, R. T., & Wui, Y.-S. (2001). The economic value of wetland services: A meta-analysis. *Ecological Economics, 37*(2), 257-270. https://doi.org/10.1016/S0921-8009(00)00276-7

WTTC. (2020). *Travel & tourism. Global economic impact and trends 2020*. World Travel & Tourism Council. file:///C:/Users/Estibaliz/Downloads/Global_Economic_Impact_&_Trends_2020.pdf

Zhang, F., Wang, X. H., Nunes, P. A. L. D., & Ma, C. (2015). The recreational value of gold coast beaches, Australia: An application of the travel cost method. *Ecosystem Services, 11*, 106-114. https://doi.org/10.1016/j.ecoser.2014.09.001

Part V
Future Challenges to the Blue Economy

Chapter 14
Monitoring Health of Oceanic Ecosystem

Aneel Salman

14.1 Blue Ecosystem and the Ocean: A Symbiosis

Blue ecosystem, as the name suggests, refers to the life forms that exist under the blue ecosystem. While the number of ecosystems that exist in the ocean will be hard to count, it is safe to say that there are thousands and thousands of ecosystems that are present in a symbiotic relationship with the oceans.

Since all life forms depend on water and 97% of the Earth's water is present in the oceans, it will not be an exaggeration to state that without healthy blue ecosystem, all life would cease to exist (Parsons and MacPherson 2016). Oceans not only support human life but also support large biospheres within them which include species of all shapes and sizes. This is not all. Since the oceans make up a huge portion of the Earth's hydrosphere, they naturally are an important element of the carbon cycle of the planet.

Simply put, the oceans are not only responsible for sustaining life forms, but they are also responsible for regulating climate and weather patterns of the planet. Thus, oceans are a very crucial and functional part of the system that runs the planet. Till now, this important source of life has been mapped 20% in total, which means of the estimated two million species that live in the blue ecosystem, only 230,000 are known (Aswani 2017). All the information given above gives an insight to the significance of this large water body for us. Without the blue ecosystem, we would not know life as we know it now.

A. Salman (✉)
Department of Economics, COMSATS University Islamabad, Islamabad, Pakistan
e-mail: aneel.salman@comsats.edu.pk

S. Hazra, A. Bhukta (eds.), *The Blue Economy*,
https://doi.org/10.1007/978-3-030-96519-8_14

14.2 Why Is Perseverance Important?

The source that supplies half the planets oxygen, provides food and employment to more than a billion people have to be preserved at all costs (Parsons and MacPherson 2016). When we in general read and talk about ocean health, even if we feel we should do something, we do not. Mostly because we feel that the blue ecosystem is not near to us, we do not need to protect it. However, even if the ocean is millions of miles away, our lives depend on them in more ways than one. If we do not protect this iota of the Earth, how will we breathe? What will we drink, and how will billions of people earn their living? The oceans are not just water storing body for humans, but they are much more than that. It is evidently clear that without the oceans, human life would be severely affected, and thus to preserve our lives, we should preserve the blue ecosystem.

An index called the Ocean Health Index ranks ocean health at 71. Even if this number tells us that the blue ecosystem may not be dying now, it shows that the ocean health is slowly deteriorating. Specifically, this number 71 guides us by letting us know how much efforts are required to get the blue ecosystem back to their original health (Ocean Health Index 2019). Even though it is very difficult to bring about changes in the blue ecosystem rapidly, it is high time that the efforts to preserve them get started. Not all blue ecosystem is equally damaged and not all is equally healthy. All the oceans of the world must be ranked separately to properly understand how much and what kind of effort is required. For this purpose, the breakdown of the oceans of the world is important.

14.3 Components of Ocean Health

There are five blue oceans in the world that are divided between continents. These are the Arctic, Atlantic, Pacific, Indian, and Southern. The Pacific Ocean separates Australasia and Asia from Americas. The Atlantic separates Americas from Europe and Africa. The Indian Ocean borders Southern Asia and separates Africa and Australia. The Southern Ocean encircles Antarctica, and the Arctic Ocean is usually known as the estuary of the Atlantic.

14.4 Ocean Health Over the Years

Since ocean health is a very wide arena, there is no one right way to measure or rank it. It requires a deep and careful study into a number of wide components to accurately judge what the blue ecosystem is telling us. The global Ocean Health Index sets specific goals to gauge the health of the blue ecosystem. The following explanation will provide an insight into the methodology chosen to measure and rank ocean health.

14.5 Methodology of Ranking Ocean Health

The goals set in measuring ocean health are there to get the maximum ecological, social, and economic benefits. Every goal measures and calculates the delivery of these benefits with respect to a certain sustainable benchmark (Ocean Health Index 2019). The goal is given a score of 100 if all of its sustainable benefits are achieved in ways that do not cause negative consequences on the abilities of the blue ecosystem to deliver those benefits in the future. If the score is lower than 100, it means that the way these benefits are gained is harming the blue ecosystem considerably.

14.6 Assessment of Regions

There is a global database which is used for yearly assessment of goals in approximately 221 regions, which includes the coastal economies, territories, and the Antarctic. The areas that are not under the jurisdiction of the countries that conduct the assessment are not assessed as regularly as the others.

14.7 Goal Setting

For the setting of goals for conducting the ocean health assessment, researchers, scientists, and economists dig deeper into the studies already conducted on ocean health and hypothesize people's expectations from the blue ecosystem in terms of benefits. These goals are then set into different categories called "goals."

There are four categories that measure and rank each goal. They are status, trend, resilience, and pressures.

Status is the current state of the ocean health in a certain region. This value is calculated by taking an average of the different values of different components for each individual country. These components are:

1. Natural products.
2. Food provision—subgoals: fisheries and marine culture.
3. Tourism and recreation.
4. Sense of place—subgoals: iconic species and lasting special places.
5. Clean water.
6. Artisanal fishing opportunities.
7. Carbon storage.
8. Livelihoods and economies—subgoals: livelihoods and economies.
9. Coastal protection.

Trend is the average status of the goal over a period of 5 years. This helps us rank the goals with accuracy since measurement of any goal related to ocean health needs a considerable period of time. **Pressures** include all the social and ecological components that decrease the current status of the goal. **Resilience** includes all the initiatives taken by the society or any community including the ecological factors that can increase the current status of the goal by eliminating unnecessary pressures.

14.8 Scores

After the goals and their dimensions are set, the goals are scored. The scores for trend, pressures, and resilience are accumulated together for the forecast of the status score for the next 5 years. The goal's score is the average or the mean calculated for the present and likely future status. In this regard, there is a breakdown of the contribution of each goal in the score setting. Status including trend makes up 83% of the score, and pressure and resilience make up 8.5% of score setting.

Even though resilience makes up a small percentage of the score, resilience plays a large role and is actually a very crucial part in making the other goals better. Resilience is the only way the pressures can be eliminated and ocean health will be improved.

14.9 Regional Scores

In a regional assessment, the goal's scores are averaged. According to each region's local conditions, score values are weighted differently which could change the score of the individual region. However, in global assessments, the goal scores are weighted equally.

14.10 Global Scores

The scores are then calculated globally. For global score calculation, the area-weighted average of all regions is taken. The EEZ score or the global regional score is the collective score that refers to all countries, territories, and Antarctica. It also includes areas beyond jurisdiction.

The present status is 50% of the goal score, and the rest is made up by the likely future status which includes trend, pressure, and resilience. The thing that helps us forecast the future likely status is trend, pressure, and resilience. Trend gives us an idea of what the future trend in ocean health might look like. If the pressures increase and surpass the resilience components, then the likely future status decreases. If the resilience factors increase and surpass the pressures, the likely future status increases.

14.11 Ocean Health Status in Southeast Asia

The focus of this chapter is on the region of Southeast Asia and the health of the blue ecosystem in that region. The following globe shows through colors the ranks of ocean health of different countries.

The different colors stand for different ranks. Colors chosen for this purpose are deep orange, light orange, yellow, sky blue, and deep blue. The ranks are then given numeric values for each color starting from 0 to 100: 0–25 is designated for deep orange, 25–50 for light orange, 50–75 for the color yellow, 75–90 for sky blue, and 90–100 for deep blue. The component or the overall score ranked at 100 shows complete health. A lower rank shows a health weakness.

14.12 South Asian Ocean Health Rank

The following countries fall under the region of South Asia, namely, Bangladesh, Pakistan, India, Sri Lanka, Afghanistan, Bhutan, and the Maldives. The Ocean Health Index designates different colors to different health ranks. The color indicated for the South Asian region as a whole is yellow. This color indicates a lower rank in terms of ocean health overall. The color yellow indicates an index rank of 50–75. These ranks show that the ocean health near the South Asian countries is lower than the global average of the ocean health. The overall global score stands at 71. That score can be broken down into the following goals. Below are given the different goals and the different scores for those goals (Ocean Health Index 2019):

- Natural products (51).
- Food provision (52)—subgoals: fisheries (55) and mariculture (6).
- Tourism and recreation (52).
- Sense of place (64)—subgoals: iconic species (66) and lasting special places (62).
- Clean waters (70).
- Artisanal fishing opportunities (77).
- Carbon storage (79).
- Livelihood and economies (82)—subgoals: livelihood (77) and economies (88).
- Coastal protection (86).
- Biodiversity (88)—subgoals: species (84) and habitats (91).

14.13 The Link Between Ocean Health and Human Health

Although there are economic consequences to the damages caused to ocean health, there are medical side effects as well. According to a few researches that have linked human health to ocean health, several algal blooms and the toxins they release cause severe health problems in humans. Algal blooms and several phytoplankton are collectively known as red tide. Red tide is an indicator of bad ocean health. These tides are considered harmful because the toxins they release are severely damaging not only to the water columns under them but also to the animals and humans that come in contact with them.

The toxins include powerful neurotoxins, derma-toxins, and hepa-toxins and in some rare cases carcinogens. In simple language, the toxins produced by these algal blooms and red tides are harmful to the liver, the skin, and the nervous system of living beings that come into contact with them (Pendleton 2020).

This exposure can be caused when people or animals eat food and drink water contaminated with these substances. The effects of these toxins are acute and prove to be fatal in some cases. As more and more information comes to light regarding red tides, it is seen that one species may be capable of producing more than one toxin. Since more than one species is capable of producing the same toxins as well, the link between the causal effects of some side effects is getting harder to find. To prevent the dangerous outcomes that these toxins cause, scientists have found ways to detect the level of these toxins in rivers, seas, and blue ecosystem. This can help the responsible authorities close off certain areas for use so that exposure is prevented (Xavier and Brandt 2016). Thus, algal blooms or red tides produce harmful toxins which can be deadly for humans and the animals alike. Since nearly a billion people come in close contact with the marine environment either through the coastal tourism activities or food consumption, it is imperative that the ocean health be preserved.

14.14 Climate Change in the South Asia

South Asia is the home to countries that are socioeconomically backward. This makes the region one of the most sensitive to changes in climate and global warming. These countries are home to about 1.5 billion people and a big chunk of the globe's poor population. What makes the situation even worse for this region is high dependence of its inhabitants on natural resources. This is why global warming will be the most deadly threat to their livelihood and sustenance. South Asia is ranked as the country's most susceptible to natural disasters due to the rising sea levels and unprecedented rainfalls. This inevitably puts the blue ecosystem of the region at a risk too. A few of the most dangerous risks are declining oxygen levels, rising sea levels, pollution, and the decrease in the ability of blue ecosystem to store carbon.

14.15 Declining Oxygen

The eventual decrease of oxygen in the ocean and coastal waters is not happening without its consequences. As a result of this decrease, the ocean waters are getting warmer and more acidic in nature. Furthermore, the marine food webs are getting altered, thus disrupting the natural habitats of the blue ecosystem. There are certain reasons for this decline in the oxygen levels. For example, the number of fisheries has increased all over the world and particularly in South Asia. Overfishing can degrade the habitat in which the fish breed and grow. Such human activities are called stressors when they result in unwanted changes in the blue ecosystem or the marine food webs. When a group of stressors result in negative consequences for the same ecological process, their combined effects can cause changes that are hard to predict and manage. The global concern that stressors are raising is that the overall resilience of the blue ecosystem is diminishing which will halt the natural regenerating processes that sustain them. The three threats to the blue ecosystem are acidification, deoxygenation, and global warming. All three of these processes are linked to each other. The warming of ocean temperatures caused by the greenhouse gases lowers the solubility of oxygen in the seawater and thereby inhibits the exchange of oxygen between surface waters that are rich in oxygen and those that are not. The individual respiration rates of the organisms also increase, further depleting the oxygen levels in the ocean and the seas. This results in further acidification which causes the blue ecosystem to degrade rapidly. The worrying fact is that the decreasing oxygen levels increase the oxygen demands by the organisms which put all the ecosystem of the oceans at risk. Moreover, the balance that is maintained by nature for the blue ecosystem to sustain is very intricate and fragile. A slight change in the oxygen concentration can make a world of difference for the different habitats. Even in the same ecosystem, different species have different tolerance for oxygen levels, which is why the deoxygenation is such a problem. The loss of habitat complexity as a result of dying coral reefs and the species that regenerate them is having a negative impact on all blue ecosystem of the planet. This effect is increasingly present in the South Asian region where destructive human activities and sensitivity to warming temperatures are depleting oxygen levels.

14.16 Blue Carbon

There is no doubt about the productivity of coastal ecosystems. However, the most important function they play is of sequestering and storing blue carbon. This blue carbon is taken from the atmosphere and the ocean and is thus an important part of the climate regulation of the planet. The biggest storage houses for blue carbon are the mangroves, tidal marshes, and seagrasses. The biggest problem these storage houses face today is the disruption in their natural habitats and their life cycles, both of which are affected by global warming, human activities, and deoxygenation of the oceans. When their ecosystem is disrupted, these storage houses release the carbon they have stored for years into the ocean and the atmosphere and become a big source of greenhouse gases. According to expert approximations, 1.02 billion tons of carbon dioxide is being released into the surrounding atmosphere and water from the coastal ecosystem which contributes 19% to the total carbon global emissions (The Blue Carbon Initiative 2019).

These coastal ecosystems are critical to the health of the coasts, the ocean, and other blue ecosystem of the world, which is why their destruction is a big threat for all. Mangroves alone provide services that are estimated to be worth US$1.6 billion each year (The Blue Carbon Initiative 2019). Mangroves store about three to four times the amount of carbon stored by a mature tropical forest (Aswani 2017). Moreover, they prevent coastal erosion and provide a breeding ground for many marine organisms. In addition to storing carbon, tidal marshes help maintain the quality of water in the coastal areas and thus are responsible to filter salts, sediment, and pollution of all kinds from entering the ocean and the sea. They help coastal communities prevent coastal erosion, storms, and floods from occurring other than providing an important habitat for healthy fisheries and coastal organisms. Their importance cannot be underestimated. Similarly, seagrass helps the oceans and the seas with the same things tidal marshes do. All blue ecosystem is the carbon sink of the planet, without which climate change cannot be controlled.

Despite their importance, the coastal ecosystem which is the most important blue ecosystem of the planet is under a grave threat. Estimations show that each year 340,000–980,000 hectares of the ecosystem is being destroyed (The Blue Carbon Initiative 2019). At this rate, nearly all coastal ecosystem will be completely wiped out in the next 100 years (The Blue Carbon Initiative 2019).[1]

14.17 Rising Sea Levels

Due to the overall increase in the greenhouse gases on the Earth's surface, the carbon cycle is severely affected. One of the key functional parts of our planet's carbon cycle is the ocean. This in turn means that the ocean health as well as the global temperature is being affected. This cycle becomes vicious, and the process of global warming keeps moving (Isensee et al. 2018). As a result of rising temperatures, the sea levels are rising all over the globe. The sea levels are predicted to cause severe damage to the people that live near the coasts or the seas. According to the estimates, approximately 250 million poor people of the South Asian countries live in low-lying river deltas. These people are at a risk of livelihood and life losses as a result of rising sea levels (ARCGIS 2020).

[1] https://www.thebluecarboninitiative.org/about

14.18 Health Threats

In addition to red tides, there are several other toxins that pose a threat to human beings and animals upon exposure to coastal waters. These threats include waterborne pathogens that are produced as a result of constant dumping of fecal matter or industrial waste into the blue ecosystem. Industrial waste also includes certain pharmaceutical products and heavy metals such as mercury which are severely damaging for the blue ecosystem's health. These threats, generated primarily by human activities, cause pathogens to grow up in ocean waters, for example, sewer flows, wastewater and industrial effluent treatment plants, oil spills, and coal power generation (Xavier and Brandt 2016). The effects of these human activities have already started to show in some parts of South Asian countries. For example, seasonal outbreaks of cholera in South Asia are linked to the phytoplankton blooms in the waters that surround the region. These blooms include pathogens such as *V. cholerae* and copepods which are the primary cause of this disease (Pendleton 2020).

14.19 Coral Reefs

Coral reefs are the life and blood of not only blue ecosystem but of the millions of organisms that live in them. They are home to the most diverse ecosystem on the planet. This makes them extremely important for humans as well. Coral reefs support millions of life cycles under the ocean and help regulate the carbon, oxygen, and nitrogen levels in the blue ecosystem as well. Even though coral reefs have the ability to sustain and regenerate, their maintenance is important.

Coral reefs are important for healthy blue ecosystem in more ways than one. They help regulate the ocean temperatures so that tropical storms and damaging waves do not harm the coastlines. This means that they are one of the primary reasons the coastal life is sustained. Following are a few reasons for the protection of coral reefs is important because:

1. They regulate the nutrient cycles of the marine organisms.
2. They protect and provide shelter for the young marine organisms.
3. They are a crucial part of the marine food chains.
4. They assist the blue ecosystem in carbon and nitrogen fixing.
5. They support the fishing industry by helping fish and other organisms spawn before making their way into the ocean. This means that the billion dollar fish industry owes its profit to the coral reefs. The Great Barrier Reef alone generates 1.5 billon dollars for the Australian fishing and tourism industry.
6. The study and research of the coral reefs of the world give us facts about climatic events from millions of years ago. It also gives us information about recent storms and effects of human impacts by showing different growth patterns (Roberson 2016).

14.20 Danger to the Coral Reefs of South Asia

The scope of this chapter is to study about the effects of human activities on the coral reefs in the South Asian region. The largest coral reefs of South Asia include those of the Maldives, Chagos, and Lakshadweep. In this region, corals also grow along the Indian subcontinent coast around Nicobar Islands, Sri Lanka, Gulf of Mannar, and Gulf of Kutch. The reefs of these regions are heavily influenced by the monsoons of the southwest and northeast. As for Bangladesh, there are no coral reefs other than the small patches offshore around St. Martin's Island. In Pakistan, a few small coral colonies are found, that too in highly turbid coastal waters. All these reefs are now deteriorating slowly given the heavy human activities that take place on and around them.

The people that inhabit the areas around these waters especially in Sri Lanka, Pakistan, Bangladesh, and India are poor people. There are hence fewer monitoring laws and even fewer concerned authorities to monitor and protect the marine life. This is why these coastal areas have severely damaged the marine life and the waters. Threats such as nonpoint pollution, runoffs, sedimentation, coral bleaching, untreated wastage disposal, and overfishing are the primary sources for the deterioration of the reefs in the region.[2]

14.21 Coral Reef Bleaching

Coral reef bleaching is a very damaging phenomenon and is linked to a number of causal factors. Presently, the increased levels of carbon dioxide are having a negative effect on the ocean's carbon content and the carbonate chemistry. The coral reefs primarily depend on the conversion of carbonates present in the seawater into a skeleton. This process is called calcification. However, the atmospheric carbon dioxide level plays an essential role in this process. If the levels of atmospheric carbon dioxide increase beyond a certain level, seawater becomes increasingly acidic, which makes calcification more difficult (Roberson 2016).

It has been observed that increased water temperatures block the photosynthetic reactions inside the zooxanthellae algae that live in a mutualistic association with coral polyps. This causes reduced amounts of carbon dioxide conversion into sugars and builds up by-products in the algae that poison it. To save itself from the algae, it starts a process called tissue sloughing (Parsons 2017). In this process, the coral spits out the zooxanthellae and a little of its own tissue leaving the coral bleached white. In tissue sloughing, another process happens as a side effect where the coral polyps separate themselves from the coral's skeleton. Without the food producing zooxanthellae, the remaining coral polyps slowly die due to a lack of food. As soon as the coral dies, the wildlife that it is home to start getting affected. This is why the

[2] https://blue ecosystems ervice.noaa.gov/facts/coralreef-climate.html

rising global temperatures are causing havoc for the corals. However, if the temperatures cool down to the desired levels, the coral may be able to recolonize the zooxanthellae and regenerate itself.

Even though coral bleaching is a natural process meant to save the coral skeleton, there is a certain point beyond which the corals start to die. When the coral reef enters the phase of destruction, the level of bleaching increases, which is the worst possible thing for the blue ecosystem and the marine wildlife. The worst coral bleaching occurred in the Great Barrier Reef after 700 years in the year 1988, which was followed by bleaching on a greater magnitude just 4 years later. This caused massive amounts of coral bleaching all over the globe. In some areas, 90% of the coral reefs were bleached from which only a few recovered (Roberson 2016).

14.22 Bangladesh

The only area in Bangladesh where corals are is the offshore island of St. Martin. Unfortunately, this is also the area that is most influenced by monsoon and cyclones due to the increasingly deteriorating health of the water. The corals in this region are not true corals. Instead, there are coral aggregations in shallow waters surrounded by soft seagrass beds and rocky habitats. This region is distinctly known because of the heavy sedimentation that takes place. This heavy sedimentation discharge from the Ganges-Brahmaputra-Meghna Delta is responsible for 6% of the total sediment that ends up in the blue ecosystem (Fischer et al. 2017).

14.23 Chagos

The archipelago in Chagos is a group of islands in the southernmost part of the Chagos-Laccadive Ridge. This area is known as the center of the Indian Ocean. This territory is British and remains uninhabited, leaving the US military base of Diego Garcia. This area is distinctly known as the largest area of undisturbed coral reefs of the Indian Ocean, boasting the highest coral reef biodiversity in the region.

14.24 India

India has four major coral reefs: Andaman and Nicobar Islands, Lakshadweep Islands, Gulf of Kutch, and Gulf of Mannar. There is a local and scattered growth on banks along the east and west coasts of the mainland. In India, like other South Asian countries, coral reefs are economically important. They are also important for providing 25% of the total fish to the communities that need fish for their livelihood and sustenance (Sarkar et al. 2019).

14.25 Maldives

The archipelago of the Maldives is located at the center of the Chagos-Laccadive Ridge having a land area of about 200 km^2. The region is rich in coral islands having 1190 coral islands and a number of sand cays making up almost 23 islands. The exclusive economic zones of the archipelago are 90,000 km^2. The elevation of these islands is very less measuring approximately 5 m (Islam et al. 2011). These elevations have a narrow fringe reef around every island of the archipelago. The reefs of the Maldives are hence very important for the protection of the shorelines of the region.

14.26 Pakistan

In Pakistan, there are a few coral reefs on which limited information is available. However, whatever scatterings of coral reef are, they are present in extremely turbid costal water conditions (Alam 2019). There is however no research or monitoring of the areas either.

14.27 Sri Lanka

Sri Lanka has a 1585 km long coastline of which 2% has fringing reefs (Shum and Kuo 2011). The region also has large reefs in the offshore areas such as Gulf of Mannar toward the northwest and east coast. The corals of Sri Lanka are grown in varying lengths on limestone, rocky reefs, and sandstone. The locations of these reefs are known albeit poorly mapped.

14.28 Loss to South Asian Economies as a Result of Rising Sea Levels

The IPCC estimates that the sea levels will rise by 4 mm annually. This means that the sea levels will rise 0.22–0.44 m more than the 1990 level by the year 2090–2099 (ARCGIS 2020). The estimated time may seem a lot, but in reality it is not. Since the rising sea levels are causing destruction at a much faster rate it may happen well before that. This means that as the year 2090 comes closer, these rising sea levels will cause destruction of unprecedented levels (Isensee et al. 2018). The South Asian region is going to fall prey to these destructions since this area is highly susceptible to storm surges, coastal erosions, floods, and inundation. All these side effects of rising sea levels will definitely cause not only losses of arable land but also displacement of millions of people in addition to loss of infrastructure. It is predicted that the people living in the following six countries of Southeast Asia and Asia—India, the Maldives, Bhutan, Nepal, Bangladesh, and Sri Lanka—will have to bear the loss of 1.8% of their GDPs as a result of the devastation caused by rising sea levels (ARCGIS 2020). This number is subject to increase by 8.8% if the current speed of using fossil fuel continues (Sivakumar and Stefanski 2011). It is pertinent to mention here that these rising sea levels are due to global warming and its adverse effects on the blue ecosystem and ice caps of the Polar Region. Thus, bad ocean health is indirectly costing the South Asian economies a great deal (ARCGIS 2020). The warnings have already been issued to these countries since, according to the IPCC's estimates, by mid-century, approximately two million people of South Asia will be displaced as a result of inundation in the region (Rajasuriya 2002). The situation is worse for some countries of the region; for example, the IPCC warns of great deal of loss to the land and lives of approximately three million people living in the Ganges-Brahmaputra-Meghna Delta by the year 2100 (Xavier and Brandt 2016). In addition to this, Bangladesh will also lose one-quarter of arable land that it had in the year 1989. This loss of land and lives is going to cause millions of people to lose their livelihood and homes (ARCGIS 2020).

Similarly, the Maldives is expected to lose 12% of its GDP annually as a result of rising sea levels by the year 2100. This is the highest loss expected for any country of this region as the lowest sea level point of the country falls below 2 m. The prediction of losses for other countries is as follows:

India: 8.7% of the GDP.
Bhutan: 6.6% of the GDP.
Sri Lanka: 6.5% of the GDP.
Nepal: 9.9% of the GDP (ARCGIS 2020).

14.29 Loss of Food

One of the largest suppliers of rice in the world is Southeast Asia and South Asia region. According to estimates, it supplies nearly 88% of the world's supply (Laal 2011). Rice is a food staple in Bangladesh and is crucial to the well-being of the country's economy. According to the WHO's estimates, Bangladesh is going to lose one million hectares of arable land which directly translates to a loss in the rice availability for domestic purposes and for export (Laal 2011).

14.30 Susceptible Populations

Coastal tourism and consumption of marine products for food is the main exposure that is at risk due to bad ocean health. The number of economies that derive a major part of their GDP from coastal tourism like the Maldives is at a risk of great loss if the ocean health is not preserved. The scale of population that stands at a risk of exposure to these toxins is significant. For example, in the USA alone, more than 62 million Americans swim in the nation's water and spend more than 800 million days at the beach in a year (Brooks et al. 2020). Thus, all these people are at a risk of exposure to deadly or damaging toxins. This then creates an opportunity for health expenditures to increase which in turn burdens the economy. Thus, indirectly, ocean health affects the economies of the globe as well.

14.31 What Are the Main Causes for This in South Asia?

The main reasons for the deterioration of the coral reefs in South Asia are (Rajasuriya 2002):

- Sedimentation.
- Coal mining.
- Destructive fishing practices.
- Excessive usage of bottom fishnets.
- Pollution from land sources.
- Nonpoint pollution.
- Boat anchoring.

In addition to this, the reef has been burdened by overfishing and excessive human activity surrounding it. This has led to elevated levels of sedimentation on the reefs which has caused an abundance of organisms alike algae, tunicates, and corallimorphs which are smothering the coral reefs especially after the bleaching event of 1988.

14.32 Pollution

Most of the pollution that is responsible for contaminating the blue ecosystem comes from the land. The primary source of ocean and sea pollution is nonpoint source pollution. This occurs as a consequence of runoff. The main sources of non-point pollution are many, including septic tanks, trucks, boats, cars, farms, ranches, and forests. Millions and millions of cars and other vehicles drop oil in small amounts each day on the roads and parking lots. This eventually makes its way to the surrounding water bodies.

Most water pollution also starts as a result of severe air pollution. The pollutants from the air settle into waterways and blue ecosystem, making them harmful for fish and wildlife habitats. Even silt from construction sites can run off into waterways.

Nonpoint pollution makes the blue ecosystem unsafe for all life. In some cases, this pollution is so severe that it becomes the cause of beaches being closed after rainstorms. Even though millions of dollars are spent each year to protect, manage, and preserve water bodies, the efforts are not enough. Authorities like NOAA work with America to protect and restore damaged water bodies. Similarly, many agencies and authorities work all over the world in collaboration with complementary authorities to manage and protect what they can. However, until the main source which is the excessive carbon footprint is lessened, the ocean health will continue to deteriorate. The only sustainable way of reducing nonpoint pollution is reducing harmful and excessive human activities; otherwise, whatever efforts we are putting in to protect our water reservoirs of the world will fall short.

14.33 How Can Blue Ecosystems' Health Be Monitored in South Asia?

The El Niño events of 1988 caused a significant amount of coral bleaching in almost all the reefs of the world and thus destroyed a major part of the blue ecosystem. The recovery of these events has been slow, and till date the coral reefs have not been able to recover. Many are completely bleached, and a few are regenerating themselves. This tragedy is exacerbated in magnitude because of the human impact in these regions and has a close link with the increasing population of the world in general. The growing population has caused an uncontrolled speed of resource exploitation and coral mining which has led to sedimentation and pollution. In addition to this, the concern of the authorities regarding coral reef health and protection has not been adequate enough to help these reefs regenerate better (Salman 2019).

There is a severe lack of research, resources, and monitoring in the management of reefs in the South Asian region, which is why the health of the blue ecosystem is at a greater risk than it was before (Alam 2019). In areas where there is room for improvement lies the management of these areas. The absence of proper infrastructure combined with the lack of awareness is wreaking havoc on the blue ecosystem and their health in general. This means that indirectly the livelihood of millions of people and the infrastructure of coastal cities stand to cause a great economic loss to the countries of these regions if they do not step up and protect the blue ecosystem. Moreover, there is an increased risk of natural disasters such as cyclones, unprecedented rainfalls, floods, and sea storms in these areas because of the bad ocean health disrupting the natural carbon cycles.

The regions are advised to monitor reef resources and research on proper management of the reefs according to their environmental conditions and requirement (Alam 2019). The concerned authorities should step up and form a proper chain of command to evaluate the present condition of marine wildlife and the impact of human activities on them. Then they should come up with an appropriate method to sustainably derive the benefits they want from the blue ecosystem without damaging their health. The overall rank of the South Asian region on the Ocean Health Index is from 51 to 55 (Ocean Health Index 2019). This rank is very far away from the global score of 71, which means that the ocean health in this region needs a lot of time to rebuild and regenerate (Ocean Health Index 2019). For this purpose, the authorities will need to reduce the human activities in these areas significantly at least for a few years. The human activities they can opt for should be sustainable for the long term so that maximum benefits from the blue ecosystem can be derived without irrevocably damaging them. If they start now, it will take a decade at least to bring a positive change in the ocean health. This gives them very few years' time before the ocean health of the region becomes harmful beyond repair. Global donor organizations such as the World Bank have helped set up multi-donor funds that help South Asia monitor its coastal ecosystem and blue ecosystem in general. These are SACEP (South Asia Co-operative Environment Programme by the World Bank) and SAWI (South Asia Water Initiative by the World Bank); other than these multi-donor organizations, local organizations are also working in separate countries for this purpose.

Strengthening the socioeconomic monitoring of the reef resources will help a great deal to these regions in addition to finding alternative livelihoods for people who are only dependent on the blue ecosystem for it. This way, a great deal of the burden will be lifted from the oceans. It will also reduce the level of human activity in the coastal regions. All this is possible only when the panning involves all levels of the government. The preferential way of deriving energy in this region is still dependent on fossil fuels. This is causing the temperatures of the ocean to rise which causes the natural system of the blue ecosystem to disrupt. Thus, these regions should shift their focus on green energy specifically solar energy since the summers last longer in South Asia with abundant sunlight. A good way of maintaining ocean health in the region would be a collaboration between all the countries of the region. This can also involve national and international NGOs dedicated for preserving the blue ecosystem. Together, these countries can raise funding and carry out important research which would be beneficial for everyone in the long run because the blue ecosystem do not belong to just one country or just one region. They are giving everyone the most important thing to live: oxygen, without which all life would seize to exist. Thus, for this purpose, a unified stand taken in this direction would prove to be in the best interests of everyone.

References

Alam, Muhammad Ammar. 2019. *Prospects of Maritime Economy For Pakistan.* MSF, Maritime Study Forum. https://www.maritimestudyforum.org/wp-content/uploads/2019/08/Prospect-of-Maritime-Economy-for-Pakistan.pdf.

ARCGIS. 2020. *Sea Level Rise : South Asia.* Accessed 12 6, 2020. https://www.arcgis.com/apps/MapTour/index.html?appid=10f767dd34d84d72bc36b016bf309342#:~:text=Sea%20Level%20Rise%3A-,Due%20to%20the%20overall%20in.

Aswani, Shankar. September 2017. "Marine resource management and conservation in the Anthropocene." *Environmental Conservation* (Cambridge University Press) 192-202.

Atanu Sarkar, Arindam Dasgupta and Suman Ranjan Sensarma. 2019. "Climate Change and Food Security in India: Adaptation Strategies and Major Challenges." In *Sustainable solutions for food security*, by Arindam Dasgupta, Suman Ranjan Sensarma, Atanu Sarkar, 497-520. Springer Link.

Islam Faisal, Hove Hilary, Parry Jo-Ellen. 2011. "Review of Current and Planned Adaptation Action: South Asia." *Adaptation Patnership/International Institute for Sustainable Development*, pp.108–118.

José C. Xavier, Angelika Brandt. 2016. "Future Challenges in Southern Ocean Ecology Research." *Frontiers in Marine Science.*

Kirsten Isensee, Francisco Chavez, Daniel Conley. 2018. *The ocean is losing its breath.* IOC-UNESCO.

Laal, Rattan. 2011. "Soil degradation and Food Security in South Asia." In *Climate change and food security in South Asia*, by M.V.K Sivakumar, S.M.A Faiz Rattan laal, 137–152. Springer Link.

Ocean Health Index. 2019. *Ocean Health Index.* Accessed December 7, 2020. http://www.ocean-healthindex.org/region-scores.

Parsons, E.C.M. September 2017. "Marine "Conservation": You Keep Using That Word but I Don't Think It Means What You Think It Means." *Frontiers in Marine Science.*

Pendleton, Linwood. May 5, 2020. "Opinion: We need a global movement to transform ocean science for a better world." *Proceedings of the National Academy of science of the United States of America* 117.

Parsons, E.C. and MacPherson, R. 2016. "Have you got what it takes? Looking at skills and needs of the modern marine conservation practitioner." *Journal of environmental studies and sciences* 515–519.

Rajasuriya, Arjan. 2002. "Status of Coral Reefs in South Asia: Bangladesh, India, Maldives and Sri Lanka." January: 95–100.

Roberson, Julia. 2016. "Restoring endangered coral reefs." *oceanconservancy.org* (Ocean Currents Blog).

Salman, Aneel. 2019. *Towards Blue Growth: A Sustainable Ocean led Development Paradigm (SODP).* Maritime Study Forum. https://www.maritimestudyforum.org/wp-content/uploads/2020/03/Towards-Blue-Growth.pdf.

Sandra Fischer, Jan Pietron and Arvid Bring. 2017. "Present to future sediment transport of the Brahmaputra River: reducing uncertainty in predictions and management." *Regional Environmental Change* 515–526.

C.K Shum and C.Y. Kuo. 2011. "Observation and Geo-Physical causes of present day sea level rise." In *Climate change and food security in South Asia*, by M.V.K Sivakumar and Faiz Rattan laal, 85–104. Springer Link.

M.V.K Sivakumar and Robert Stefanski 2011. *Climate change and food security in South Asia.* Springer Link.

The Blue Carbon Initiative. 2019. *Mitigating Climate Change through Coastal Ecosystem Management.*

Willa R. Brooks, Morgan E. Rudd, Samantha H. Cheng. May 2020. "Social and ecological outcomes of conservation interventions in tropical coastal marine ecosystems: a systematic map protocol." *Environmental Evidence.* https://doi.org/10.1186/s13750-020-00193-w.

Chapter 15
Potential Future Challenges and Impacts on Fisheries and Coastal Economies

Ignacio Cazcarro, Iñaki Arto, Jose A. Fernandes-Salvador, and Valentina Lauriad

15.1 Introduction

The concept of Anthropocene captures the transition to human dominance of the Earth, with a more clear dominant influence on the climate and environment (e.g. see Donges et al. 2017; Ribot 2014; Steffen et al. 2011). Coastal areas, and especially deltas, represent clear hotspots where natural processes and intense and growing human activity intersect (Fernandes 2018; Renaud et al. 2013; Tittensor et al. 2010). Many relevant processes of environmental, economic and social change are progressing faster and more intensely in deltas than their global averages (Nicholls et al. 2019). For example, it was estimated that more than 500 million people (around 7% of world human population lives in 1% of global land area) live in and around deltas (Ericson et al. 2006; Woodroffe et al. 2006) and in some cases areas close to rapidly growing mega-cities such as Dhaka, Kolkata, Shanghai or Cairo. Still, most large delta populations are located in the so-called Global South, and especially in Asia, usually largely being rural areas and livelihoods relying on agriculture and fishing.

I. Cazcarro (✉)
ARAID (Aragonese Agency for Research and Development), Agrifood Institute of Aragon (IA2), University of Zaragoza, Zaragoza, Spain

Basque Centre for Climate Change Research (BC3), Leioa, Spain
e-mail: icazcarr@unizar.es

I. Arto
Basque Centre for Climate Change Research (BC3), Leioa, Spain

J. A. Fernandes-Salvador
AZTI, Basque Research and Technology Alliance (BRTA), Sukarrieta (Bizkaia), Spain

V. Lauriad
IRBIM-CNR Institute of Biological Resources and Marine Biotechnologies, National Research Council, Mazara del Vallo (TP), Italy

S. Hazra, A. Bhukta (eds.), *The Blue Economy*,
https://doi.org/10.1007/978-3-030-96519-8_15

In this chapter, we will focus on two aspects related to the future of the coastal economies. First, we look at one of the main activities upon which many coastal areas, especially in developing countries, rely on—fisheries. We discuss the present state and future estimates, reflecting on the effects that this may have on those coastal, mostly rural areas mainly depending on them. Second, we look in detail to particularly relevant areas of coastal economies, the deltas. This second aspect extends from the multidisciplinary work of the DECCMA ('Deltas, Vulnerability and Climate Change: Migration and Adaptation') project (Das et al. 2020; Hossain et al. 2018; Kebede et al. 2018; Lauria et al. 2018; Nicholls et al. 2019; Nicholls et al. 2017). According to many of these and other studies, agriculture and particularly fishing are very important in the livelihoods in the deltas, which furthermore it extends along the supply chain to other industrial sectors of fish transformation, selling and export. Still, as we will see and discuss, historically, some of the large ones have generated what today are large or mega-cities, where obviously the importance of fisheries gets relatively reduced. We explore the expected socioeconomic future of those considered key areas in terms of the blue economy and vulnerabilities to global changes.

15.2 Fishing and Aquaculture in Asia

15.2.1 Present State

The following section shows a nice perspective and summary of other FAOSTAT data (e.g. from the FishStatJ—Software for Fishery and Aquaculture Statistical Time Series) on the world fishing and aquaculture evolution (FAO 2020). Fishing production has been increasing over decades (see Fig. 15.1, Panel a). Of this, aquaculture represented 46% of the total, being inland aquaculture 62% of the total aquaculture (hence 29% of the total). The increase in these types of activities then has notably led the increase, while captures have remained relatively stable or on slight decrease (Pauly and Zeller 2019).

Fish and fishery products still are also some of the most traded food commodities in the world (67 million tonnes in 2018, i.e. 38% of total fisheries and aquaculture production). Overall, from 1976 to 2018, the monetary value of global fish exports increased from US$7.8 billion to peak at US$164 billion (yearly nominal growth rate of 8%, in real terms, adjusted for inflation, 4%, when global export quantities increased at a real growth rate of 3%). This slower growth rate in fishing export quantities shows the increasing fish prices and the larger share of processed products in trade, revealing some reliance of developing countries on fish export for foreign income (Fernandes 2018; Lauria et al. 2018).

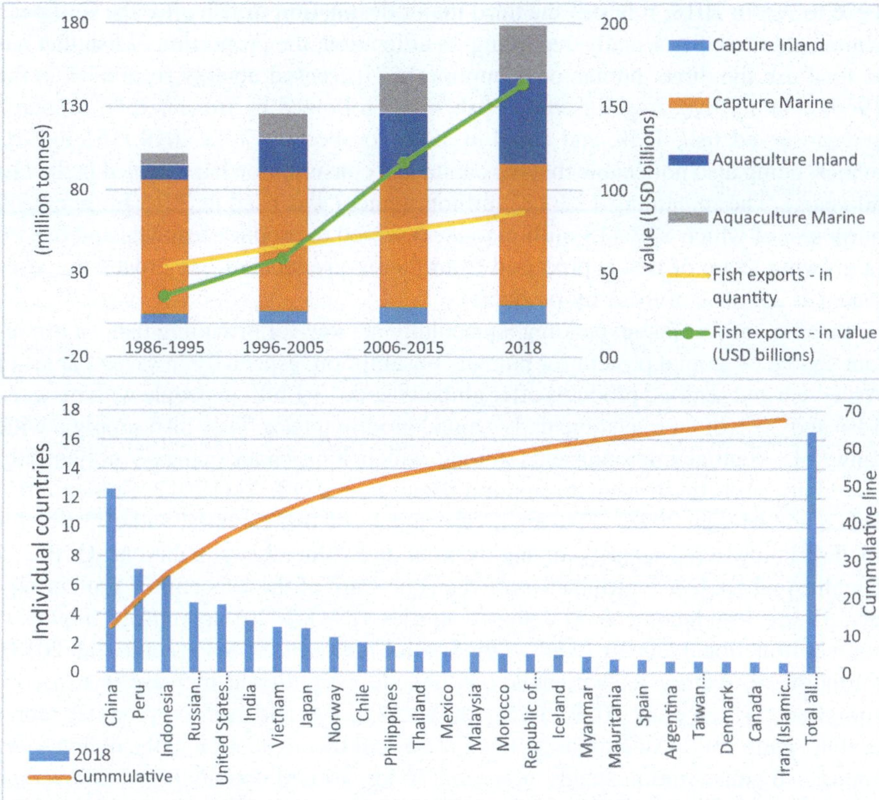

Fig. 15.1 Panel (**a**) Evolution of capture fish and aquaculture (inland and marine). (**b**) Marine catch in the most recent year (2018) by country (million tonne, live weight). (Source: Own elaboration from the FishStatJ—Software and (FAO 2020)).

In addition, Asia, the focus of this book, has a predominant role in the fishing production and trade. China has the major fish catch[1] (more than doubling that of Indonesia and Peru, which are followed by India, the Russian Federation, the USA, Vietnam and Japan) and exporter since 2002, but also since 2011, it is the third major importing country by value. Figure 15.1 (Panel b) shows only marine capture (not inland, and hence the figures are smaller for the countries with some relevant captures there, such as India, with about 35% of the total catch, China with close to 15%, Indonesia ~7%, Russia and Vietnam <3%).

Also, after China and Norway, the third and fourth major exporters are Vietnam (since 2014) and India (since 2017), followed by Chile and Thailand. In terms of imports, although Japan has reduced its share in global fish imports from 21% in

[1] While the estimates of total catches for China in the FAO database are generally considered to be complete, improvements are needed to more accurately assign China's distant water fishery catches by area and the disaggregation of catches by species.

1976 to 9% in 2018, it is still the third major destination of fish after the European Union and the USA. Finally, regarding its utilization, the proportion of fish that has as final use the direct human consumption has increased notably from 67% in the 1960s (mainly live, fresh or chilled fish, 44%, followed by frozen, 35%, prepared and preserved fish, 11%, and cured at 10%) to about 88% in 2018 (156 million tonnes, being also noticeable that per capita fish consumption has doubled in the last 50 years). The remaining 12% (22 million tonnes) was used in 2018 for non-food purposes, of which 82% (18 million tonnes) served to produce fishmeal and fish oil (a growing share of this is produced from the by-products of fish processing, estimated at 25–35% of these by-products).

Fishing and the fast-expanding aquaculture are key for providing fish, an important source of animal protein for human consumption, especially in several areas of Asia. Fish and seafood products offer globally about 3.1 billion people with between 17% and 20% of their average daily animal protein intake. They also supply a wide range of essential micronutrients among which long-chain omega-3 polyunsaturated fatty acids for direct consumption (Béné et al. 2015; FAO 2012; Golden 2016; Tacon and Metian 2018). This makes the supply of fish essential and effective for food and nutritional security among the poor and vulnerable populations. Of the 34 countries where fish contributes more than one-third of the total animal protein supply, 18 are low-income food-deficit countries (FAO 2020) apart from important sources of national income (e.g. see for South China Sea, SCS, Pernetta et al. 2013). Furthermore, dietary recommendations include a significant increase in fish consumption (Willett et al. 2019). In Southeast Asia, fish and seafood products represent the main source of animal protein for most of the population in the region—per capita fish consumption stands at around 36 kg, around double the world average and accounts for about 42% of total animal protein intake for individuals (FAO 2017; OECD-FAO 2017).

15.2.2 The Future

The world population is expected to reach nine billion by 2050, implying increases in the needs of food and nutrition, under the limits of natural resources (land, water, energy, etc.), furthermore in the context of global change. For example, in the ASEAN (Association of Southeast Asian Nations) region, fish consumption is projected to rise from 24.5 million tonnes in 2015 (38.4 kg per capita) to 36.9 million tonnes in 2030 (51.5 kg per capita) and then reach 47.1 million tonnes in 2050 (61.5 kg per capita) (Chan et al. 2017).

In what concerns fish availability, 34% of assessed fish stocks currently are fished at levels that exceed biological sustainability (FAO 2020). The status of fish stocks for developed countries is considered to be improving, but worsening for many developing countries in terms of overcapacity, production per unit of effort and stock status. This situation will need ad hoc management plans especially in the context of the expected impacts of climate change (Jose A. Fernandes et al. 2013,

2016; Lauria et al. 2018; Lotze et al. 2019). However, to put in place long-lasting resource management plans, it is necessary to have a full understanding of the catches (sometimes unreported, see Watson 2017), local practices in targeting specific fish, based on what are economically important fish species (e.g. see for the case of India, Das et al. 2019; Raman et al. 2017; Tarun Kumar and Shivani 2014, and for SCS in Pernetta et al. 2013), the marine ecosystem dynamics and the effects due to the role of those species in the food web (Das et al. 2018). Neither one can ignore the aspects that we examine in the second part, related to human development and alternative livelihoods for coastal poor households, since the lack of them is often related to overfishing or unsustainable practices (Ahmed et al. 2010; Das et al. 2019; Dutta et al. 2012; Hoq 2007; Teh et al. 2017; Trajano et al. 2017).

Looking, for example, at the projected economic impact of climate change on marine capture fisheries in the Philippines (Suh and Pomeroy 2020), the contribution of marine capture to GDP is expected to decrease between 9% and 18% by 2060 depending on the scenarios, notably affecting income of the fishermen, considered among the poorest groups, with high incidence of poverty. Prospects for Fisheries and Aquaculture for 2030 have been performed with the IMPLAN model (WB. 2013). A more recent study using FAOSTAT fish (Cheung et al. 2019) explained that across exclusive economic zones (EEZs), the 'business as usual' (BAU) scenario and without any high seas protected area, maximum catch potential, the indicator of the theoretical productivity of fish stocks and proxy for maximum sustainable yield, was projected to decrease throughout the twenty-first century between 1% (RCP2.6) and 4% (RCP8.5), with some buffering of the decline (around 25%) with protected area scenarios. Without any protected area, maximum catch potential in all the EEZs was projected to decline between 1% and 4% relative to the 2000s. Specifically, in low-income countries, mean species abundance remains relatively constant on average under SSP1 and RCP2.6 by 2050. However, it is projected to decrease by around 7% by 2050 under SSP3 and SSP5 under RCP8.5. This leads to projections on the highest losses in terms of profits for low-income countries, especially under SSP3 (Cheung et al. 2019). Furthermore, some specific regions may even find reductions in potential catch also for the sustainable fishing scenarios (e.g. see on tuna, Indian oil sardine and hilsa in Das et al. 2020), so it is strongly advocated that proper management plans to sustain the existing fish stocks should be in place and consider the studied possible futures (Das et al. 2019; Dutta et al. 2012; Pitcher et al. 2009; Pomeroy 2012; Trajano et al. 2017).

This needs to be also put in perspective taking into account several studies on the future effects of global change on the deltas, where multiple and interlinked drivers of change are examined (Tessler et al. 2015; Kebede et al. 2018). Deltas are linked with the threat of climate-induced sea-level rise and subsidence which may affect economic development (De Souza et al. 2015; Hallegatte et al. 2016) and to changes in water levels (due to their low elevation) which may have strong hydrological and livelihood effects, including inundation, salinity and waterlogging with severe impacts on livelihoods. The issue of forced migration from deltas due to sea-level rise was raised a long time ago (Milliman et al. 1989) and keeps being signalled

(Gemenne 2011), although there is little systematic scientific investigation of current and future likely demographic movements and settlements.

According to Tessler et al. (2015), of the 48 deltas into a risk space defined by each delta's specific anthropogenic, geophysical and socioeconomic characteristics (proxying hazard, exposure and vulnerability), we may highlight the high values of the Krishna delta, the Ganges-Brahmaputra (especially due to high vulnerability), Brahmani, Godavari (having these 4 Indian deltas the highest risk of the 48), but also the Indus in Pakistan, the Shatt al Arab in Iraq/Iran/Kuwait, the Hong in Vietnam, Mahanadi in India and Irrawaddy in Myanmar (Burma) rank seventh to 11th in risk (after the cited, Limpopo and Sebou). The Chinese Pearl, Yangtze and Yellow also rank at a medium-high risk level. The delta systems such as the Yangtze are though at clear risk both from RSLR, leading to increased exposure to flooding, and from reduced effectiveness of risk reduction strategies that may not be sustainable on the century scale. This leads to conclude that under future scenarios in which the role of GDP is seen or estimated as less positive, or less capable to compensate or reduce risk (e.g. based on increased costs of energy, etc.), this delta is much more vulnerable, being the Pearl also strongly affected, while deltas in low-GDP regions, such as the Irrawaddy (also Tana and Fly in other regions), are the least sensitive to these scenarios.

Worldwide, the fishing industry consumes 21% more fuel per tonne of catch than 20 years ago (Parker et al. 2018), whereas catches are forecasted to remain the same or up to 30% lower due to climate change (Lotze et al. 2019). Europe's fishing fleet's GHG emissions represent more than 10% of the global fleet's emissions (Parker et al. 2018). Furthermore, these estimations might be an underestimation since they focus on larger fishing vessels over 24 m when the high proportion of fishing vessels are smaller, particularly in Asian countries (Taconet et al. 2019). Several studies suggest that significant fuel reductions (up to 10%) are possible with vessel retrofits, more efficient gears and better antifouling (Basurko et al. 2013; Notti et al. 2019). However, higher reductions (over 15%) could be achieved with good fishing and routing decision systems using climate and environmental data services (Basurko et al. 2013; Jafarzadeh et al. 2016; Vettor et al. 2016). These decision systems will make the capture fisheries carbon efficient and economically resilient. Reduction of fishing vessel fuel consumption and consequent climate change mitigation and adaptation is possible through better use of environmental data, as the fishers get better info about where and when to venture on their fishing trips. This data will also help to adapt to the changes in species distribution due to climate change that are already affecting fisheries (Baudron et al. 2020; Dufour et al. 2010) since fishermen's fishing decision is biased towards past successful fishing grounds (Jennings and Lee 2012; Maina et al. 2016).

15.3 Deltas in Asia: Present and Future Socioeconomic Estimates

We have highlighted how agriculture and particularly fishing have been highlighted in the literature as key activities for the livelihoods in the deltas because of the interplay there of sediment delivery and reworking, destructive marine processes and subsidence, extensive ecosystem services and accessible transport links which have favoured human settlements, etc. Furthermore, the importance does not remain only for the primary activities but tend to have cascading effects along the supply chain to other industrial sectors of fish transformation, selling, export or consumption (e.g. see Cai et al. 2016; Cazcarro et al. 2018). In addition to climatic drivers, external drivers such as changes in upstream catchments (Dunn et al. 2018) may notably affect deltas, including changes in river nutrient levels (Whitehead et al. 2018)—something that can be hazardous to fishing, aquaculture and agriculture, as well as potential human health (Syvitski 2003; Syvitski et al. 2005).

Low-income households in deltas may have very crucial bottlenecks to sustain their livelihoods due to the particular combination of hazards, exposure and vulnerability, which provides important risks. Notably, the fishing activities are very prone to those risks, which may trigger outmigration from coastal areas but also likely to some coastal mega-cities. Historically, some of the large deltas have generated what today are large or mega-cities, where obviously the importance of fisheries gets relatively reduced in the overall economy. As shown in Tessler et al. (2015) and in the analysis of the following subsection (with their spatial delimitation of deltas), this is the case in the Ganges-Brahmaputra (Kolkata, Dhaka, Khulna), Yangtze (Shanghai), Pearl (Guangdong-Hong Kong-Macao Greater Bay Area), Chao Phraya (Bangkok) and Hong (Hanoi). Many others, on the contrary, are truly areas without major cities, which tend to be also relatively more rural and poorer. Dependence on fisheries nutrition has been emphasized, for example, on the Mekong River Delta, as one of the most productive fishing zones with wild fishing and extensive fish farming (Uyen 2017). The Government of Vietnam would be aiming for turning the country into a global leading seafood exporter with a country's fisheries development strategy plan 2020 in which the seafood industry is expected to contribute 30–35% to the country's agroforestry-fisheries GDP.

15.3.1 Estimates of the Future Socioeconomics of the Deltas: Gridded GDP, GDP per Capita, Population and HDI

The increasing role of geographical information systems (GIS) to study the biophysical sphere has extended in the last decades also to the socioeconomic data, with further downscaling of traditional measures such as the nationally obtained gross domestic product (GDP). This has derived into metrics of gross regional product (GRP) as a monetary measure of the market value of all final goods and services produced in a region or subdivision of a country in a period (typically yearly) of time but also some kind of gridded GDP/GRP under concepts such as gross cell product (GCP) (Nordhaus 2006). In this regard, we may find, for example, the gridded GDP data set with its association with night-time lights satellite imagery, a good proxy to assess economic activity, is much more useful (Ghosh et al. 2010). The Global Risk Data Platform of the United Nations also allows for obtaining historical GDP raster distribution model and expanding from 1975 to 2007 (Deichmann 2013).[2] The Nordhaus (2006) and Nordhaus and Chen (2016) database clearly had an economic orientation, using microeconomic information, such as rural/urban population data, income/labour (regional by industry), land area and some estimates of mineral production, etc., from several statistical agencies. In Fig. 15.2 (Panel a), we show how Asia looked based on this data, but apart from the relatively outdate of this data (up to 2005),[3] we have in this work a perspective more into the future which takes us to use some other more recent and prospective data.

Figure 15.2 shows the differences in resolution and scope of some of the databases, as well as some key aspects we want to highlight with different metrics. For example, we may observe how some Chinese deltas with very high GCP (in absolute terms, Panel a), generally, lie in relatively (with respect to the world) medium or low GCP per capita (hence in those the effect of huge populations living there play a role). Deltas that we may highlight as falling, at least partly in areas of high GCP, are the Chinese Yangtze and Pearl, the Han between the Koreas, the Ganges-Brahmaputra-Meghna and the Shatt al Arab.

Gridded data as the ones cited on GDP and GDP per capita has been used (together with government effectiveness measure from) in Tessler et al. (2015) to obtain an Investment Deficit Index (IDI), as a proxy of a vulnerability index of key deltas, which are our focus now in this chapter. They also assessed the risk sensitivity to future investment capacity reducing the weight of the GDP indicators in the IDI. Using here their layers on deltas (hence having clear-cut different definitions of deltas for the Ganges and Mahanadi than in the DECCMA work, which include high urban areas now such as Dhaka, Kolkata and Bhubaneshwar) and the SSP projections of Murakami and Yamagata (2016)[4] in purchasing power parity (PPP, billion USD in 2005 year rate), we obtain here the estimated future GDP, population and GDP per capita of each of those main Asian deltas for their SSPs (Table 15.1 shows mainly SSP2 and SSP3, which seem to be so far the closer trends for most countries in the last decades). SSP1, which refers to the sustainable scenario, yields a compact population distribution relative to SSP3, which denotes the

[2] As well as population extrapolation from 1975 to 2007 (Dao 2013) and socioeconomic data in general (Dao et al. 2013)

[3] Also, some authors find challenges in the resolution (e.g. for more localized studies), of 1 degree longitude/latitude (e.g. see Figure 2, Panel a), vs. the resolution, e.g. of 30 arc seconds of some of these other data sets.

[4] As a disclaimer, it should be noted that we find several relatively extreme values (particularly for SSP2, in low populated areas), with outliers, grids where GDP per capita shows relatively unreasonable data.

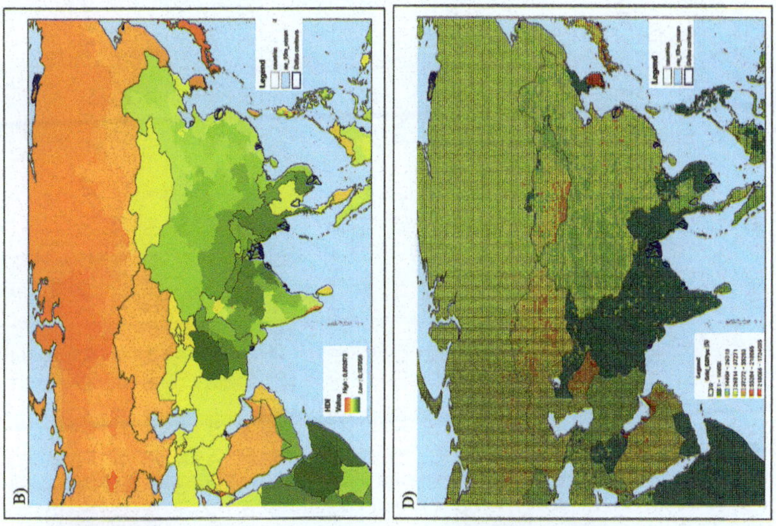

Fig. 15.2 Analysed deltas with blue contour. Panel (**a**) Absolute GCP (PPP, billion US$) from the Nordhaus (2006) database. Panel (**b**) (Kummu et al. 2018, 2020). Panel (C) GDP (PPP, US$2005/year) per capita in 2050 under SSP2 from Murakami and Yamagata (2016). Panel (D) GDP per capita in 2050 under SSP3 (same source, both 'natural breaks'). (Source: Own elaboration using data from the above-cited sources (Nordhaus and Chen 2016; Kummu et al. 2018, 2020; Murakami and Yamagata 2016) with the Arc GIS 10.5.1 software)

Table 15.1 Estimates from different sources on metrics of the main Asian deltas identified in Tessler et al. (2015)

		Area[a]	HDI[b]		GDP PPP $/pc deltas[c]				Population (1000 people) deltas[d]			
		Deltas	Deltas	Country/ies	SSP2	SSP2	SSP3	SSP3	SSP2	SSP2	SSP3	SSP3
Delta	Country	Area_ km^2	1990–2015	1990–2015	2020	2050	2020	2050	2020	2050	2020	2050
Amur	Russia	4037	0.71	0.75	8852	24,834	8796	24,799	21	22	21	20
Ganges	India	108,908	0.40	0.52	2342	8994	2275	6002	151,681	190,900	152,838	207,463
Brahmani	India	7047	0.34	0.52	3572	13,515	3488	8422	4167	3980	4520	5779
Chao Phraya	Thailand	25,066	0.57	0.67	9141	27,233	9145	19,833	22,413	30,683	20,349	23,958
Godavari	India	4328	0.44	0.52	2869	12,023	2797	7905	5555	8230	5415	7943
Han	South Korea/North Korea	4039	0.68	0.83/–	16,010	34,185	15,723	32,285	2712	2644	2705	2409
Hong	Vietnam	6459	0.48	0.59	4395	13,968	4353	10,294	6904	6433	7222	7967
Indus	Pakistan	6952	0.41	0.47	1911	6920	1897	4488	251	278	272	425
Irrawaddy	Myanmar	36,019	0.35	0.45	1820	5237	1879	3678	14,213	15,753	13,822	15,436
Krishna	India	2619	0.44	0.52	3432	13,765	3294	8878	1002	1164	1045	1398
Mahanadi	India	7733	0.34	0.52	4149	15,006	4022	9442	5916	7065	6122	8343
Mekong	Vietnam/Cambodia	52,066	0.45	0.59/0.46	3427	11,779	3409	8848	24,597	26,021	25,238	29,798
Shatt al Arab	Iraq/Iran/Kuwait	6309	0.57	0.61/0.69/0.77	6617	15,178	6676	14,413	739	927	737	1028
Tone	Japan	1559	0.81	0.86	32,646	47,578	32,124	44,703	444	355	447	341
Yangtze	China	50,505	0.55	0.62	19,916	50,903	19,682	34,322	56,887	56,950	55,876	55,184
Yellow	China	10,188	0.52	0.62	8363	28,185	8224	23,196	899	623	956	804
Pearl	China	13,540	0.52	0.62	17,046	45,915	16,764	33,380	30,666	32,835	29,598	30,305
Lena[e]	Russia	235,633	0.74	0.75	6060	21,604	6074	23,128	0	0	0	0
Mahakam	Indonesia	5532	0.53	0.62	5832	19,774	5877	13,208	781	718	835	931
Nile	Egypt (Africa)	37,603	0.55	0.62	6916	21,358	6899	14,733	66,161	81,157	68,894	99,838
Fly	Papua New Guinea (Oceania)	3595	0.36	0.46	2293	8745	2796	5700	7	7	7	7

		Area[a]	HDI[b]		GDP PPP $/pc deltas[c]				Population (1000 people) deltas[d]			
		Deltas	Deltas	Country/ies	SSP2	SSP2	SSP3	SSP3	SSP2	SSP2	SSP3	SSP3
Delta	Country	Area_ km^2	1990–2015	1990–2015	2020	2050	2020	2050	2020	2050	2020	2050
Danube	Ukraine/Romania (Europe)	9922	0.69	0.70/0.74	7070	19,945	6937	15,357	197	174	198	178
Dnieper	Ukraine (Europe)	1710	0.71	0.70	8575	23,952	8485	16,607	354	357	343	315

[a] Area is obtained from the original data of Tessler et al. (2015)

[b] Using the shapefiles from [a], we obtain the HDI from Kummu et al. (2018, 2020)

[c] GDP (PPP, US$2005/year) per capita is estimated from Murakami and Yamagata (2016)

[d] We also evaluate population futures from Merkens et al. (2016). It should be noted that these population projections differ from those of Murakami and Yamagata (2016), e.g. for the Lena delta

[e] The Lena delta has some uncoherent results from the GDP per capita data, probably related to marginal populations captured in some scenarios, and hence we do not give much importance to it

fragmentation scenario. Their results show that GDP growth in major metropolitan areas changes significantly depending on the scenarios.

A very interesting recent work is also that of Kummu et al. (2018, 2020), since for the period 1990–2015 (so in this case, no projections), they obtain not only the GDP and GDP per capita but also the Human Development Index (HDI), which is a summary measure of average achievement in key dimensions of human development: a long and healthy life (life expectancy), being knowledgeable (education) and have a decent standard of living (the gross national income plays also a role). Having also its limitations (Jahan 2019; Lind 1992, 2004; Morse 2003; UNDP 2019), still GDP has been often more criticized as single measure of development (Constanza et al. 2013; Kubiszewski et al. 2013; Stiglitz et al. 2010), and HDI has been shown useful and recognized as better proxy of human development (Khodabakhshi 2012; Lind 1992). Importantly, some studies have supported the assumption made by the HDI that higher levels of GDP are associated with greater levels of development, but at a decreasing rate (Cahill 2002).

We make use of this data as well to evaluate more specifically the climate-induced and socioeconomic scenarios for the deltas. Table 15.1 summarizes the estimates of different metrics for the deltas identified in Tessler et al. (2015) placing the focus on the Asian ones (as well as some very nearby ones which allow having also some points of comparison). We may highlight how the Human Development Index is typically lower in the delta areas than the general figures of the country or countries they fall in. It also highlights the idea on deltas where due to relatively low development (there only measured in terms of economic growth) people may face higher vulnerabilities. In this regard, especially Indian, Burmese (Myanmar) and Pakistani deltas, with some Vietnamese/Cambodian, Thai or Chinese deltas illustrate this idea of populations with high social vulnerability which may suffer from relatively low capacity to build infrastructure, capacity or health assistance in case of emergencies.

Furthermore, Merkens et al. (2016) extended the SSP narratives to the coastal zone, with gridded subnational population projections until 2100 for each SSP. They concluded that coastal population is up to 35% higher compared to the original IIASA projections, finding projecting to be >1 billion by 2050 in all SSPs in low elevation coastal zone (LECZ). In that regard, as reviewed in Murakami and Yamagata (2016), since the work of Gaffin et al. (2004), there has also been several works which have downscaled and geospatial gridded socioeconomic projections, e.g. from the IPCC special reports, especially on population. In Murakami and Yamagata (2016), we find the downscaling of the Shared Socioeconomic Pathways (SSPs) (O'Neill et al. 2014, 2017) projections that describe future socioeconomic conditions under various scenarios, including SSP1–3.

SSP1 makes relatively good progress towards sustainability under an open and globalized world. SSP2 is a middle-of-the-road scenario assuming that the typical trends in the last decades will continue, and in SSP3, the world is closed and fragmented into regions, but it fails to achieve sustainability (quite unfortunately in our opinion, what the world is looking more like currently). The estimated GDP per capita is for almost all areas higher in SSP2 than SSP3, being for population levels a more mixed case, even with higher populations in the delta in more cases in SSP3 than in SSP2

Additionally, departing from the data of Kummu et al. (2018, 2020), the projections of Cuaresma and Lutz (2015) on education (and gender and age) and HDI themselves, inequality projections of Rao et al. (2019) and other available projections at the national level, HDI towards 2050 can be examined and potentially projected. For example, for Pakistan, they show that under SSP2 (what has been considered by many scholars as a kind of BAU), the HDI could reach 0.56, and in SSP5 0.64. Still this seems to provide an image of the fast economic growth scenario as the most positive, but the sustainable one (environmentally with smaller fertility and mortality rates) offers a very close value of 0.63, while more disruptive scenarios, based on the fragmented and stalled social development SSP3 provides the value of 0.40 and the inequal world of SSP4 of 0.41. In this regard, it is clear, for example, that a delta such as the Indus but similarly also several other deltas in the area would rarely benefit from unsustainable and unequal scenarios (SSP4, SSP3, SSP2, in this order, tend to be the most dramatic in this regard), as we have also seen for the case of fish catch.

We may think that those scenarios would present further challenges in terms of capabilities (Sen 1999) of those vulnerable populations, which require more human-focused approaches to understanding environmental change (Brown and Westaway 2011). For example, also in the Chinese Yellow delta, it was shown a wide typology of respondents (of comprehensive survey concerning environmental awareness and vulnerability) with low levels of education, earnings and awareness of global processes of climate variability and sea-level rise (Wolters et al. 2016; Fernandes 2018). Vulnerability itself affects people's ability to adapt in deltas (Tompkins et al. 2019), i.e. natural processes may be amplified by more recent human interference with the delta systems (poorly maintained engineering dams, navigation, flood control works, etc., demographic pressures, changes in land use, etc.). It is recently highlighted then that adaptation strategies need to consider long-term planning (Suckall et al. 2018), management policies enforcement (Monirul Islam et al. 2014), finances and risk management literacy (Ahsan 2011), corruption (Bennett et al. 2014), gambling (Lincoln 2014), disaster preparedness and sustainable development programs, but also feasibility and practicality in communities (Smit and Wandel 2006). Coming back, for example, to the case of fisheries, in a delta one may find several dozens of adaptation options, but implementation costs usually vary importantly (Arto et al. 2019) being very important to understand then costs and benefits and what actions can be undertaken by fishing households and/or other institutions.

Even more, the fact that some deltas are defined in a wide sense including some large fast-growing cities may be even hiding some additional human development effects of migration among/from/to coastal areas, probably especially in inequalitarian, more fragmented and/or unsustainable futures. Still, as we extend below, these aspects require much more study since, to start with, we find a variety of cases, especially in Southeast Asia. Typically, rural coastal areas are seen as places of out-migration when the biophysical events may occur; however, the expected by many scholars increases in the share of populations living in urban areas (greater than the current ~55% globally), especially with increasing concentrations in mega-cities, should be weighted with the fact that many are in coastal and delta areas, especially in Asia (de Campos et al. 2019).

15.3.2 Further Interplay of Expected Impacts of Climate Change on Fisheries and Deltas

We have seen that the fishing sector tends to play an important role in Asian countries, especially in coastal zones. In the past decade, the contribution to GDP has been estimated (Lymer et al. 2008; Sugiyama et al. 2004) in India around 0.5% for capture fisheries and 0.5% for aquaculture, 1% and 2–5% in China, around 2% for each of them in Bangladesh, and in Thailand (with some decreasing share in the second case), 1.5% and 4–5%, respectively, for Laos, 2% and 1.5% for Indonesia (with some decreasing trend), 10–11% and 1%, respectively, for Cambodia, or around 4% each in Vietnam. These estimates vary among sources and more recently tend to have a smaller share in GDP (Subasinghe 2017), often due to higher GDP growth levels than those of fisheries. Also, obviously, the share of aquaculture is surpassing that of capture fisheries, except notably for Thailand and especially Cambodia. As regards the deltas, for example, in several Bay of Bengal deltas (of India and Bangladesh), the fishing sector share was estimated around 4–5% of GDP (e.g. see Cazcarro et al. 2018; Lauria et al. 2018). It is easy to estimate then just in GDP terms what expected future losses as the ones cited in several studies of between 10% and 20% of this may damage the economies. Still, vulnerability differs highly between nations according to the contribution of such fisheries to their economy (Cooley et al. 2012; see also Sections 6.4.1, 7.3.2.4, 7.4.2 of IPCC 2014).

Higher temperatures, rising extreme weather events and damage to ecosystems are expected to have other effects (not just fishing) on the coastal areas and deltas. For example, coastal flooding has been estimated (Kirezci et al. 2020) that it will threaten assets in 2050~15% more than today and 32–46% more in 2100 than today, being worth in this final moment up to 12–20% of global GDP. Schinko et al. (2020) specifically simulate GDP losses by 2050 in different scenarios, which strongly depend on the level of adaptation and the assumed degree of ice melting (without further adaptation measures, 0.4% with high ice melting and ~0.2% with low, while full adaptation leaves impacts to less than 0.1% of global GDP in all cases). Hence, adaptation is found to be highly economically efficient, with adaptation costs being much lower than the corresponding benefits from avoided damages. The highest levels of GDP loss are projected for China (0.8–0.9% under RCP26-SLR and 0.9–1.0% under RCP45-SLR) and India (0.5–0.6% under both scenarios), followed by Canada and Indonesia (0.2–0.3% under both scenarios). These are also the countries with the highest direct impacts according to DIVA (Dynamic Interactive Vulnerability Assessment) modelling framework projections.

If global temperatures rise by more than 1.5 °C above pre-industrial levels, the Asian and Pacific region will face climate-linked disasters, estimated (ESCAP 2019) with an annual average loss of $675 billion (equivalent to 2.4% of the region's GDP in 2018). This could hinder economic and social development gains through negative impacts on infrastructure, health and education attainment and on income distribution (Alisjahbana et al. 2020). Economic growth undermined by climate disasters could increase the region's Gini coefficient by 0.24, increase under-five mortality rates by 0.3 and decrease education rates by 0.26 percentage points, respectively. This means that the projected HDI and ultimately the development path could be also highly affected in several of the Asian rural coastal areas and deltas.

As reviewed in Anika and De Cassandra (2016) in the fifth IPCC report (IPCC 2014), it was predicted that climate change will adversely affect food security in Asia by the middle of the twenty-first century, with South Asia being most affected (WGII (B), p. 1343). The AR5 stated that—after Indonesia—Cambodia and the Socialist Republic of Vietnam are the most vulnerable countries to climate impacts on marine fisheries (ibid.). One major factor behind this statement is the vulnerability of coral reefs to both warming and ocean acidification (WGII (B), p. 1355).

As highlighted in Pörtner et al. (2015), a 2 °C global temperature increase by 2050 is estimated to cause global losses in landed value of US$17 to 41 billion annually (in 2005 value), with an estimated cost of adaptation for the fisheries of US$7 to 30 billion annually over a 40-year time frame between 2010 and 2050. The largest loss in landed value is projected to occur in East Asia and the Pacific (Sumaila and Cheung 2010). Overall impacts and the regional manifestations will partially depend on the flexibility and response capacities of food production systems (Elmqvist et al. 2003; Planque et al. 2011). Hence, in order to avoid those negative effects, deltas and in general coastal zones necessarily will have to adapt. According to M. Ahmed and Suphachalasai (2014), to avoid the damage and economic losses from climate change under the BAU scenario, South Asia needs to provide an average adaptation expenditure of 0.48% of GDP per annum ($40 billion) by 2050 and 0.86% of GDP per annum ($73 billion) by 2100, being slightly lower under global actions to reduce climate change.

15.4 Conclusions and Discussion

We have seen in the chapter then the importance of fishing activities in coastal areas and how especially low-income households may suffer to sustain their livelihoods due to the particular important risks involved. Apart from the role of climate change, as shown empirically in several of the cited studies and theoretically summarized with scenarios by Attila N. Lázár et al. (2019), there are policy-driven trade-offs for fishing in which short- and long-term economic gains may be switched depending on the fishing efforts and more or less intrusive farming practices which exploit or not the environment.

Notably, the fishing activities are highly sensitive to all those changes, which may trigger outmigration from coastal areas but also likely to some coastal megacities. Migration brings benefits to those moving both in terms of avoiding risks in source areas and, more importantly, providing economic, social and educational opportunities in destination areas (as further studied in depth, e.g. by de Campos et al. 2019), but also as most population health research indicates (e.g. see Whitmee et al. 2015, and as many people have suggested regarding climate change, diseases like COVID-19, etc.), significant time lags tend to appear between ecological degradation and the effects on the overall health of populations, which suggest that forthcoming challenges and illnesses may appear in the near future related to pollution, toxic effluents, heat, etc. (Whitmee et al. 2015). Furthermore, outmigration may also have counteracting effects. Also for example aquaculture is providing rural employment and income (e.g. in Vietnam more than ½ million are employed, more than in capture fisheries) and is planned to keep expanding offering rural diversification and providing jobs and an alternative to urban migration (Uyen 2017).

We have also hinted how especially the population grids of Merkens et al. (2016) for coastal societies as well as the different indicators for deltas of Tessler et al. (2015) can be used in coastal impact, adaptation and vulnerability (IAV) research to assess exposure of population to climate change impacts and natural hazards in coastal economies. The two main aspects studied strongly interact. For example, coastal areas with rural economies are enormously dependent on fisheries catch, which are the main sources of protein intake (mainly from less valued species for food) for poorer households. The possible additional income from selling them though comes more from high valued species (which are typically demand more at close by urban canters) than low valued ones. Such income can be obtained especially by those capable of profiting from the increasing value added of aquaculture, of processed seafood (escalating in the ladder of value added, vertical specialization, etc.), and of domestic and foreign international trade to higher purchasing power destinations. The vicious and virtuous circles though are often present, especially given that as we have seen the capability of managing sustainable marine areas often depends on the development itself, on the possibilities of having alternative sources of income to allow fisheries to recover and even of other political aspects such as being able to enforce only local/domestic fishing so that other foreign commercial fishing does not exhaust the local resources.

For these reasons, although often difficult in modelling, due to the simplicity of having projections from several institutions on GDP and population, but not so frequently on education and health attainment likely futures, we have tried to find ways in which future scenarios may be affected with a more pluralistic way on development. This goes in line with the shifting narrative of climate policy evaluation from one of costs/benefits or economic growth (which tended to conclude that climate agreements were too costly) to a message of improving social welfare as a criterion for judging the effects of climate policy (van den Bergh and Botzen 2018).

Acknowledgements This work was developed thanks to the DEltas, vulnerability and Climate Change: Migration and Adaptation (DECCMA) project which is part of Collaborative Adaptation Research Initiative in Africa and Asia (CARIAA), with financial support from the UK Government's

Department for International Development (DfID) and the International Development Research Centre (IDRC), Canada. The views expressed in this work are those of the creators and do not necessarily represent those of DfID and IDRC or its Board of Governors. Ignacio Cazcarro and Iñaki Arto also thank the support of the Spanish Ministry of Science, Innovation, and Universities, through the project MALCON, RTI 2018-099858-A-I00. Their research is also possible thanks to the María de Maeztu excellence accreditation 2018-2022 (Ref. MDM-2017-0714), funded by MCIN/AEI/10.13039/501100011033/; and by the Basque Government through the BERC 2022-2025 program.

References

Ahmed, M., & Suphachalasai, S. (2014). *Assessing the Costs of Climate Change and Adaptation in South Asia*. Retrieved from http://www.adb.org/sites/default/files/pub/2014/assessing-costs-climate-change-and-adaptation-south-asia.pdf

Ahmed, N., Troell, M., Allison, E. H., & Muir, J. F. (2010). Prawn postlarvae fishing in coastal Bangladesh: Challenges for sustainable livelihoods. *Marine Policy*, *34*(2), 218–227. https://doi.org/10.1016/j.marpol.2009.06.008

Ahsan, D. A. (2011). Farmers' motivations, risk perceptions and risk management strategies in a developing economy: Bangladesh experience. *Journal of Risk Research, 14*(3): 325–349. https://doi.org/10.1080/13669877.2010.541558.

Alisjahbana, A. S., Hahm, H., & Malik, H. A. (2020). *Economic and social survey of Asia and the Pacific 2020. Towards sustainable economies*. Bangkok, Thailand: United Nations Economic and Social Commission for Asia and the Pacific (ESCAP), United Nations 2020.

Anika, S., & De Cassandra, Y. (2016). *Climate Change Implications for Fisheries and Aquaculture. Summary of the findings of the IPCC Fifth Assessment Report. FAO Fisheries and Aquaculture Circular* (Vol. FIAP/C1122).

Arto, I., García-Muros, X., Cazcarro, I., González-Eguino, M., Markandya, A., & Hazra, S. (2019). The socioeconomic future of deltas in a changing environment. *Science of The Total Environment*, *648*, 1284–1296. https://doi.org/10.1016/j.scitotenv.2018.08.139

Attila N. Lázár, Hanson, S. E., Nicholls, R. J., Allan, A., Hutton, C. W., Salehin, M., & Kebede, A. S. (2019). Choices: Future Trade-Offs and Plausible Pathways. In *Deltas in the Anthropocene*. Palgrave-McMillian.

Basurko, O. C., Gabiña, G., & Uriondo, Z. (2013). Energy performance of fishing vessels and potential savings. *Journal of Cleaner Production*, *54*, 30–40. https://doi.org/10.1016/j.jclepro.2013.05.024

Baudron, A. R., Brunel, T., Blanchet, M. A., Hidalgo, M., Chust, G., Brown, E. J., … Fernandes, P. G. (2020). Changing fish distributions challenge the effective management of European fisheries. *Ecography*, *43*(4), 494–505. https://doi.org/10.1111/ecog.04864

Béné, C., Barange, M., Subasinghe, R., Pinstrup-Andersen, P., Merino, G., Hemre, G. I., & Williams, M. (2015). Feeding 9 billion by 2050 – Putting fish back on the menu. *Food Security*, *7*(2), 261–274. https://doi.org/10.1007/s12571-015-0427-z

Bennett, N., Dearden, P., Murray, G. & Kadfak, A. (2014). The capacity to adapt?: communities in a changing climate, environment, and economy on the northern Andaman coast of Thailand. Ecology and Society, 19(2): art: 5. http://dx.doi.org/10.5751/ES-06315-190205.

Brown, K., & Westaway, E. (2011). Agency, Capacity, and Resilience to Environmental Change: Lessons from Human Development, Well-Being, and Disasters. *Annual Review of Environment and Resources*, *36*(1), 321–342. https://doi.org/10.1146/annurev-environ-052610-092905

Cahill, M. B. (2002). Diminishing returns to GDP and the human development index. *Applied Economics Letters*, *9*(13), 885–887. https://doi.org/10.1080/13504850210158999

Cai, J., Huang, H., & Leung, P. (2016). *Understanding the Contribution of Aquaculture and Fisheries to GDP. Proceedings of the Eighteenth Biennial Conference of the International Institute of Fisheries Economics and Trade*.

Campos, R. S. de, Codjoe, S. N. A., Adger, W. N., Mortreux, C., Hazra, S., Siddiqui, T., … Abu, M. H. R. M. (2019). Where People Live and Move in Deltas. In R. J. Nicholls, W. N. Adger, C. W. Hutton, & S. E. Hanson (Eds.), *Deltas in the Anthropocene*. Palgrave Macmillan. https://doi.org/10.1007/978-3-030-23517-8

Cazcarro, I., Arto, I., Hazra, S., Bhattacharya, R., Osei-Wusu Adjei, P., Ofori-Danson, P., … Hossen, Z. (2018). Biophysical and Socioeconomic State and Links of Deltaic Areas Vulnerable to Climate Change: Volta (Ghana), Mahanadi (India) and Ganges-Brahmaputra-Meghna (India and Bangladesh). *Sustainability*. https://doi.org/10.3390/su10030893

Chan, C. Y., Tran, N., Dao, D. C., Sulser, T. B., Philips, M. J., Batka, M., … Preston, N. (2017). *Fish to 2050 in the ASEAN Region. Working Paper: 2017-01* (No. 2017–01). Penang, Malaysia: WorldFish and Washington DC, USA.

Cheung, W. W., Lam, V. W., & Cc Wabnitz, C. (2019). *Fisheries; Climate change Future scenarios and projections for fisheries on the high seas under a changing climate Produced by IIED's Shaping Sustainable Markets Group Partner organisation*. Retrieved from http://pubs.iied.org/16653IIEDwww.iied.org@iiedwww.facebook.com/theIIED

Constanza, R., Alperovitz, G., Daly, H. E., Farley, J., Franco, C., Jackson, T., … Victor, P. (2013). *Building a Sustainable and Desirable*. Retrieved from https://www.jstor.org/stable/j.ctt5hgz53

Cooley, S. R., Lucey, N., Kite-Powell, H., & Doney, S. C. (2012). Nutrition and income from molluscs today imply vulnerability to ocean acidification tomorrow. *Fish and Fisheries*, *13*(2), 182–215. https://doi.org/10.1111/j.1467-2979.2011.00424.x

Cuaresma, J. C., & Lutz, W. (2015). The demography of human development and climate change vulnerability: A projection exercise. *Vienna Yearbook of Population Research*, *13*(1), 241–262. https://doi.org/10.1553/populationyearbook2015s241

Dao, H. (2013). Population extrapolation from 1975 to 2007.

Dao, H., Debono, A., & Deichmann, U. (2013). Socio-economical data.

Das, I., Hazra, S., Das, S., Giri, S., Chanda, A., Maity, S., & Ghosh, S. (2018). Trophic-level modelling of the coastal waters of the northern Bay of Bengal, West Bengal, India. *Fisheries Science*, *84*(6), 995–1008. https://doi.org/10.1007/s12562-018-1246-x

Das, I., Hazra, S., Das, S., Giri, S., Maity, S., & Ghosh, S. (2019). Present Status of the Sustainable Fishing Limits for Hilsa Shad in the northern Bay of Bengal, India. *Proceedings of the National Academy of Sciences India Section B - Biological Sciences*, *89*(2), 525–532. https://doi.org/10.1007/s40011-018-0963-3

Das, I., Lauria, V., Kay, S., Cazcarro, I., Arto, I., Fernandes, J. A., & Hazra, S. (2020). Effects of climate change and management policies on marine fisheries productivity in the north-east coast of India. *Science of the Total Environment*, *724*, 138082. https://doi.org/10.1016/j.scitotenv.2020.138082

De Souza, K., Kituyi, E., Harvey, B., Leone, M., Murali, K. S., & Ford, J. D. (2015). Vulnerability to climate change in three hot spots in Africa and Asia: key issues for policy-relevant adaptation and resilience-building research. *Regional Environmental Change*, *15*(5), 747–753. https://doi.org/10.1007/s10113-015-0755-8

Deichmann, U. (2013). GDP raster distribution model and expanding from 1975 to 2007. Retrieved from https://preview.grid.unep.ch/index.php?preview=data&events=socec&evcat=1&lang=eng

Donges, J. F., Winkelmann, R., Lucht, W., Cornell, S. E., Dyke, J. G., Rockström, J., … Schellnhuber, H. J. (2017). Closing the loop: Reconnecting human dynamics to Earth System science. *Anthropocene Review*, *4*(2), 151–157. https://doi.org/10.1177/2053019617725537

Dufour, F., Arrizabalaga, H., Irigoien, X., & Santiago, J. (2010). Climate impacts on albacore and bluefin tunas migrations phenology and spatial distribution. *Progress in Oceanography*, *86*(1), 283–290. https://doi.org/10.1016/j.pocean.2010.04.007

Dunn, F. E., Nicholls, R. J., Darby, S. E., Cohen, S., Zarfl, C., & Fekete, B. M. (2018). Projections of historical and 21st century fluvial sediment delivery to the Ganges-Brahmaputra-Meghna, Mahanadi, and Volta deltas. *Science of The Total Environment*, *642*, 105–116. https://doi.org/10.1016/j.scitotenv.2018.06.006

Dutta, S., Maity, S., Chanda, A., & Hazra, S. (2012). Population Structure, Mortality Rate and Exploitation Rate of Hilsa Shad (Tenualosa ilisha) in West Bengal Coast of Northern Bay of Bengal, India.

Elmqvist, T., Folke, C., Nyström, M., Peterson, G., Bengtsson, J., Walker, B., & Norberg, J. (2003). Response diversity, ecosystem change, and resilience. *Frontiers in Ecology and the Environment*, *1*(9), 488–494.

Ericson, J., Vorosmarty, C., Dingman, S., Ward, L., & Meybeck, M. (2006). Effective sea-level rise and deltas: Causes of change and human dimension implications. *Global and Planetary Change*, *50*(1–2), 63–82. https://doi.org/10.1016/j.gloplacha.2005.07.004

ESCAP. (2019). People's Republic of China: 2019 Article IV Consultation. Washington, DC. Retrieved from imf.org/~/media/Files/Publications/CR/2019/1CHNEA2019003.ashx.

FAO. (2012). The State of the World Fisheries and Aquaculture 2012. Rome: FAO.

FAO. (2017). Global Fishery and Aquaculture Production (database). Retrieved from www.fao.org/fishery/statistics/global-production/en

FAO. (2020). *The State of World Fisheries and Aquaculture 2020. Sustainability in action*. Rome: Food and Agriculture Organization (FAO). https://doi.org/10.4060/ca9229en

Fernandes, J.A. (2018). Chapter 13: Southern Asian fisheries in the Arabian Sea, Bay of Bengal and East Indian Ocean. In M. Barange, T. Bahri, M. Beveridge, K. Cochrane, S. Funge-Smith, & F. Poulain (Eds.), *Impacts of Climate Change on fisheries and aquaculture: Synthesis of current knowledge, adaptation and mitigation options* (pp. 281–303). Rome.

Fernandes, Jose A., Cheung, W. W. L., Jennings, S., Butenschön, M., De Mora, L., Frölicher, T. L., … Grant, A. (2013). Modelling the effects of climate change on the distribution and production of marine fishes: Accounting for trophic interactions in a dynamic bioclimate envelope model. *Global Change Biology*, *19*(8), 2596–2607. https://doi.org/10.1111/gcb.12231

Fernandes, Jose A., Kay, S., Hossain, M. A. R., Ahmed, M., Cheung, W. W. L., Lazar, A. N., & Barange, M. (2016). Projecting marine fish production and catch potential in Bangladesh in the 21st century under long-term environmental change and management scenarios. *ICES Journal of Marine Science*, *73*(5), 1357–1369. https://doi.org/10.1093/icesjms/fsv217

Gaffin, S. R., Rosenzweig, C., Xing, X., & Yetman, G. (2004). Downscaling and geo-spatial gridding of socio-economic projections from the IPCC Special Report on Emissions Scenarios (SRES). *Global Environmental Change*, *14*(2), 105–123. https://doi.org/10.1016/j.gloenvcha.2004.02.004

Gemenne, F. (2011). Why the numbers don't add up: A review of estimates and predictions of people displaced by environmental changes. *Global Environmental Change*, *21*, S41–S49. https://doi.org/10.1016/j.gloenvcha.2011.09.005

Ghosh, T., Powell, R., Elvidge, C. D., Baugh, K. E., Sutton, P. C., & Anderson, S. (2010). Shedding light on the global distribution of economic activity. *The Open Geography Journal*, *3*, 148–161.

Golden, C. (2016). Are farmed fish just for the wealthy? Golden et al. reply. *Nature*, *538*(7624), 171. https://doi.org/10.1038/538171e

Hallegatte, S., Bangalore, M., Bonzanigo, L., Fay, M., Kane, T., Narloch, U., … Vogt-Schilb, A. (2016). Shock Waves: managing the Impacts of Climate Change on Poverty. International Bank for Reconstruction and Development / The World Bank. Retrieved from http://www.journals.cambridge.org/abstract_S0022112081001535

Hoq, M. E. (2007). An analysis of fisheries exploitation and management practices in Sundarbans mangrove ecosystem, Bangladesh. *Ocean & Coastal Management*, *50*(5), 411–427. https://doi.org/10.1016/j.ocecoaman.2006.11.001

Hossain, M. A., Ahmed, M., Ojea, E., & Fernandes, J. A. (2018). Impacts and responses to environmental change in coastal livelihoods of south-west Bangladesh. *Science of the Total Environment*, *637*, 954–970.

IPCC. (2014). Climate Change 2014 – Impacts, Adaptation and Vulnerability.

Jafarzadeh, S., Ellingsen, H., & Aanondsen, S. A. (2016). Energy efficiency of Norwegian fisheries from 2003 to 2012. *Journal of Cleaner Production*, *112*, 3616–3630. https://doi.org/10.1016/j.jclepro.2015.06.114

Jahan, S. (2019). The Human Development Index – what it is and what it is not. Retrieved from http://hdr.undp.org/en/hdi-what-it-is

Jennings, S., & Lee, J. (2012). Defining fishing grounds with vessel monitoring system data. *ICES Journal of Marine Science*, *69*(1), 51–63. https://doi.org/10.1093/icesjms/fsr173

Kebede, A. S., Nicholls, R. J., Allan, A., Arto, I., Cazcarro, I., Fernandes, J. A., … Whitehead, P. W. (2018). Applying the global RCP–SSP–SPA scenario framework at sub-national scale: A multi-scale and participatory scenario approach. *Science of the Total Environment, 635*. https://doi.org/10.1016/j.scitotenv.2018.03.368

Khodabakhshi, A. (2012). Relationship between GDP and Human Development Indices in India. *SSRN Electronic Journal* (2010), 1–9. https://doi.org/10.2139/ssrn.1867887

Kirezci, E., Young, I. R., Ranasinghe, R., Muis, S., Nicholls, R. J., Lincke, D., & Hinkel, J. (2020). Projections of global-scale extreme sea levels and resulting episodic coastal flooding over the 21st Century. *Scientific Reports*, *10*(1), 1–12. https://doi.org/10.1038/s41598-020-67736-6

Kubiszewski, I., Costanza, R., Franco, C., Lawn, P., Talberth, J., Jackson, T., & Aylmer, C. (2013). Beyond GDP: Measuring and achieving global genuine progress. *Ecological Economics*, *93*, 57–68. https://doi.org/10.1016/j.ecolecon.2013.04.019

Kummu, M., Taka, M., & Guillaume, J. H. A. (2018). Gridded global datasets for Gross Domestic Product and Human Development Index over 1990-2015. *Scientific Data*, *5*, 1–15. https://doi.org/10.1038/sdata.2018.4

Kummu, M., Taka, M., & Guillaume, J. H. A. (2020). Gridded global datasets for Gross Domestic Product and Human Development Index over 1990-2015.

Lauria, V., Das, I., Hazra, S., Cazcarro, I., Arto, I., Kay, S., … Fernandes, J. A. (2018). Importance of fisheries for food security across three climate change vulnerable deltas. *Science of the Total Environment (under Review)*.

Lincoln, Z. (2014). Socio-Economic and Cultural Impacts of Tourism in Bangladesh. *European Scientific Journal*, *9*(10), 326–331. https://doi.org/10.19044/esj.2013.v9n10p%25p

Lind, N. C. (1992). Some thoughts on the human development index. *Social Indicators Research*, *27*(1), 89–101. https://doi.org/10.1007/BF00300511

Lind, N. C. (2004). Values reflected in the human development index. *Social Indicators Research*, *66*(3), 283–293. https://doi.org/10.1023/B:SOCI.0000003587.32655.b7

Lotze, H. K., Tittensor, D. P., Bryndum-Buchholz, A., Eddy, T. D., Cheung, W. W. L., Galbraith, E. D., … Worm, B. (2019). Global ensemble projections reveal trophic amplification of ocean biomass declines with climate change. *Proceedings of the National Academy of Sciences of the United States of America*, *116*(26), 12907–12912. https://doi.org/10.1073/pnas.1900194116

Lymer, D., Funge-Smith, S., Clausen, J., & Miao, W. (2008). Status and potential of fisheries and aquaculture in Asia and the Pacific 2008. Contributions of fisheries and aquaculture in Asia and the Pacific region. Bangkok, Thailand. Retrieved from http://www.fao.org/3/i0433e/I0433E04.htm

Maina, I., Kavadas, S., Katsanevakis, S., Somarakis, S., Tserpes, G., & Georgakarakos, S. (2016). A methodological approach to identify fishing grounds: A case study on Greek trawlers. *Fisheries Research*, *183*, 326–339. https://doi.org/10.1016/j.fishres.2016.06.021

Merkens, J. L., Reimann, L., Hinkel, J., & Vafeidis, A. T. (2016). Gridded population projections for the coastal zone under the Shared Socioeconomic Pathways. *Global and Planetary Change*, *145*, 57–66. https://doi.org/10.1016/j.gloplacha.2016.08.009

Milliman, J. D., Broadus, J. M., & Gable, F. (1989). Environmental and economic implications of rising sea level and subsiding deltas: the Nile and Bengal examples. *Ambio*, *18*(6), 340–345. https://doi.org/10.2307/4313605

Monirul Islam, M., Sallu, S., Hubacek, K., & Paavola, J. (2014). Limits and barriers to adaptation to climate variability and change in Bangladeshi coastal fishing communities. *Marine Policy*, *43*, 208–216. https://doi.org/10.1016/j.marpol.2013.06.007

Morse, S. (2003). Greening the United Nations' Human Development Index? *Sustainable Development*, *198*(July), 183–198.

Murakami, D., & Yamagata, Y. (2016). Estimation of gridded population and GDP scenarios with spatially explicit statistical downscaling.

Nicholls, R. J., Adger, W. N., Hutton, C. W., & Hanson, S. E. (Eds.). (2019). *Deltas in the Anthropocene*. Palgrave Macmillan. https://doi.org/10.1007/978-3-030-23517-8

Nicholls, R., Kebede, A., Allan, A., & Al, E. (2017). *The DECCMA integrated scenario framework: A multi-scale and participatory approach to explore migration and adaptation in deltas*.

Nordhaus, W. D. (2006). Geography and macroeconomics: new data and new findings. *Proceedings of the National Academy of Sciences of the United States of America, 103*(10), 3510–3517. https://doi.org/10.1073/pnas.0509842103

Nordhaus, W. D., & Chen, X. (2016). Global Gridded Geographically Based Economic Data (G-Econ), v4 (1990, 1995, 2000, 2005). Palisades, NY: NASA Socioeconomic Data and Applications Center (SEDAC). Retrieved from https://sedac.ciesin.columbia.edu/data/set/spatialecon-gecon-v4/metadata

Notti, E., Figari, M., Sala, A., & Martelli, M. (2019). Experimental assessment of the fouling control coating effect on the fuel consumption rate. *Ocean Engineering, 188*, 106233. https://doi.org/10.1016/j.oceaneng.2019.106233

O'Neill, B. C., Kriegler, E., Ebi, K. L., Kemp-Benedict, E., Riahi, K., Rothman, D. S., … Solecki, W. (2017). The roads ahead: Narratives for shared socioeconomic pathways describing world futures in the 21st century. *Global Environmental Change, 42*, 169–180. https://doi.org/10.1016/j.gloenvcha.2015.01.004

O'Neill, B. C., Kriegler, E., Riahi, K., Ebi, K. L., Hallegatte, S., Carter, T. R., … van Vuuren, D. P. (2014). A new scenario framework for climate change research: the concept of shared socioeconomic pathways. *Climatic Change, 122*(3), 387–400. https://doi.org/10.1007/s10584-013-0905-2

OECD-FAO. (2017). Chapter 2. Southeast Asia: Prospects and challenges. In *Agricultural Outlook 2017-2026*. OECD-FAO. Retrieved from https://www.oecd-ilibrary.org/sites/agr_outlook-2017-5-en/index.html?itemId=/content/component/agr_outlook-2017-5-en

Parker, R. W. R., Blanchard, J. L., Gardner, C., Green, B. S., Hartmann, K., Tyedmers, P. H., & Watson, R. A. (2018). Fuel use and greenhouse gas emissions of world fisheries. *Nature Climate Change, 8*(4), 333–337. https://doi.org/10.1038/s41558-018-0117-x

Pauly, D., & Zeller, D. (2019). Agreeing with FAO: Comments on SOFIA 2018. *Marine Policy, 100*(December 2018), 332–333. https://doi.org/10.1016/j.marpol.2018.12.009

Pernetta, J. C., Ong, J.-E., Padilla, N. E. O., Rahim, K. A., & Chinh, N. T. (2013). Determining regionally applicable economic values for coastal habitats and their use in evaluating the cost effectiveness of regional conservation actions: the example of mangroves, in the South China Sea. *Ocean & Coastal Management, 85*, 177–185. https://doi.org/10.1016/j.ocecoaman.2013.04.001

Pitcher, T., Kalikoski, D., Pramod, G., & Short, K. (2009). Not honouring the code. *Nature, 457*(7230), 658–659. https://doi.org/10.1038/457658a

Planque, B., Bellier, E., & Loots, C. (2011). Uncertainties in projecting spatial distributions of marine populations. *Journal of Marine Science, ICES, 68*(6), 1045–1050.

Pomeroy, R. S. (2012). Managing overcapacity in small-scale fisheries in Southeast Asia. *Marine Policy, 36*(2), 520–527. https://doi.org/10.1016/j.marpol.2011.10.002

Pörtner, H. O., Karl, D. M., Boyd, P. W., Cheung, W. W. L., Lluch-Cota, S. E., Nojiri, Y., … Wittmann, A. C. (2015). Ocean systems. *Climate Change 2014 Impacts, Adaptation and Vulnerability: Part A: Global and Sectoral Aspects*, 411–484. https://doi.org/10.1017/CBO9781107415379.011

Raman, R. K., Sathianandan, T. V., Sharma, A. P., & Mohanty, B. P. (2017). Modelling and Forecasting Marine Fish Production in Odisha Using Seasonal ARIMA Model. *National Academy Science Letters, 40*(6), 393–397. https://doi.org/10.1007/s40009-017-0581-2

Rao, N. D., Sauer, P., Gidden, M., & Riahi, K. (2019). Income inequality projections for the Shared Socioeconomic Pathways (SSPs). *Futures, 105*(June 2018), 27–39. https://doi.org/10.1016/j.futures.2018.07.001

Renaud, F. G., Syvitski, J. P. M., Sebesvari, Z., Werners, S. E., Kremer, H., Kuenzer, C., … Friedrich, J. (2013). Tipping from the Holocene to the Anthropocene: How threatened are major world deltas? *Current Opinion in Environmental Sustainability*, *5*(6), 644–654. https://doi.org/10.1016/j.cosust.2013.11.007

Ribot, J. (2014). Cause and response: vulnerability and climate in the Anthropocene. *The Journal of Peasant Studies*, *41*(5), 667–705. https://doi.org/10.1080/03066150.2014.894911

Schinko, T., Drouet, L., Vrontisi, Z., Hof, A., Hinkel, J., Mochizuki, J., … Lincke, D. (2020). Economy-wide effects of coastal flooding due to sea level rise: a multi-model simultaneous treatment of mitigation, adaptation, and residual impacts. *Environmental Research Communications*, *2*(1), 015002. https://doi.org/10.1088/2515-7620/ab6368

Sen, A. (1999). *Development as Freedom*. Oxford University Press.

Smit, B., & Wandel, J. (2006). Adaptation, adaptive capacity and vulnerability. *Global Environmental Change*, *16*(3), 282–292. https://doi.org/10.1016/j.gloenvcha.2006.03.008

Steffen, W., Persson, Å., Deutsch, L., Zalasiewicz, J., Williams, M., Richardson, K., … Svedin, U. (2011). The Anthropocene: From Global Change to Planetary Stewardship. *AMBIO*, *40*(7), 739. https://doi.org/10.1007/s13280-011-0185-x

Stiglitz, J. E., Sen, A., & Fitoussi, J. P. (2010). *Mismeasuring our lives: Why GDP doesn't add up* (The New Pr). New York.

Subasinghe, R. (2017). *Regional review on status and trends in aquaculture development in Asia-Pacific, 2015. FAO Fisheries and Aquaculture Circular* (Vol. 5).

Suckall, N., Tompkins, E. L., Nicholls, R. J., Kebede, A. S., Lázár, A. N., Hutton, C., … Mensah, A. (2018). A framework for identifying and selecting long term adaptation policy directions for deltas. *Science of The Total Environment*, *633*, 946–957. https://doi.org/10.1016/j.scitotenv.2018.03.234

Sugiyama, S., Staples, D., & Funge-Smith, S. (2004). Status and potential of fisheries and aquaculture in Asia and the Pacific 2004. Contributions of fisheries and aquaculture in the Asia-Pacific region. Bangkok, Thailand.

Suh, D., & Pomeroy, R. (2020). Projected Economic Impact of Climate Change on Marine Capture Fisheries in the Philippines. *Frontiers in Marine Science*, *7*(April), 1–14. https://doi.org/10.3389/fmars.2020.00232

Sumaila, U. R., & Cheung, W. W. L. (2010). Development and Climate Change: Cost of Adapting Fisheries to Climate Change. Washington, DC, USA.

Syvitski, J. P. M. (2003). Supply and flux of sediment along hydrological pathways: research for the 21st century. *Global and Planetary Change*, *39*(1), 1–11. https://doi.org/10.1016/S0921-8181(03)00008-0

Syvitski, J. P. M., Vörösmarty, C. J., Kettner, A. J., & Green, P. (2005). Impact of Humans on the Flux of Terrestrial Sediment to the Global Coastal Ocean. *Science*, *308*(5720), 376–380. https://doi.org/10.1126/science.1109454

Tacon, A. G. J., & Metian, M. (2018). Food matters: A comparative analysis of fish, income and food supply.

Taconet, M., Kroodsma, D., & Fernandes, J. A. (2019). *Global Atlas of AIS-based fishing activity - Challenges and opportunities*. Rome: FAO.

Tarun Kumar, S., & Shivani, P. (2014). Marine Fisheries; Its Current Status, Sustainable Management and Socio-Economic Status of the Marine Fishers of Odisha, Through Indian Marine Policy: A Case Study. *Res. J. Animal, Veterinary and Fishery Sci. International Science Congress Association*, *2*(7), 10–19. Retrieved from www.isca.me

Teh, L. S. L., Witter, A., Cheung, W. W. L., Sumaila, U. R., & Yin, X. (2017). What is at stake? Status and threats to South China Sea marine fisheries. *Ambio*, *46*(1), 57–72. https://doi.org/10.1007/s13280-016-0819-0

Tessler, Z. D., Vörösmarty, C. J., Grossberg, M., Gladkova, I., Aizenman, H., Syvitski, J. P. M. M., & Foufoula-Georgiou, E. (2015). Profiling risk and sustainability in coastal deltas of the world. *Science*, *349*(6248), 638–643. https://doi.org/10.1126/science.aab3574

Tittensor, D. P., Mora, C., Jetz, W., Lotze, H. K., Ricard, D., Berghe, E. Vanden, & Worm, B. (2010). Global patterns and predictors of marine biodiversity across taxa. *Nature*, *466*(7310), 1098–1101. https://doi.org/10.1038/nature09329

Tompkins, E. L., Vincent, K., Suckall, N., Rahman, R., Ghosh, T., Mensah, A., … Price, V. (2019). Adapting to Change: People and Policies. In R. J. Nicholls, W. N. Adger, C. W. Hutton, & S. E. Hanson (Eds.), *Deltas in the Anthropocene*. Palgrave Macmillan. https://doi.org/10.1007/978-3-030-23517-8

Trajano, J. C., Gong, L., Sembiring, M., & Astuti, R. (2017). Marine Environmental Protection in the South China Sea: Challenges and Prospects Part I By. *NCT Insight* (December), 1–16.

UNDP. (2019). Human Development Index (HDI). Retrieved from http://hdr.undp.org/en/content/human-development-index-hdi

Uyen, N. T. (2017). Fisheries Country Profile: Viet Nam. Retrieved from http://www.seafdec.org/fisheries-country-profile-viet-nam/

van den Bergh, J. C. J. M., & Botzen, W. J. W. (2018). Global impact of a climate treaty if the Human Development Index replaces GDP as a welfare proxy. *Climate Policy*, *18*(1), 76–85. https://doi.org/10.1080/14693062.2016.1227954

Vettor, R., Tadros, M., Ventura, M., & Guedes Soares, C. (2016). Route planning of a fishing vessel in coastal waters with fuel consumption restraint. *Proceedings of 3rd International Conference on Maritime Technology and Engineering, MARTECH 2016*, *1*(July), 167–174. https://doi.org/10.1201/b21890-24

Watson, R. A. (2017). A database of global marine commercial, small-scale, illegal and unreported fisheries catch 1950-2014. *Scientific Data*, *4*, 1–9. https://doi.org/10.1038/sdata.2017.39

WB. (2013). FISH TO 2030. Prospects for Fisheries and Aquaculture.

Whitehead, P. G., Jin, L., Macadam, I., Janes, T., Sarkar, S., Rodda, H. J. E., … Nicholls, R. J. (2018). Modelling impacts of climate change and socio-economic change on the Ganga, Brahmaputra, Meghna, Hooghly and Mahanadi river systems in India and Bangladesh. *Science of The Total Environment*, *636*, 1362–1372. https://doi.org/10.1016/j.scitotenv.2018.04.362

Whitmee, S., Haines, A., Beyrer, C., Boltz, F., Capon, A. G., De Souza Dias, B. F., … Yach, D. (2015). Safeguarding human health in the Anthropocene epoch: Report of the Rockefeller Foundation-Lancet Commission on planetary health. *The Lancet*, *386*(10007), 1973–2028. https://doi.org/10.1016/S0140-6736(15)60901-1

Willett, W., Rockström, J., Loken, B., Springmann, M., Lang, T., Vermeulen, S., … Murray, C. J. L. (2019). Food in the Anthropocene: the EAT–Lancet Commission on healthy diets from sustainable food systems. *The Lancet*, *393*(10170), 447–492. https://doi.org/10.1016/S0140-6736(18)31788-4

Wolters, M. L., Sun, Z., Huang, C., & Kuenzer, C. (2016). Environmental awareness and vulnerability in the Yellow River Delta: Results based on a comprehensive household survey. *Ocean & Coastal Management*, *120*, 1–10. https://doi.org/10.1016/j.ocecoaman.2015.11.009

Woodroffe, C. N., Nicholls, R. J., Saito, Y., Chen, Z., & Goodbred, S. L. (2006). Landscape variability and the response of Asian megadeltas to environmental change. In N. Harvey (Ed.), *Global change and integrated coastal management: The Asia-Pacific region* (pp. 277–314). New York, NY, USA.

Part VI
Ocean Related Policies

Chapter 16
Present Status of Ocean and International Maritime Regulations and Securities

Sanjay Upadhyay

16.1 Background

The United Nations in its Concept Paper in 2014 gave a general definition of the pre-existing idea of 'Blue Economy' as an ocean economy that aims at 'the improvement of human well-being and social equity, while significantly reducing environmental risks and ecological scarcities'.[1] The concept has evolved over the years finding relevance in significant international commitments of the global community at large, including the Sustainable Development Goals, specifically SDG 14 which aims to 'conserve and sustainably use the oceans, seas and marine resources for sustainable development'. Having recognized the potential of Blue Economy as a driver for regional economic development, the South Asian nations have been continuously initiating dialogues among themselves, either multilaterally or bilaterally, through several regional organizations such as the Association of Southeast Asian Nations (ASEAN), South Asian Association for Regional Cooperation (SAARC),

[1] Blue Economy Concept Paper: Sustainable Development Knowledge Platform. (2014, August 15). Sustainable Development Goals. https://sustainabledevelopment.un.org/index.php?page=view&type=111&nr=2978&menu=35

This chapter has been greatly benefitted by research inputs from Ms. Fizza Zaidi and Mr. Prannoy Joe Sebastian, Senior Associate and Former Associate Respy of Enviro Legal Defence Firm. The author is deeply indebted to them.

S. Upadhyay (✉)
Enviro Legal Defence Firm, Delhi, India
e-mail: sanjay@eldfindia.com

S. Hazra, A. Bhukta (eds.), *The Blue Economy*,
https://doi.org/10.1007/978-3-030-96519-8_16

Asia-Pacific Economic Cooperation (APEC), Indian Ocean Rim Association (IORA) and Bay of Bengal Initiative for Multi-Sectoral Technical and Economic Cooperation (BIMSTEC). In its draft Policy Framework on India's Blue Economy (Draft Policy Framework on India's Blue Economy 2020), the Government of India has emphasized certain aspects such as offshore sovereignty; economically valuable resources in water, as well as on and under the seabeds, onshore infrastructure like seaports, maritime routes connected with domestic and international trade and offshore energy resources, be they fossil based or renewable; scientific marine technologies; production of goods and services from fisheries, marine manufacturing, shipping and tourism that is connected with the sea and the oceans; and ocean governance, maritime security, sustainability concerns and adherence to international treaties and commitments. To this end, the Government of India has made budgetary allocations in its 2021–2022 budget wherein it seeks to further the mission of Blue Economy by launching a Deep Ocean Mission for deep ocean survey exploration and conservation of deep-sea biodiversity.[2] One of the key elements to ensure that such an economy unfolds without any major impediments is to examine and understand the legal, regulatory and institutional framework which governs our oceans and coasts within which such an economy has to thrive.

16.2 International Framework on Oceans Governance

The international scenario relating to governance of oceans dates back to the early 1600s. In 1609, the *principle of mare liberum* formulated that the sea was an international territory and provided for all nations to use the seas for free trade (Martinez 1980). This principle provided for a part of the most basic definition of sea and the evolution of the definition took place in the coming years. In contrast, the *mare clausum*, a contrasting principle, was given by John Selden which states that the certain parts of water bodies could be claimed to be a country's exclusive jurisdiction (Martinez 1980). It was understood that oceans could be appropriated as land. A combination of both these principles gave rise to the modern definition of ocean governance (as enunciated under the UNCLOS) wherein certain parts of oceans (up to 12 nautical miles in the ocean as territorial waters, etc.) can be claimed to be exclusive jurisdiction of a country while the area beyond these 12 nautical miles cannot be claimed to be exclusive jurisdiction and belong to all the countries equally (United Nations Convention on the Law of the Sea 1982a, b).

Later, the *Baltic and International Maritime Council* (BIMCO) of 1905 and the *International Convention for the Prevention of Pollution of the Sea by Oil* of 1954 were the first formally formulated international conventions and councils relating to the issues of governance of oceans. In the early years, the basic issues that were

[2]Budget 2021-2022: Speech of Nirmala Sitharaman, Minister of Finance (2021, February 1). Government of India. https://www.indiabudget.gov.in/doc/budget_speech.pdf

faced were oil spills, those relating to high seas fishing, etc.; therefore, the early treaties dealt with these issues itself. In 1948, the *International Maritime Organization*, which was originally called the Inter-Governmental Maritime Consultative Organization, (International Maritime Organization, 1948) was set up because a need was felt to establish a body that could bring international consensus between countries.

Currently, the treaties, conventions and the laws relating to oceans deal with myriad issues such as maritime satellite, prevention of pollution from ships, safety of fishing vessels, suppression of unlawful acts, recycling of ships, among others. The latest in the list of international treaties are the *Hong Kong International Convention for the Safe and Environmentally Sound Recycling of Ships* (Hong Kong International Convention for the Safe and Environmentally Sound Recycling of Ships 2009), the *Athens Convention relating to the Carriage of Passengers and their Luggage by Sea* (Athens Convention relating to the Carriage of Passengers and their Luggage by Sea 1974), the *International Convention for the Control and Management of Ships' Ballast Water and Sediments* (International Convention for the Control and Management of Ships' Ballast Water and Sediments 2004), etc. Other prominent ones in this arena include the *International Whaling Commission* dealing with the governance and protection of whales in aquatic ecosystems (International Whaling Commission 1946), *International Convention for the Prevention of Pollution from Ships* ('MARPOL') which covers the aspect of prevention of pollution of marine environment by ships from operational or accidental causes (International Convention for the Prevention of Pollution from Ships 1973), *International Convention for the Safety of Life at Sea* which specifies the minimum standards for the construction, equipment and operation of ships, compatible with their safety (International Convention for the Safety of Life at Sea 1974), *London Dumping Convention* (Convention on the Prevention of Marine Pollution by Dumping of Wastes and Other Matter 1972) whose objective is to promote the effective control of all sources of marine pollution and to take all practicable steps to prevent pollution of the sea by dumping of wastes and other matter, etc.

The biggest step in improving the governance of the oceans internationally was taken with the adoption of the *United Nations Convention on the Law of the Sea* ('UNCLOS') which was adopted and signed in 1982 (United Nations Convention on the Law of the Sea 1982a, b). This convention replaced and merged the four Geneva conventions which dealt with the territorial sea and the contiguous zone, the continental shelf, the high seas, fishing and conservation of living resources on the high seas separately (United Nations Convention on the Law of the Sea 1982a, b). The UNCLOS has set up three international organizations that are provided with different functions, i.e. the International Tribunal for the Law of the Sea ('ITLOS'), the International Seabed Authority and the Commission on the Limits of the Continental Shelf. The UNCLOS and the International Union for Conservation of Nature and Natural Resources (IUCN) work together towards filling gaps in the governance of oceans (United Nations Convention on the Law of the Sea 1982a, b). A positive result would provide a measure of protection and conservation of Areas Beyond National Jurisdiction (ABNJ) where there is none at present. The UNCLOS states that "area of the seabed and ocean floor and the subsoil thereof, beyond the limits of national jurisdiction, as well as its resources, are the common heritage of mankind, the exploration and exploitation of which shall be carried out for the benefit of mankind as a whole, irrespective of the geographical location of States" (United Nations Convention on the Law of the Sea 1982a, b). This states the nature of oceans in relation to their ownership. That ocean and all its parts are a common heritage of mankind and do not allow any country to claim ownership beyond a certain limit shows the international intention to make sure that States are not able to overpower other States on the basis of their financial and technological advancement. This is in consonance with the principles of *mare liberum* and *mare clausum* of the 1600s that provide for different kinds of jurisdiction over oceans (Martinez 1980). They provide for individual exclusive jurisdictions to countries while making sure that no country is unfairly treated because of such exclusive jurisdiction. However, as per the assessment done by the *International Union for Conservation of Nature, World Commission on Environmental Law – Ocean Specialist Group Global Marine and Polar Programme & Environmental Law Centre*, the UNCLOS misses out on a clear obligation of the countries to make sustainable use of marine resources and to also ensure that any activity under any countries' jurisdiction does not cause significant harm to the resources in the ABNJ.[3] A need is felt to state the obligation more clearly and explicitly.

International treaties governing the environment are not without their own challenges. *First*, international law is considered to be ineffective in its working and implementation because of the limitations in their redressal mechanism and dispute resolution, among others. Many treaties lack the effectiveness of law to make sure that they achieve the aim they were meant to achieve. *Second*, international law governs only a small part of the nations' duties and obligations, therefore playing a very limited role. A very clear example of that would be scenarios where countries do not sign or ratify the treaties as per their need and convenience, therefore removing themselves from the jurisdiction of that treaty. In such scenarios, there is little that international law or international organizations can do. *Third*, since there are multiple treaties working in the field of oceans and seas, sometimes there are overlapping (and contradictory) provisions governing the same issue. This leads to contradiction and confusion in implementation. It is therefore important that outdated, contradictory treaties and treaties governing the same issue are reassessed. There are newer issues and challenges that now need to be integrated more succinctly such as use of technology in committing crimes in the oceans, human trafficking and waste management, among others. Thus, for example, with evolving technology in the use of artificial intelligence and cyberattacks, it has become important to

[3] International Union for Conservation of Nature, World Commission on Environmental Law - Ocean Specialist Group Global Marine and Polar Programme & Environmental Law Centre (2020, February 20). *International legally binding instrument under the United Nations Convention on the Law of the Sea on the conservation and sustainable use of marine biological diversity of areas beyond national jurisdiction.* https://www.iucn.org/sites/dev/files/content/documents/iucn_comments_on_revised_bbnj_draft_text_february_2020.pdf

incorporate them in the working of laws. *Fourth*, another way of improving the international law governing the oceans and seas is through capacity building for greater compliance and enforcement by port States, coastal States and flag States (Seijo and Salas 2014). Countries have a special obligation in the implementation of the law in oceans because of lack of clarity in fixing responsibility. To improve the capacity building, it is important that there is increased information sharing and cooperation between countries as well as with international organizations (Seijo and Salas 2014). International law is based on cooperation between countries. It is only with mutual understanding between countries that international law will work.

To the already existing issues that exist in the oceans, newer security issues are coming up. Apart from the original issues of security including piracy, IUU fishing, etc., new issues of marine terrorism, use of drones and other technology and people smuggling have also evolved (Premarathna 2021). Most of the laws, both domestically and internationally, do not take into account the evolving nature of crimes, which has a direct bearing on the economic activity on the ocean space. Responding to these challenges will need cohesion between countries so that they can play an active role in the prevention of such crimes.

16.3 Regional Framework on Oceans Impacting Blue Economy: The South Asian Region

At the regional level, there are multiple regional and sub-regional international organizations which have recognized the potential of the Blue Economy as a driver for the economic development of States in the South Asian region.

These regional and sub-regional groups have significant impact on the bilateral and multilateral relations among the states. Some of the important regional organizations and their initiative with respect to the Blue Economy have been discussed below.

In the Southeast Asian context, the Association of Southeast Asian Nations (ASEAN) was established in 1967, followed by the South Asian Association for Regional Cooperation (SAARC) in 1985, Asia-Pacific Economic Cooperation (APEC) in 1989, Indian Ocean Rim Association in 1977 and Bay of Bengal Initiative for Multi-Sectoral Technical and Economic Cooperation (BIMSTEC) in 1977, among others.

ASEAN: The Association of Southeast Asian Nations (ASEAN) was established on 8 August 1967 in Bangkok, Thailand, with the signing of the ASEAN Declaration. The member states include Brunei Darussalam, Cambodia, Indonesia, Lao PDR, Malaysia, Myanmar, the Philippines, Singapore, Thailand and Vietnam. ASEAN was formed with the aim of 'fostering economic growth, social progress, and cultural development' in the Southeast Asian region through joint endeavours and maintaining close and beneficial cooperation with existing international and regional organizations with similar aims (ASEAN 1967). ASEAN builds its relations with other countries through its ASEAN Plus summits such as ASEAN Plus Three which includes China, Republic of Korea and Japan and ASEAN Plus Six which includes Australia, India and New Zealand. ASEAN regularly initiates dialogues on the concept of 'Blue Economy.' In the first 'ASEAN-India Workshop on Blue Economy: from Concepts to Action' held during November 2017, a number of dimensions were examined ranging from maritime connectivity, renewable energy, marine resource conservation and coastal management.

ASEAN-India cooperation recognizes the scope in the field of Blue Economy through 'efficient harnessing of blue resources, including through use of new and emerging technologies, the issues of poverty, food insecurity, unemployment and ecological imbalance can be effectively tackled. Research and innovations in marine biotechnology, higher access to seabed resources, investment in marine information and communication technologies and proper integration of coastal tourism and other services can play a crucial role in injecting stimulus to create additional economic activities for both ASEAN and India.'[4]

SAARC: The South Asian Association for Regional Cooperation (SAARC) was established in 1985 with the objective 'to promote welfare of people of South Asia and improve their quality of life; to accelerate economic growth, social progress and cultural development in the region and to provide all individuals the opportunity to live in dignity and to realize their full potentials; to promote and strengthen collective self-reliance among the countries of South Asia; to contribute to mutual trust, understanding and appreciation of one another's problems; to promote active collaboration and mutual assistance in the economic, social, cultural, technical and scientific fields; to strengthen cooperation with other developing countries; to strengthen cooperation among themselves in international forums on matters of common interests; and to cooperate with international and regional organizations with similar aims and purposes.'[5] The SAARC comprises eight member states – Afghanistan, Bangladesh, Bhutan, India, the Maldives, Nepal, Pakistan and Sri Lanka—which represent a combined population of approximately 1.9 billion.

The 18th SAARC Summit held in November 2014 in Kathmandu recognized the manifold contributions of ocean-based 'Blue Economy' in the SAARC region and the need for collaboration and partnership in this area.[6]

The concept of Blue Economy can be implemented in its true spirit by SAARC countries if the member countries forge political alliances for sustainable

---

[4] *Second ASEAN-India Blue Economy Workshop Keynote Address by Secretary (East)*. (2017, July 18). Ministry of External Affairs. https://www.mea.gov.in/Speeches-Statements.htm?dtl/30097/2nd_ASEANIndia_Blue_Economy_Workshop_Keynote_Address_by_Secretary_East

[5] SAARC Charter. (2020, July 12). Retrieved from https://www.saarc-sec.org/index.php/about-saarc/saarc-charter

[6] *18th SAARC Summit Declaration*. (2014, November 27). South Asian Association for Regional Cooperation. https://www.saarc-sec.org/index.php/press-release/106-18th-saarc-summit-declaration

management of ocean resources in order to achieve economic stability, ensuring food security and meeting SDG 14 (Alharthi and Hanif 2020, Conclusion).

APEC: The Asia-Pacific Economic Cooperation (APEC) is a regional economic forum established in 1989 to leverage the growing interdependence of the Asia-Pacific with the aim to create greater prosperity for the people of the region by promoting balanced, inclusive, sustainable, innovative and secure growth and by accelerating regional economic integration. The fourth APEC Oceans Ministerial Meeting held on 28 August 2014 led to issuing the Xiamen Declaration which recognizes 'the potential linkages between Blue Economy, sustainable development and economic growth' and called 'for cooperation on Blue Economy in the Asia-Pacific region.'[7] Subsequently, the 22nd APEC Economic Leaders' Meeting in November 2014 encouraged the Ocean and Fisheries Working Group to work with APEC for advance Blue Economy cooperation.[8] In 2015, the APEC members endorsed the APEC High-Level Policy Dialogue on Food Security and Blue Economy Plan of Action. It further welcomed the Oceans and Fisheries Working Group (OFWG) Food Security Action Plan and efforts for sustainable use and management of marine resources.[9]

IORA: The Indian Ocean Rim Association is an intergovernmental organization established in 1997 for strengthening regional cooperation and sustainable development within the Indian Ocean region through the States of Indian Ocean Rim. Presently, there are 23 member states which include India, Sri Lanka, Australia, Bangladesh, France, the Maldives, Singapore, Oman and United Arab Emirates, among others. The Indian Ocean Rim is one of the most vital sea lanes in the world with over 80% of world's seaborne trade in oil and around 100,000 commercial vessels passing through the region. The IORA began its focus on Blue Economy since the 14th IORA Ministerial Meeting in Perth in 2014 due to its growing global interest and potential for 'generating employment, food security, poverty alleviation and ensuring sustainability in business and economic models in the Indian Ocean.'[10]

Capacity building programmes have been carried out since 2014 which cover a wide range of areas which have been identified in six priority pillars by IORA Secretariat and include inter alia 'fisheries and aquaculture; renewable ocean energy; seaports and shipping; offshore hydrocarbons and seabed minerals; marine

---

[7] *APEC ocean-related Ministerial Meeting Joint Statement*. (2014, August 28). APEC. https://www.apec.org/Meeting-Papers/Sectoral-Ministerial-Meetings/Ocean-related/2014_ocean

[8] Leaders' Declaration. (2014, November 11). APEC. https://www.apec.org/Meeting-Papers/Leaders-Declarations/2014/2014_aelm

[9] Plan of Action in APEC High-Level Policy Dialogue on Food Security and Blue Economy. (2015). APEC. https://www.apec.org/Meeting-Papers/Annual-Ministerial-Meetings/2015/2015_amm/annexb

[10] *14th Meeting of the COM – Indian Ocean Rim Association – IORA*. (2014). Indian Ocean Rim Association. https://www.iora.int/en/events-media-news/events/structures-and-mechanisms/council-of-ministers/2014/14th-meeting-of-the-com

biotechnology, research and development; and tourism.'[11] The First IORA Ministerial Blue Economy Conference (BEC) was held in Mauritius on 2–3 September 2015 wherein the Blue Economy Declaration was adopted to 'harness oceans and maritime resources to drive economic growth, job creation and innovation, while safeguarding sustainability and environmental protection' (IORA Blue Economy 2015). Indonesia hosted the Second Ministerial Blue Economy Conference on 'Financing the Blue Economy' in 2017 where the Jakarta Declaration on the Blue Economy was adopted with the aim to 'ensure sustainable management and protection of marine and coastal ecosystems to avoid adverse impacts, including by strengthening their resilience, and taking action for their restoration in order to maintain healthy and productive oceans, and achieve inclusive economic growth in the Indian Ocean region.'[12] It further declared to undertake sustainable development of the Blue Economy in accordance with the 1982 United Nations Convention on the Law of the Sea (UNCLOS) as well as strengthening maritime cooperation for peace and stability in Indian Ocean, among others (Declaration of the IORA on the Blue Economy in the Indian Ocean Region 2017). It further established the Blue Economy Working Group as part of IORA Action Plan 2017–2021.[13]

Recognizing that a safe and secure Indian Ocean is important for socio-economic development, the IORA assigned Maritime Safety and Security (MSS) as the top priority area of focus in 2011. The IORA has been addressing MSS in the Indian Ocean through a broad range of activities and has established the IORA Working Group for MSS which finalized a regional Work Plan in 2019 for a period of 2 years (2019–2021) to enhance international cooperation in security and governance to successfully tackle the challenges faced by the region.

BIMSTEC: The Bay of Bengal Initiative for Multi-Sectoral Technical and Economic Cooperation (BI-STEC) came into being on 6 June 1997 with Bangladesh, India, Sri Lanka and Thailand as its original members. The organization later added Myanmar, Nepal and Bhutan into the group and is now called BIMSTEC. The BIMSTEC as per its Declaration aims to accelerate economic growth and social progress among members across multiple sectors such as *trade, technology, energy, transport, tourism and fisheries, agriculture, public health, poverty alleviation, counterterrorism, environment, culture, people to people contact and climate change* (Bangkok Declaration 1997). The Bay of Bengal covers over 2.2 million square km of maritime space. During the Third BIMSTEC Summit in 2014, the member states agreed and declared to cooperate in 'fisheries management, sustainable utilization of marine resources, environmental protection, and disaster

[11] *Blue Economy – Indian Ocean Rim Association – IORA*. (2015). Indian Ocean Rim Association. https://www.iora.int/en/priorities-focus-areas/blue-economy

[12] Declaration of the Indian Ocean Rim Association on the Blue Economy in the Indian Ocean Region, 08.05.2017, https://www.iora.int/media/8218/jakarta-declaration-on-blue-economy-final.pdf

[13] Declaration on the establishment of Bangladesh-India-Sri Lanka-Thailand Economic Cooperation. (1997, June 6). BIST-EC. https://drive.google.com/file/d/0B8Fv9wDGJqx2NkJTVzlZek5va0U/view?resourcekey=0-7YYfgTO3IbbjSNaDhv2XRQ

management' (Third BIMSTEC Summit Declaration 2014). The fourth BIMSTEC Summit Declaration held in Kathmandu, Nepal, emphasized the importance of Blue Economy and agreed to cooperate with an Intergovernmental Expert Group to develop an action plan on Blue Economy, keeping in mind the specific needs and circumstances of landlocked member states.[14]

16.4 National Legal and Policy Framework on Oceans Governance: The Driver of Blue Economy

The current legal regime on the marine environment in India can be divided into three categories: laws pertaining to the conservation of marine areas, laws pertaining to the use of marine areas and laws pertaining to the legal classification of various zones of the marine environment. The first two are functional classifications, while the third is a jurisdictional classification (Upadhyay and Upadhyay 2002, p. 76). A classification of this nature is necessary to assess the long-term viability of the development of a Blue Economy. It aids in understanding that, in addition to the extractive framework, the conservation framework, and thus sustainability, is equally important.

Before getting into the details of the legislation that impacts marine systems specifically, it is crucial to examine the overarching national regime, i.e. the constitutional mandate and the Ocean Policy Statement.

Constitutional mandate—Marine resources have been of prime concern under the Constitution of India. The Union of India has sovereignty/jurisdiction over valuable resources such as land, minerals, etc., which lie beneath the ocean in various zones such as territorial waters, continental shelf and exclusive economic zone. The types of resources included in these territorial zones have evolved over time, culminating into a comprehensive legislation in 1976. The territorial zones have been expanded by the constitutional mandate to include not only the territorial waters, the continental shelf and the exclusive economic zones (i.e. non-living resources) but also all marine living resources of India's exclusive economic zone as specified by the Maritime Zones Act, 1976. Furthermore, Article 48A of the Constitution requires the State to protect, safeguard and improve the marine environment, while Article 51-A(g) establishes all citizens' fundamental duty to conserve and enhance the natural environment. It has long been customary to invoke provisions of the Directive Principles of State Policy to ensure that fundamental governance, including environmental protection, is upheld, although the above two Articles relate to the Directive Principles of State Policy, which means that they are not enforceable in a court of law. The courts are liberal in interpreting the above-mentioned Articles in

[14] Fourth BIMSTEC Summit Declaration, Kathmandu, Nepal (August 30–31, 2018). (2018, August 31). Ministry of External Affairs. https://mea.gov.in/bilateral-documents.htm?dtl/30335/Fourth_BIMSTEC_Summit_Declaration_August_3031_2018

protecting environment of India. (Constitution of India, 1950) "This clearly arms the State to take all the necessary legal steps not only to protect, but also to improve marine environment" (Upadhyay and Upadhyay 2002, p. 77). It is based on larger governance goals of the country as a part of her overall economic development.

Ocean Policy Statement—The Ocean Policy Statement recognizes the coastal States' economic jurisdiction, stating that the nation has the sole right to use biotic and abiotic resources within the maritime zones. It also recognizes that the ocean environment's complexity and uncertainty necessitate a coordinated, centralized response based on adequate knowledge of marine space, including the seabed, water and air columns, in order to manage, maintain and utilize the sea's rich and diverse natural resources (Upadhyay and Upadhyay 2006, pp. 306). It was issued by the Department of Ocean Development, Ministry of Earth Sciences in November 1982 post the United Nations Conference on the Law of the Sea. The core of the policy statement, in order to go beyond optimal utilization of biotic marine resources as well as abiotic resources, should include, among other things, the use of renewable ocean energy sources. While the policy statement throws light on 'marine development,' which is related to sci-tech inputs, it also requires measures for the preservation of marine environments and resources through an inclusive legal framework with punitive measures (Upadhyay and Upadhyay 2002, pp. 77–78).

16.4.1 Laws Relating to the Conservation of Marine Habitats

The laws that govern the conservation of marine areas essentially consist of those laws that protect the marine habitat and environment. It is imperative to have a holistic understanding of these laws because each one of them has their own relevance and significance in safeguarding the vast coastal stretches of the country and their unique environment.

Territorial Waters, Continental Shelf, Exclusive Economic Zone and Other Maritime Zones Act, 1976—It is India's most important piece of legislation, governing the country's territorial waters, continental shelf, exclusive economic zone and other marine zones. Aside from actual jurisdiction over maritime zones, the Union of India has the sole authority over the preservation and protection of the marine environment, and also the prevention and management of marine pollution. India's territorial waters and the 'seabed and subsoil underpinning the air space over such seas' are under the sovereignty of the Indian government.[15] It has the authority to impose contiguous zone restrictions, full and exclusive rights to the continental shelf and exclusive control over India's exclusive economic zone (EEZ). A 'designated area' can be declared anywhere on the continental shelf or in the EEZ. Furthermore, 'historic waters' refers to internal waterways that would otherwise fall under the authority of the relevant state government if a historic title did

[15] Territorial Waters, Continental Shelf, Exclusive Economic Zone and Other Maritime Zones Act, § 3(1) (1976). https://legislative.gov.in/sites/default/files/A1976-80_0.pdf

not exist. The central government's sovereignty extends to the historic seas (Upadhyay and Upadhyay 2002, pp. 78–79).

Coast Guard Act, 1978—This Act constitutes and regulates an Armed Force of the Union for ensuring the security of the maritime zones of India in order to protect maritime and other national interests. The Act defines 'maritime zones of India'[16] as the 'territorial waters, the contiguous zone, the continental shelf, the exclusive economic zone or any other maritime zone of India, which in turn have respective meanings assigned to them under the Territorial Waters, Continental Shelf, Exclusive Economic Zone and Other Maritime Zones Act, 1976.'.[17] The Coast Guard's responsibilities and functions include, but are not limited to, ensuring the safety and protection of artificial islands, offshore terminals, installations and other structures and devices in any maritime zone; protecting fishermen, by providing assistance at sea when they are in distress; and taking measures to ensure the safety of life and property at sea (Upadhyay and Upadhyay 2002, p. 79).

Coastal Regulation Zone Notification, 2011—Activities along India's coastline which spans across a length of 7500 km have been governed by the Coastal Regulation Zone Notification, 2011, for over a decade. This Notification issued under the Environment (Protection) Act, 1986, allows different extents of land utilization in the demarcated coastal areas. It seeks to conserve and protect coastal stretches and their unique environment and marine area and promote development in a sustainable manner based on scientific principles taking into account the dangers of natural hazards in the coastal areas. It restricts the setting up and expansion of any industry, operations or processes and manufacture or handling or storage or disposal of hazardous substances in the Coastal Regulation Zone (CRZ). All permissible activities in the CRZ shall be considered for CRZ clearance as per the procedure laid down. For the purpose of implementation and enforcement of the provisions of this Notification, the powers, either original or delegated, are available under the Environment (Protection) Act, 1986, with the Ministry of Environment, Forest and Climate Change (MoEF&CC), State Government, National Coastal Zone Management Authority (NCZMA) and the UT/State Coastal Zone Management Authorities (UT/SCZMA).[18]

The interpretation of the CRZ Notification has been further developed through myriad judicial decisions. Such interpretations have mostly taken place as a consequence of the Court ruling against coastal violations and mandating adherence to the provisions of the CRZ Notification. In *Indian Council for Enviro Legal Action v. Union of India & Ors*, besides directing the central and state governments to finalize the Coastal Zone Management Plans (CZMPs), the Court held that the amendments which reduced the No Development Zone were illegal (Indian Council for Enviro Legal Action v. Union of India and Ors 1996). During the course of adjudicating on the adverse impacts of coastal pollution caused by non-traditional and unregulated

---

[16] Coast Guard Act, § 2(m) (1978) (Ind). https://legislative.gov.in/sites/default/files/A1978-30.pdf

[17] Coast Guard Act, § 2(y) (1978). (Ind). https://legislative.gov.in/sites/default/files/A1978-30.pdf

[18] Coastal Regulation Notification 2011 (Ind). http://www.indiaenvironmentportal.org.in/files/CRZ-Notification-2011.pdf

prawn farming, the Supreme Court in *S. Jagannath v. Union of India* held that prawn farming industries were prohibited in the coastal regulation zones under the CRZ Notification 1991, excluding traditional systems of aquaculture (S. Jagannath v. Union of India 1997). The issue before the Supreme Court in *Vaamika Island v. Union of India & Ors* pertained to the categorization of certain properties in the Vembanad Backwaters of Kerala as CRZ-1 in the Coastal Zone Management Plan of the State of Kerala. The Court affirmed that "islands could be coastal stretches of river or backwater islands in Kerala are clearly covered by CRZ I". In view of the violation of the CRZ Notification and in the larger public interest to save Vembanad lake, the Court confirmed the decision of the High Court directing the demolition of illegal constructions (Vaamika Island v. Union of India and Ors 2013).

Island Protection Zone Notification, 2011—This Notification declares the coastal stretches of Middle Andaman, North Andaman, South Andaman and Greater Nicobar and the entire area of the other islands of Andaman and Nicobar and the Lakshadweep and their water area up to territorial water limit as the Islands Protection Zone and thereby restricts the areas from the setting up and expansion of any industry, operations or processes and manufacture or handling or storage or disposal of hazardous substances specified in the Hazardous Waste Rules, 2016, except in the manner provided in the Island Coastal Regulation Zone and Integrated Islands Management Plan. This Notification following a similar scheme as the CRZ Notification permits varying degrees of land use in demarcated coastal areas, albeit limited to the aforesaid islands (Island Protection Zone Notification of 2011).

Coastal Regulation Zone Notification, 2019—This Notification has superseded the CRZ Notification, 2011. Although Para 6(i) states that until and unless the Coastal Zone Management Plans (CZMPs) are revised or updated as per the 2019 Notification, provisions of this notification shall not apply and the CZMP as per the provisions of the CRZ Notification, 2011, shall continue to be followed for appraisal and CRZ clearance to such projects (Coastal Regulation Zone Notification of 2019).

Island Coastal Regulation Zone Notification, 2019—This Notification has superseded the IPZ Notification, 2011. Although Para 5(i) states that until and unless the Island Coastal Regulation Zone Plan (ICRZP) are revised or updated as per the 2019 Notification, provisions of this notification shall not apply and the ICRZP as per the provisions of the IPZ Notification, 2011, shall continue to be followed for appraisal and CRZ clearance to such projects (Island Coastal Regulation Zone Notification of 2019).

16.4.2 Laws Relating to Use of Marine Areas

The regulations relating to marine habitat conservation that have been listed above must be viewed in light of the laws that expressly regulate the usage of maritime regions. Historically, rules governing the use of maritime areas have taken precedence over laws governing the protection of marine habitat, which is a relatively new phenomenon. The regulation of the usage of maritime areas and marine products can be split into two categories: one that governs the commercial aspect of marine goods, including their export, and the other that facilitates their trade by allowing movement of vessels and ships. The former has a direct impact on marine resource protection, whereas the latter has an impact on ecosystems due to potential pollution from oil spills or other accidents (Upadhyay and Upadhyay 2002).

Marine Products Export Development Authority Act (MPEDA), 1972—One of the important legislations under the 'use' category is the Marine Products Export Development Authority Act of 1972 (MPEDA) which aims to streamline the marine products industry by creating a central body to regulate, organize and promote the industry on a commercial basis. The body has the authority to take actions to promote the growth of the marine products business, such as encouraging exports, registering fishing vessels and constructing processing units, among other things. 'All varieties of fisheries goods known commercially as shrimp, prawn, lobster, crab, fish, shellfish, other aquatic creatures or plants or parts thereof, and any other products which the Authority classifies as marine products' are included in the definition of marine products[19] (Upadhyay and Upadhyay 2002, pp. 82–83).

Indian Ports Act, 1908—The Indian Ports Act, 1908, governs the minor ports in India. Due to minor ports falling under the Concurrent List, they are also governed by the respective state governments. The term 'port' under the Act is inclusive of any part of a river or channel. This Act deals with various aspects of ports and port charges, viz. powers and duties of port officials; safety of shipping and conservation of ports; and port dues, fees and charges, among others. An essential provision includes the rule making power under this Act. Section 6 of the Act states that the government may enact rules and regulations to control the activities in ports such as the manner in which oil or water mixed with oil shall be discharged in any port for disposal[20]; the use of piers, jetties, landing places, etc.; vessels while taking in or discharging ballast or cargo, usage of fires and lights within any port. Section 21 of the Act particularly forbids the inappropriate discharge of ballast and/or trash into the sea, and it is penalized if the ballast or trash is likely to produce a shoal (Upadhyay and Upadhyay 2002, p. 83).

Merchant Shipping Act 1958—The Merchant Shipping Act of 1958 aims to 'foster the development and ensure the efficient maintenance of an Indian mercantile marine in a manner best suited to serve the national interests.' The Act also establishes a National Shipping Board for the said purpose. Besides this, the Act provides for the registration, certification, safety and security of Indian ships, along with governing merchant shipping (Merchant Shipping Act of 1958). Several rules for specific purposes have been issued under the Act such as the Merchant Shipping (Carriage of Cargo) Rules, 1995; Merchant Shipping (Civil Liability for Oil Pollution Damage) Rules, 2008; Merchant Shipping (Control of Pollution by Noxious Liquid Substances in Bulk) Rules, 2010; Merchant Shipping (Prevention of Pollution by Harmful Substances Carried by Sea in Packaged Form) Rules, 2010;

---

[19] Marine Products Export Development Authority Act 1972, § 3(h) (Ind). https://legislative.gov.in/sites/default/files/A1972-13.pdf

[20] Indian Ports Act 1908, § 6(1) (ee) (Ind). https://legislative.gov.in/sites/default/files/A1908-15.pdf

Merchant Shipping (Prevention of Pollution by Sewage from Ships) Rules, 2010; Merchant Shipping (Prevention of Pollution by Garbage from Ships) Rules, 2010; Merchant Shipping (Prevention of Pollution by Oil from Ships), 2010; Merchant Shipping (Wrecks and Salvage) Rules, 1974; and the Merchant Shipping (Prevention of Collisions at Sea) Rules, 1975.

Marine Insurance Act, 1963—This Act deals with marine insurance, wherein 'marine insurance' is an agreement 'whereby the insurer undertakes to indemnify the assured, in the manner and to the extent thereby agreed, against marine losses, that is to say, the losses incidental to marine adventure.'[21] [Sec. 3 of the Act of 1963].

Maritime Zones of India (Regulation of Fishing by Foreign Vessels) Act, 1981—The enactment of Maritime Zones of India (Regulation of Fishing by Foreign Vessels) Act, 1981, strengthened the objectives of the Territorial Waters, Continental Shelf, Exclusive Economic Zones and Other Maritime Zones Act, 1976. A separate legislation was required to address the regulation, exploration and exploitation of marine resources found across various zones (Upadhyay and Upadhyay 2002, p. 84). This legislation was enacted with a view to regularize activities such as fishing by vessels from international waters within particular maritime zones of India, where the term fishing includes 'catching, taking, killing, attracting or pursuing fish by any method and includes the processing, preserving, transferring, receiving, and transporting of fish.'[22] [Sec. 2(c) of the Act of 1981].

Coastal Aquaculture Authority Act, 2005—The Coastal Aquaculture Authority Act, 2005, establishes the Coastal Aquaculture Authority for regulating coastal aquaculture activities in coastal areas. Coastal aquaculture entails 'culturing, under controlled conditions in ponds, pens, enclosures or otherwise, in coastal areas, of shrimp, prawn, fish or any other aquatic life in saline or brackish water, but does not include freshwater aquaculture.'[23] [Sec. 2(c) of the Act of 2005].

Recycling of Ships Act, 2019—This Act regulates the recycling of ships by setting certain standards and laying down the statutory mechanism for their enforcement. The Act restricts and prohibits the use or installation of hazardous materials, which apply irrespective of whether a ship is meant for recycling or not. Such restriction or prohibition will apply immediately on new ships. Ship recycling facilities are required to be authorized. The Act imposes a statutory duty on ship recyclers to 'ensure safe and environmentally sound removal and management of hazardous materials from a ship.'[24] [Sec. 21 of the Act of 2019].

Draft National Fisheries Policy, 2020—The objective of the Draft National Fisheries Policy, 2020, is to secure the overall development of capture fisheries and aquaculture in the country. The Policy aims to 'develop, harness, manage and

[21] Marine Insurance Act 1963, § 3 (Ind). https://legislative.gov.in/sites/default/files/A1963-11.pdf

[22] Maritime Zones of India (Regulation of Fishing by Foreign Vessels) Act, § 2(c) (1981), (Ind). https://www.indiacode.nic.in/bitstream/123456789/1817/1/198142.pdf

[23] Coastal Aquaculture Act, § 2(c) (2005), (Ind). https://legislative.gov.in/sites/default/files/A2005-24.pdf

[24] Recycling of Ships Act, § 21 (2019). (Ind). https://www.indiacode.nic.in/bitstream/123456789/15690/1/AAA2019___49.pdf

regulate capture and culture of fisheries in a sustainable and responsible manner'[25] (Ministry of Fisheries 2020). The Policy encompasses the entire land and EEZ of the country and is set within a time frame of 10 years (2021–2030).

Major Port Authorities Act, 2021—The Major Port Authorities Act, 2021, has superseded the Major Port Trusts Act, 1963, for governing the regulation, operation and planning of major ports in India. One of the major shifts that the new Act has introduced is that the Board of Trustees under the old Act has now been replaced with the Board of Major Port Authority,[26] albeit consisting of a lower number of representatives from the concerned state governments, port employees and labour as compared to earlier. By designating the Board of Major Port Authority as the successor of the Board of Trustees, the new Act brings about a huge structural overhaul by converting the existing 'trustees' set-up into an 'authorities' set-up. The new Act predominantly aims to reorient the governance model in central ports by changing it to the landlord port model. The publicly governed port authority, in the landlord port model, acts both as a regulatory body and as a landlord, while private companies carry out other activities, namely cargo handling activities. The ownership of the port, in the new model, therefore, enjoyed by the port authority, whereas the infrastructure is leased to private firms for providing and maintaining their own superstructure and installing their own equipment to handle cargo. The landlord, i.e. the port authority, in this case, gets, in return, a fair share of the revenue from the private entity (Manoj 2016).

16.5 Response to National Maritime Security Challenges

The geo-strategic and geo-economic linkages within the Indian Ocean are affected both by traditional maritime security conflicts among States and threats by non-State actors (e.g. maritime terrorism and piracy) as well as by non-traditional threats that affect the environment (Khan 2021). In this background, it becomes imperative to discuss some of the noteworthy strategic responses taken by India to such national maritime security challenges.

Project Mausam — The Ministry of Culture has initiated a project titled Project 'Mausam' with Archaeological Society of India (ASI), New Delhi, as the nodal agency and Indira Gandhi National Centre for the Arts (IGNCA), New Delhi, as its research unit. 'Mausam or Arabic "Mawsim" refers to the season when ships could sail safely. This distinctive wind system of the Indian Ocean region follows a regular pattern: southwest from May to September; and northeast from November to March. The endeavour of Project "Mausam" is to position itself at two levels: at the macro level, it aims to re-connect and re-establish communications between

---

[25] Ministry of Fisheries, Animal Husbandry and Dairying. (2020) Draft National Fisheries Policy. Hyderabad: National Fisheries Development Board

[26] Major Port Authorities Act, § 3 (2021), (Ind). https://egazette.nic.in/WriteReadData/2021/225265.pdf

countries of the Indian Ocean world, which would lead to an enhanced understanding of cultural values and concerns; while at the micro level the focus is on understanding national cultures in their regional maritime milieu. The central themes that hold Project "Mausam" together are those of cultural routes and maritime landscapes that not only linked different parts of the Indian Ocean littoral but also connected the coastal centres to their hinterlands. More importantly, shared knowledge systems and ideas spread along these routes and impacted both coastal centres, and also large parts of the environs.' It has been touted as India's answer to China's Maritime Silk Road (Project Mausam, Indira Gandhi National Centre for the Arts, Ministry of Culture).[27]

Necklace of Diamonds Strategy—'Through its String of Pearls Strategy, China is expanding its footprints to contain Indian hold in the Indian Ocean. It is creating a ring around India through strategically placed nations such as at Chittagong (Bangladesh), at Karachi & Gwadar Port (Pakistan), at Colombo & Hambantota (Sri Lanka) and other facilities. As a counter-action, India has started working on the 'Necklace of Diamonds' strategy. This strategy aims at garlanding China or in simple words, the counter encirclement strategy. India is expanding its naval bases and is also improving relations with strategically placed countries to counter China's strategies.' India's strategic bases include those at Changi Naval Base (Singapore), Sabang Port (Indonesia), Duqm Port (Oman), Assumption Island (Seychelles) and Chabahar Port (Iran). Besides gaining direct access to the strategic naval bases, India is also developing new naval bases, developing the old bases and is enhancing relations with other nations to garland China such as Mongolia, Japan, Vietnam and other Central Asian countries.[28]

16.6 Concluding Remarks

The above makes it clear that a robust framework exists within which the Blue Economy in the region as well as in India can thrive rapidly. However, the competitiveness between the Big Two needs to be more cooperative than the present. The race to establish might result in decisions that harm both nations. It is also clear from the above that while there is a huge thrust on the development of the Blue Economy, especially since 2014, both internationally and regionally, India has had a reasonably strong economic development framework since the early 1970s. The

[27] Culture, M. O. (n.d.). Project Mausam. Retrieved July 13, 2021, from https://www.indiaculture.nic.in/project-mausam Also see, *About The Project | IGNCA*. (n.d.). Project 'Mausam'- Mausam/Mawsim: Maritime Routes and Cultural Landscapes. Retrieved July 13, 2021, from http://ignca.gov.in/about-the-project/

[28] *What is Necklace of Diamonds Strategy?* (n.d.). What Is Necklace of Diamonds Strategy? Retrieved July 13, 2021, from https://ibteducation.blogspot.com/2020/06/what-is-necklace-of-diamonds-strategy.html

Also see, Vidhi Bubna and Sanjna Mishra. (2020, July 14). String of Pearls vs Necklace of Diamonds. Retrieved July 13, 2021, from https://asiatimes.com/2020/07/string-of-pearls-vs-necklace-of-diamonds/

Merchant Shipping Act, the Maritime Zones Act, the MPEDA, the Regulation of Foreign Fishing Vessels Act and the Major Ports Act laid a strong foundation for blue economic development without stating as such. It is therefore important that the latest regional developments for economic cooperation build on the existing strong legal framework and ensure that neighbouring cooperating countries undertake significant legal reforms to improve their oceans and coastal governance framework so that a level playing field is created which is based on transparency, equity and fairness.

References

About The Project | IGNCA. (n.d.). Project 'Mausam'- Mausam/Mawsim: Maritime Routes and Cultural Landscapes. Retrieved July 13, 2021, from http://ignca.gov.in/about-the-project/

Alharthi, M. and Hanif, I. (2020), "Impact of blue economy factors on economic growth in the SAARC countries", *Maritime Business Review*, Vol. 5 No. 3, pp. 253-269. https://doi.org/10.1108/MABR-01-2020-0006

Athens Convention relating to the Carriage of Passengers and their Luggage by Sea. (1974). https://www.imo.org/en/About/Conventions/Pages/Athens-Convention-relating-to-the-Carriage-of-Passengers-and-their-Luggage-by-Sea-(PAL).aspx

Baltic and International Maritime Council. (1905). https://www.bimco.org

Blue Economy – Indian Ocean Rim Association – IORA. (2015). Indian Ocean Rim Association. https://www.iora.int/en/priorities-focus-areas/blue-economy

Constitution of India 1950 (Ind). https://legislative.gov.in/sites/default/files/COI_1.pdf

Convention on the Prevention of Marine Pollution by Dumping of Wastes and Other Matter. (1972). https://www.imo.org/en/OurWork/Environment/Pages/London-Convention-Protocol.aspx

Declaration of the Indian Ocean Rim Association on the Blue Economy in the Indian Ocean Region, 08.05.2017, https://www.iora.int/media/8218/jakarta-declaration-on-blue-economy-final.pdf

Declaration on the Establishment of Bangladesh-Sri Lanka-Thailand Economic Cooperation. (1997, June 6). Bangkok Declaration. https://drive.google.com/file/d/0B8Fv9wDGJqx2NkJTVzlZek5va0U/view?resourcekey=0-7YYfgTO3Ibbj SNaDhv2XRQ

Hong Kong International Convention for the Safe and Environmentally Sound Recycling of Ships. (*2009*). https://www.imo.org/en/About/Conventions/Pages/The-Hong-Kong-International-Convention-for-the-Safe-and-Environmentally-Sound-Recycling-of-Ships.aspx

India's Blue Economy - A Draft Policy Framework. (2020, September). Ministry of Earth Sciences. https://moes.gov.in/writereaddata/files/BlueEconomyPolicy.pdf

Indian Council for Enviro Legal Action vs Union of India & Ors (April 18, 1996) (1996) 5 SCC 281.

International Convention for the Prevention of Pollution from Ships (MARPOL). (1973). https://www.imo.org/en/About/Conventions/Pages/International-Convention-for-the-Prevention-of-Pollution-from-Ships-(MARPOL).aspx

International Convention for the Prevention of Pollution from Ships. (n.d.) https://www.imo.org/en/KnowledgeCentre/ConferencesMeetings/Pages/Marpol.aspx

International Convention for the Safety of Life at Sea. (1974). https://www.imo.org/en/About/Conventions/Pages/International-Convention-for-the-Safety-of-Life-at-Sea-(SOLAS),-1974.aspx

International Maritime Organisation (1948, March 17). https://www.imo.org/en

International Whaling Commission, (1946, December 2). https://iwc.int/home

Island Protection Zone Notification 2011 (Ind). http://www.indiaenvironmentportal.org.in/files/IslandProtectionZoneNotification2011.pdf

Khan, S. E., (2021). New Maritime Security Challenges in the Indian Ocean: Shared Concerns and Opportunities from Bangladesh Perspective, Retrieved July 13, 2021, from https://nesa-center.org/new-maritime-security-challenges-in-the-indian-ocean-shared-concerns-and-opportunities-from-bangladesh-perspective/

Manoj, P. (2016). *Move to landlord port model means privatizing state-run cargo berths*. Mint. https://www.livemint.com/Industry/TrIxBCGekevmqGS8dbGo0I/Move-to-landlord-port-model-means-privatizing-staterun-carg.html

Martinez, A. D. (1980). Conflicting Law of the Sea Principles: Mare Liberum versus Mare Clausum. *Townson State Journals of International Affairs,* 14(2), 93-101. https://cpb-us-w2.wpmucdn.com/wp.towson.edu/dist/b/55/files/2018/04/Spring-1980-Martinez.pdf

Merchant Shipping Act 1958 (Ind). https://legislative.gov.in/sites/default/files/A1958-44.pdf

Ministry of Fisheries, Animal Husbandry and Dairying. (2020) Draft National Fisheries Policy. Hyderabad: National Fisheries Development Board.

Premarathna, Isuru. (2021). Maritime Security Challenges in the Indian Ocean: Special Reference to Sri Lanka. *International Journal of Research and Innovation in Social Science*, 5(1), 158-173. https://doi.org/10.47772/IJRISS.2021.5106.

S. Jagannath vs Union of India (December 11, 1996) (1997) 2 SCC 87.

Seijo, J. C., & Salas, Silvia (2014). The role of capacity building for improving governance of fisheries and conservation of marine ecosystems. In Garcia, S. M., Rice, Jake, & Charles, Anthony (Eds.), *The role of capacity building for improving governance of fisheries and conservation of marine ecosystems*. https://doi.org/10.1002/9781118392607.ch26

ASEAN. (1967). The ASEAN Declaration (Bangkok Declaration) Bangkok, 8 August 1967 - ASEAN: One Vision One Identity One Community. https://asean.org/the-asean-declaration-bangkok-declaration-bangkok-8-august-1967/.

The International Convention for the Control and Management of Ships' Ballast Water and Sediments. (2004). https://www.imo.org/en/About/Conventions/Pages/International-Convention-for-the-Control-and-Management-of-Ships%27-Ballast-Water-and-Sediments-(BWM).aspx

Third BIMSTEC Summit Declaration. (2014, March 4). Ministry of External Affairs. https://www.mea.gov.in/in-focus-article.htm?23029/Third+BIMSTEC+Summit+Declaration

United Nations. (1982a, December 10). United Nations Convention on Law of Seas https://www.un.org/depts/los/convention_agreements/texts/unclos/unclos_e.pdf

United Nations. (1982b, December 10). United Nations Convention on Law of Seas https://www.iucn.org/theme/marine-and-polar/our-work/international-ocean-governance/unclos

Upadhyay, S., & Upadhyay, V. (2002). International and National Instruments and Marine Turtle Conservation in India. *Journal of International Wildlife Law and Policy*; p. 82.

Upadhyay, S., & Upadhyay, V. (2006). Laws and Marine Turtle Conservation in India. In *Marine Turtles of the Indian Subcontinent*. Chennai: Universities Press.

Vaamika Island vs Union of India & Ors (August 8, 2013) (2013) 8 SCC 760.

Chapter 17
Ocean Governance and Integrated Ocean Management

Heman Das Lohano and Muhammad Bilal Maqbool

17.1 Introduction

Oceans are valuable natural resources providing numerous ecosystem services. Oceans have great importance not only from an economic perspective but also from a geographical and political perspective. However, oceans are becoming increasingly more vulnerable over time due to population growth and achievement of economic growth. By anthropogenic activities such as dumping untreated sewage and industrial wastewater, plastic, and other pollutants, oil spills, and climate change, ocean ecosystems and their biodiversity are rapidly changing and are under threat. As oceans are transboundary in nature, these issues need to be resolved at the country level as well as the regional and international levels. Furthermore, keeping in view the importance of oceans, the conservation and sustainable use of oceans has been included in the United Nations Sustainable Development Goals (SDGs).

SDG 14 states this goal as "conserve and sustainably use the oceans, seas and marine resources." The United Nations has specified the following targets under this goal:

1. Index of coastal eutrophication and floating plastic debris density.
2. Proportion of national exclusive economic zones managed using ecosystem-based approaches.
3. Average marine acidity (pH) measured at agreed suite of representative sampling stations.
4. Proportion of fish stocks within biologically sustainable levels.
5. Coverage of protected areas in relation to marine areas.

H. D. Lohano (✉) · M. B. Maqbool
Institute of Business Administration, Karachi, Pakistan
e-mail: hlohano@iba.edu.pk; mbmaqbool@iba.edu.pk

S. Hazra, A. Bhukta (eds.), *The Blue Economy*,
https://doi.org/10.1007/978-3-030-96519-8_17

6. Progress by countries in the degree of implementation of international instruments aiming to combat illegal, unreported, and unregulated fishing.
7. Sustainable fisheries as a proportion of GDP.
8. Proportion of total research budget allocated to research in the field of marine technology.
9. Progress by countries in the degree of application of a legal/regulatory/policy/institutional framework which recognizes and protects access rights for small-scale fisheries.
10. Number of countries making progress in ratifying, accepting, and implementing through legal, policy, and institutional frameworks, ocean-related instruments that implement international law, as reflected in the United Nations Convention on the Law of the Sea (Ritchie et al. 2018).

17.2 Contribution of Oceans

Oceans provide ecosystem services including provisioning, supporting, regulating, and cultural services, also covered in the previous chapters. Environmental and other economists have attempted to measure the value of these ecosystem services. In this section, we focus on the contribution of oceans in terms of GDP and major economic activities.

Three-fourths of earth's area is covered by the oceans; therefore, oceans have the largest and also complicated ecosystem. WWF (2015) reported that the estimated value of ocean assets around the world is about USD 24 trillion that entails an annual output of around USD 2.4 trillion. If the ocean economy is measured separately, this would have ranked seventh among the top 10 global economies (WWF 2015).

Ocean economy's share in GDP varies from country to country, ranging from 1% to 30%. However, there are some miscalculations or underestimation while calculating its share in GDP. Many do not consider the indirect role of oceans; they just calculate direct impact on GDP such as fisheries, aquaculture, and tourism only. There are many economies in this region where it plays a key role in their GDP such as India (4.10%), China (9.6%), Vietnam (18.80%), and Indonesia (14.85%) (Juneja et al. 2021).

Asia is surrounded by the Indian Ocean to the south, Pacific to the east, Arctic Ocean to the north, and Red and Black Sea to the southwest. Among the world's oceanic divisions, the Indian Ocean is the third largest, covering an area of more than 70 million sq. km that includes extensive exclusive economic zones (EEZ) of different countries and large "high seas." These countries are home to one-third of the world's population that rely extensively on marine resources for livelihood and food security. The Indian Ocean is projected to become a dominant global geopolitical and economic force in the twenty-first century. The contribution of Indian ocean nations to global GDP has significantly increased over the last four decades, from an average 6–7% in 1980 to 10% or USD 78 trillion in 2014 (Roy 2019).

The Arabian Sea covers a total area of about 3,862,000 square km and forms part of the principal sea route between Europe and India. Pakistan has a long coastline of 1050 km and the exclusive economic zone covering about 240,000 sq. km. It is bounded to the Horn of Africa, the Arabian Peninsula, Iran, Pakistan, and India. The Arabian Sea, with its strategic location vis-à-vis the Red Sea (including the Suez Canal) and the Persian Gulf, contains some of the world's busiest shipping lanes; and the chief routes originate in those two extensions. Persian Gulf shipping largely consists of tankers, some of immense capacity, that transit the Arabian Sea en route to destinations in East Asia, Europe, and North and South America. Besides that, it has an abundance of natural resources in the form of oil and gas along with seafood.

Therefore, it has become more challenging to enhance cooperation not only at the international level but also among local-level different administrative units within the country for the conservation and long-term use of riverbeds, deltas, seas, oceans, and other marine resources. Conservation and sustainability of these resources has also been given due attention in Sustainable Development Goals (SDGs).

Blue Economy Blue Economy is defined as the "sustainable use of ocean resources for economic growth, improved livelihood and jobs, and ocean ecosystem health" (World Bank, 2017). Thus, blue economy can potentially help nations to achieve SDGs.

In the present study, we will focus on the Asian region for analysis. Most of the Asian region is covered by Asia and Pacific region. Many of them lie in developing and underdeveloped countries that are directly or indirectly linked with ocean economy. There are 13 least developing countries in this region out of 47 around the world. Names of these countries are Afghanistan, Bangladesh, Bhutan, Cambodia, Kiribati, Lao People's Democratic Republic, Myanmar, Nepal, Solomon Islands, Timor-Leste, Tuvalu, Vanuatu, and Yemen.

Asia-Pacific contains a large area of both land and coastline in the world. This region includes East Asia, South Asia, Southeast Asia, and Oceania. The area includes two oceans, the Indian Ocean (the third largest ocean) and the Pacific Ocean (the largest ocean), as well as several seas like Bay of Bengal and other water bodies. It possesses some of the most ecologically and economically important sea areas of the world which provide a rich array of services that directly and indirectly contribute to human survival and quality of life, supporting local coastal communities and their economies. The scope of Blue Economy is thus large in Asia as several states in this region have a significant share of marine economy in their gross domestic product (GDP).

There are some economic sectors that are dependent on oceans such as port and shipping, tourism, and fishing. The Asian region contributes a significant part in global trade. In terms of global trading volume (loaded and unloaded), Asia has the largest share as depicted from the figure below (Fig. 17.1).

Asian economies have witnessed a miraculous growth over the last decades. Most of the goods are manufactured in other countries while assembled somewhere else. Therefore, transportation of such good is heavily dependent on the shipping

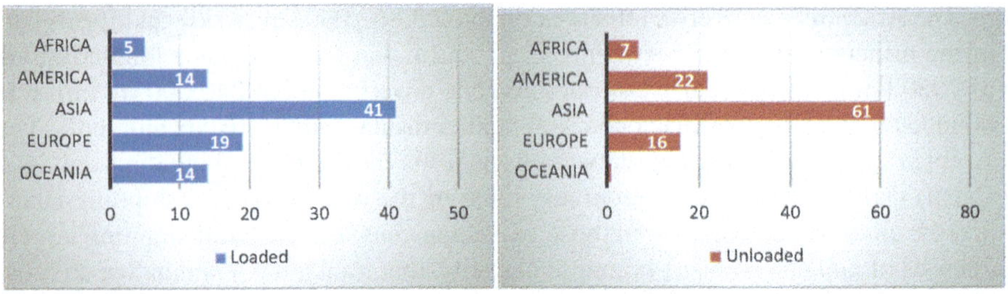

Fig. 17.1 International Maritime Trade by region in 2018 (% share in world tonnage). (Source: UNCTAD 2019)

industry. Interestingly, worldwide shipping industry is ruled by three Asian countries, namely, China, Korea, and Japan.

Asia excluding China captures 34% of fishing and aquaculture marked to the world. Since the last 20 years, its production has doubled. Asia's share (excluding China) in the global international trade market has reached up to 42% in 2019. In terms of volume of production, Asia's share has reached to 89% of global fish production (UNESCO 2017).

Tourism plays a key role in Asian economies. Southeast Asian economies are most dependent and have much contribution in their respective GDP through tourism. Indonesia, Malaysia, Thailand, Singapore, and the Philippines are among the leading countries in this sector. However, South Asian economies are far behind in this sector; now, India, Sri Lanka, and Pakistan are trying their best to attract tourist from across the globe as they are rich in natural spots along with historical places. In Asian economies, South Asian countries have just 13% share, while Eastern Asia-Pacific countries have 68% share in tourism.

17.3 Blue Economy and Challenges

Blue Economy is defined as the "sustainable use of ocean resources for economic growth, improved livelihood and jobs, and ocean ecosystem health" (World Bank 2017). It is used in a broader term, sometimes Blue Economy also mentioned as "marine economy," "coastal economy," or "ocean economy" in the literature (Mohanty et al. 2015).

To ensure sustainable use of ocean resources and provision of ecosystem services, oceans need to be protected from degradation. Oceans are facing some major threats and challenges including pollution (untreated sewage, industrial wastewater, plastic, and other waste material), oil spills, unsustainable fishing, climate change, reduction of mangroves, and many others.

One of the major threats faced by the ocean is plastic pollution which is increasing day by day and is expected to triple by 2050. Plastic pollution is affecting environment in two ways. Firstly, burning plastic increases carbon emissions, and secondly, plastic pollution affects badly to biodiversity and other ocean habitants. Half of the plastic pollution is produced by the Asia-Pacific region which needs to be taken seriously. In this regard, all stakeholders at each level, domestic, national, and regional levels, should coordinate in policy design and then most importantly its implementation. Another issue is the usage of coastal land for industrial and commercial purposes. These industries are flowing their waste in terms of garbage and industrial water without any treatment. Many big and industrial cities are located nearby coastal areas. These cities especially in developing countries are pumping untreated effluent into the oceans. This has resulted in a staggering drop in marine life populations by 40% (Moeen 2019).

Oil spillovers are another threat to the ocean. Such incidents are rare but still causing damage to the ecosystem and biodiversity in the oceans. To reduce the risk of oil spillover, there is a need of clean and renewable energy so that the risk of such incident could be minimized. In this regard, developed countries and other international agencies should step forward in technology transfer and manufacturing to such plants.

Another threat is unsustainable fishing in the Asia-Pacific Ocean which has caused the stock to dwindle over the time. In this region, as per UN reports, about 20% of fishing is done by unregulated, illegal, and unreported manners. This situation is even more worse in Indian, Northern Pacific, and Western Central Pacific Oceans where this percentage goes up to 30% (UNESCAP 2020).

Climate change is one of the major threats to the ocean life. Since the 1980s, 20–30% of human-induced carbon emissions are absorbed by the ocean. Therefore, small islands in Asia along the ocean are more vulnerable to these changes. Besides this, increased rainfall, glacier melting with more intensity, sea surface temperature, and warmer air are affecting seagrasses, mangroves, and other ocean habitants. Climate change causes the rise of sea level and to some extent intrusion of agricultural land into the sea. Due to climate change, rapid melting of glaciers and changing pattern and intensity of rainfall have resulted in frequent floods. Due to lack of water preserving capacity, floodwater run toward the sea, hence causing destruction of coastal areas.

Alongside the coastal area, mangrove trees and forests are of vital importance for sea ecosystem especially for fisheries and built-in barrier for various disastrous threats. It is a pool of biodiversity and provides habitat for a diverse community of organisms, ranging from bacteria, fungi, fish, shrimps, birds, and mammals. But with the passage of time, there is a decline in the area of mangroves. Global warming, arid conditions, prolong drought spells, inadequate supply of fresh water from rivers and other canals, industrial and thermal pollution, dumping of untreated effluents, overexploitation of mangroves for fuelwood/fodder, and population pressure are the main causes of degradation of Indus delta mangroves.

To achieve economic growth equally, small economies, especially underdeveloped ones, must be connected in the shipping industry as most of the world trade is done by oceans. In spite of this step, there is a need of greening the oceans to overcome the impact of pollution generated by these heavy cargo ships.

17.4 Integrated Ocean Management Policy

Since the last four decades, Blue Economy is increasing continuously to meet the needs of humankind in terms of food, transportation, energy, and tourism. At the same time, new industries are opening along with the expansion of existing ones. Therefore, new challenges and issues are emerging due to climate change, pollution, and environmental degradation. In this regard, there is a need of both short- and long-term policies to make the oceans more prosperous, healthy, and resilient against these unfavorable natural and anthropogenic shocks. All these benefits from the oceans and losses to the ocean's ecosystem are resulted from different industries. Therefore, there is a need of an integrated ocean management (IOM) policy. Keeping in view the importance of oceans, the need of an integrated management policy for oceans was observed many years ago (Underdal 1980). To discuss further details of IOM, we describe two approaches, namely, ecosystem-based management and integrated coastal zone management.

Ecosystem-based management (EBM) is defined as management of natural resources mainly by focusing on their health and productivity along with resilience of a specific ecosystem both at individual and at group level (Winther et al. 2020). It is a management approach which emphasizes the full range of interaction in the ecosystem including human. It has the following features: (a) emphasis on the protection of ecosystem, its structure, and functions; (b) recognition of linkages between target species and systems such as air, land, and sea; and (c) focused on individual ecosystem and account for human activities affecting it (Winther et al. 2020).

Integrated coastal zone management (ICZM) is the process of managing the coast and nearshore waters in an integrated and comprehensive manner with the goal of achieving conservation and sustainable use (Katona et al. 2017). Sometimes it is also named as integrated coastal management. The main purpose of this is to collect information, planning, decision making, management, and implementing with the help of all relevant stakeholders. Further, it also includes marine spatial planning and adaptive ocean management along with their features (Katona et al. 2017).

Coastal countries' jurisdiction is over 200 nautical miles, also called exclusive economic zone (Agardy et al. 2011). There are often multiple authorities to manage and oversee. Therefore, it is sometimes difficult for different agencies or departments to manage efficiently due to lack of coordination. Keeping in view all of this, it is a need of time to harmonize and coordinate among these different agencies for better and efficient management.

With the passage of time, its importance has been accepted not only globally but also at the regional, national, and domestic levels. Even then, there are many challenges regarding policy framework, implementation, new laws, capacity deficiency in knowledge, and lack of coordination among different departments and ministries and between their mandates as well. The main purpose of integrated ocean management is to protect oceans (here ocean includes both marine and coastal areas) for sustainable development and long-term usage and make it healthier and more resilient against the unhealthy activities such as pollution and environmental changes. In this regard, IOM aims to bring all the stakeholders from international to domestic, from government to civil society, all those who are directly or indirectly linked with the ocean economy.

The World Ocean Assessment in its first report under the UN General Assembly to examine the status of the marine environment concludes as (United Nations 2015):

> The sustainable use of the ocean cannot be achieved unless the management of all sectors of human activities affecting the ocean is coherent. Human impacts on the sea are no longer minor in relation to the overall scale of the ocean. A coherent overall approach is needed. This requires considerations of the effects on ecosystems of each of the many pressures, what is being done in other sectors and the way that they interact.

The UN General Assembly in 1999 passed a resolution on oceans and law of the sea to address the issues and challenges; again, in 2018, the General Assembly passed a resolution whose preamble states that.

> [T]he problems of ocean space are closely interrelated and need to be considered as a whole through an integrated, interdisciplinary and intersectoral approach, and reaffirming the need to improve cooperation and coordination at the national, regional and global levels, in accordance with the Convention, to support and supplement the efforts of each State in promoting the implementation and observance of the Convention and the integrated management and sustainable development of the oceans and seas.. .

Further, the UN adopted Sustainable Development Goals as part of Agenda 2030. Many of these goals are relevant to the ocean, but specifically Goal 14 is related to life below water. It addresses marine issues specifically.

17.5 Ocean Governance

17.5.1 Ingredients of Successful Ocean Governance

Today, in the era of modern and complex world, there is more need of better coordination among different sectors and departments for improved ocean governance (Klinger et al. 2018). In most of the countries, especially in underdeveloped countries, legal and institutional mechanisms are divided among different agencies or departments both at national and domestic or at local level. That makes economic development along with other issues such as environmental and marine ecosystem difficult to manage. This needs to be resolved, and this can be done only by political will. In general, governance is defined by the UNDP as "the exercise of economic, political and administrative authority to manage a country's affairs at all levels. It comprises the mechanisms, processes and institutions through which citizens and groups articulate their interests, exercise their legal rights, meet their obligations and mediate their differences" (UNDP 1997). Rotberg (2004) perceives governance as "the term used to describe the tension-filled interaction between citizens and their rulers and various means by which governments can either help or hinder their constituents' ability to achieve satisfaction and material prosperity." Keeping all this in view, some aspects of better ocean governance are described below in detail.

All over the world, issues and problems vary across the country and region. Besides these issues and challenges, some ingredients must be followed: (a) a detailed examination of all relevant agencies, their structure, powers, and obligations either they have overlapping or shortcomings in their responsibilities; (b) an assessment of the present condition, variability, and expected future trends in climate and ecosystem with the help of all available data using modern scientific techniques (also maintain a data center to observe these changes and trend continuously); (c) an examination of information regarding human activities and interest along with their conflicts; and (d) include all relevant stakeholders in this exercise on a regular basis.

As discussed above, the main aspects of a successful ocean management now need some detail.

A legal structure for IOM can provide a baseline for better ocean management for sectoral and long-term economic growth. In this regard, there is a need of new law and also provisions in the existing laws at the domestic and international levels just to harmonize the polices and for better coordination among different stakeholders. The European Union in 2008 formulated their marine policy at the continent level, and on the other side, at the state level, USA state Massachusetts introduced new laws for ocean in 2008. Besides legal regime, there is also a need of check and balance on illicit activities, capacity constraints, and gap between reality and legal framework.

Another issue is the deficiency of scientific data and its usage for better ocean management. Many countries are facing this problem as they do not have enough data for implementing and then monitoring the international governance framework (UNESCO 2017). Availability of new technologies has increased the government's capability to oversee what is happening in the oceans. For example, Global Fishing Watch offers tracking of fishing activities in the ocean. National level cooperation helps to pool up resources for better management and monitoring.

Finally, the issue in this regard is the involvement of all stakeholders from common persons to international community. Planning at the domestic level requires inclusion of all groups related to oceans. These are common persons, businesses, local agencies, and domestic and national governments. Participation of all these actors is the most effective approach in ocean management. Even with compilation of resource from all these stakeholders, resource constraints are still there especially in developing countries and small islands. Keeping this in view, there is a need of international cooperation among different agencies and countries. There are different actors or sectors that are directly or indirectly involved in ocean economy that are fisheries, recreation, tourism, aquaculture, mining, renewable energy, petroleum, and shipping.

17.5.2 Future Policy for Ocean Governance

For sustainable use of oceans, there is a need of effective governance with cooperation not only at the international level but also among local-level different administrative units within the country aiming at conservation and long-term use of riverbeds, deltas, seas, oceans, and other marine resources. Conservation and sustainability of these resources has also been given due attention in Sustainable Development Goals (SDGs).

There are more than 100 agreements at the regional level for ocean governance. The scope of these agreements is related to biodiversity, fisheries, pollution, and climate change. Some agreements cover some sections of biodiversity, fisheries, pollution and climate change, while others include all sections (Mahon et al. 2015). Many of these agreements are still in papers. However, some agreements are fully implemented while others are partially implemented. In a governance perspective, many agreements are overlapping in their jurisdictions and mandates. Therefore, according to Mahon et al. (2015), there is a need of "one ocean one policy" for better and effective governance. Mahon et al. (2015) called this approach as a global to regional issue-based network. At the regional level, all agreements should be in line with each other. Conflicts and overlapping in jurisdiction and mandate of these agreements must be removed. This can be achieved through better coordination and linkages among different regional stakeholders.

17.6 Conclusion and Way Forward

Keeping in view large interests in the ocean, there is a need of transformative actions to make ocean life better and sustainable. As discussed above, the UN adopted SDGs for Agenda 2030 that also include ocean named as "life below water.". Johansen (2020) estimated that there is a need of $174 billion on annual basis to achieve this goal, and most of these funds should be spent to avoid marine and coastal pollution. Oceans are transboundary, therefore, the flow of polluted water and plastic move from one country to the other. Thus, these problems call for mutual agreement and cooperation among countries in the Asian region.

Climate change is one of the major threats to the ocean life. Since the 1980s, 20–30% of human-induced carbon emissions are absorbed by the ocean. Therefore, small islands in Asia along the Asia-Pacific and Indian Oceans are more vulnerable to these changes. Besides this, increased rainfall, glacier melting with more intensity, sea surface temperature, and warmer air are affecting seagrasses, mangroves, and other ocean habitants.

Another threat to the ocean is plastic pollution which is increasing day by day and is expected to triple by 2050. It has double affects, one carbon emission and second it affects badly to biodiversity and other ocean habitants. In Asia, half of the plastic pollution is produced by the Asia-Pacific region alone, which depicts the gravity of the situation, and therefore, it must be taken seriously. In this regard, all stakeholders at each level, domestic, national, and regional levels, should coordinate in policy design and then most importantly its implementation.

To achieve economic growth equally, small economies, especially underdeveloped ones, must be connected in the shipping industry as most of the world trade is done by oceans. In spite of this step, there is a need of greening the oceans to overcome the impact of pollution generated by these heavy cargo ships.

Another threat is unsustainable fishing in the Asia-Pacific Ocean which has caused the stock to dwindle over the time. In this region, as per UN reports, about 20% of fishing is done by unregulated, illegal, and unreported manners. This situation is even more worse in Indian, Northern Pacific, and Western Central Pacific Oceans where this percentage goes up to 30%.

Oil spillovers are another threat to the ocean. Although such incidents are rare but still causing damage to the ecosystem and biodiversity in the oceans. To reduce the risk of oil spillover, there is a need of clean and renewable energy so that the risk of such incident could be minimized. In this regard, developed countries and other international agencies should step forward in technology transfer and manufacturing to such plants.

Currently, there are many organizations that are involved in ocean governance at different levels that make it difficult to make a unified approach for ocean protection. These are performing their duties at different layers such as international and domestic. These include the International Maritime Organization, which exclusively deals with oceans, and the Food and Agriculture Organization (FAO), which basically deals with food by partly dealing with the sea. Other international organizations deal with oceans at the regional level such as the Oslo-Paris convention that deals with Northeast Atlantic Ocean. Some other nongovernment organizations are also working; besides these, most of the countries have their own institutions for ocean governance. All that need is the establishment of single body or commission to regulate at each level of governance including the national, regional, and international levels just to have a unified policy for better ocean governance to protect this asset for future generations.

References

Agardy, T., Di Sciara, G. N., & Christie, P. (2011). Mind the gap: addressing the shortcomings of marine protected areas through large scale marine spatial planning. Marine Policy, 35(2), 226-232.

Moeen H. (2019, October 14) Karachi's polluted sea. DAWN. Available at: https://www.dawn.com/news/1510742

Johansen, D. F., & Vestvik, R. A. (2020). The cost of saving our ocean-estimating the funding gap of sustainable development goal 14. Marine Policy, 112, 103783.

Juneja, M., De Souza, C., Giriyan, A. L., & Ganeshan, S. (2021). Contextualising Blue Economy in Asia-Pacific Region.

Katona, S., Polsenberg, J., Lowndes, J., Halpern, B. S., Pacheco, E., Mosher, L., ... & Carr, S. (2017). Navigating the seascape of ocean management: waypoints on the voyage toward sustainable use.

Klinger, D. H., Eikeset, A. M., Davíðsdóttir, B., Winter, A. M., & Watson, J. R. (2018). The mechanics of blue growth: management of oceanic natural resource use with multiple, interacting sectors. Marine Policy, 87, 356-362.

Mahon, R., Fanning, L., Gjerde, K. M., Young, O., Reid, M., & Douglas, S. (2015). Transboundary Waters Assessment Programme (TWAP) Assessment of Governance Arrangements for the Ocean Volume 2: Areas Beyond National Jurisdiction. Intergovernmental Oceanographic Commission Technical Series, 119.

Ritchie, H., Roser, M., Mispy, J., & Ortiz-Ospina, E. (2018). Measuring progress towards the sustainable development goals. SDG-Tracker. org, website.

Rotberg, R. I. (2004). Strengthening governance: Ranking countries would help. The Washington Quarterly, 28(1), 71-81.

Roy A. (2019) "Blue Economy in the Indian Ocean: Governance Perspectives for Sustainable Development in the Region". Observer Research Foundation. Available at: https://www.orfonline.org/wpcontent/uploads/2019/01/ORF_Occasional_Paper_181_Blue_Economy.pdf

Mohanty, S. K., Dash, P., Gupta, A., & Gaur, P. (2015). Prospects of blue economy in the Indian Ocean. Research and Information System for Developing Countries, New Delhi.

World Bank. (2017). What is the Blue Economy? Infographic, June 6. Available at: http://www.worldbank.org/en/news/infographic/2017/06/06/blue-economy

WWF (2015). WWF 2015 Annual Report. Available at https://www.worldwildlife.org/publications/wwf-2015-annual-report

UNCTAD (United Nations Conference on Trade and Development) (2019). Review of Maritime Transport 2019. Available at: https://unctad.org/system/files/official-document/rmt2019_en.pdf

Underdal, A. (1980). Integrated marine policy: what? why? how?. Marine Policy, 4(3), 159-169.

UNESCO (United Nations Educational, Scientific and Cultural Organization) (2017). Global Ocean Science Report: The Current Status of Ocean Science around the World. Available at: https://unesdoc.unesco.org/ark:/48223/pf0000250428.locale=en

UNESCAP. (2020). "The Changing Sails: Accelerating Regional Actions for Sustainable Oceans in Asia and the Pacific."

Winther, J. G., Dai, M., Rist, T., Hoel, A. H., Li, Y., Trice, A., ... & Whitehouse, S. (2020). Integrated ocean management for a sustainable ocean economy. Nature ecology & evolution, 4(11), 1451-1458.

UNDP. (1997). Governance for sustainable human development. UNDP.

United Nations. (2015). "The First Global Integrated Marine Assessment." Regular Process for Global Reporting and Assessment of the State of the Marine Environment, Including Socioeconomic Aspects. https://www.un.org/regularprocess/content/first-world-ocean-assessment.

Word Notes

Blue Carbon

Blue carbon is the carbon stored in the carbon sinks. Carbon captured from the atmosphere by the ocean and coastal ecosystems helps in reducing global warming. Mangroves, seagrasses, salt marshes, etc. sequester atmospheric carbon at a much faster rate and store them under the seabed and ground. Degradation of ocean and coastal ecosystems may release this carbon again into the atmosphere and create extreme danger to climate.

Bioprospecting

Bioprospecting indicates the exploration of new species of plants, animals or micro-organisms, which have prospective social and commercial value.

Carbon Sequestration

Carbon sequestration describes the process of storing carbon into any natural pool, like plants, soil, ocean, biota, geological formations, etc., other than atmosphere. By carbon sequestration, the danger of global warming can thus be mitigated. Carbon often gets sequestrated naturally by biotic processes like photosynthesis by plants and shrubs in the terrestrial ecosystem and by plants and phytoplankton in aquatic pools like the ocean. On the other hand, engineering techniques are sometimes applied to capture and store carbon in the natural pools through an abiotic process. These engineering techniques are used mainly for restraining industrial emissions in

S. Hazra, A. Bhukta (eds.), *The Blue Economy*,
https://doi.org/10.1007/978-3-030-96519-8

reaching the atmosphere. Generally, 30% of anthropogenic emissions of carbon are absorbed by the oceans, 45% are retained by the atmosphere, and the rest are incorporated into terrestrial ecosystems.

Carbon Sinks

The natural reservoirs, like plants, soil, ocean, biota and different geological formations, that store carbon and restrain it from entering the atmosphere are known as carbon sinks.

Cultural Services

Cultural services are non-material, intangible benefits obtainable from ecosystems by the cognitive development, spiritual enrichment, recreation and aesthetic experiences of the people. Since use values offered by the cultural services are non-consumptive in nature, their quantification and monetary valuation, except the recreational and aesthetic value they provide, is a little bit difficult.

Ecosystems

An ecosystem is a functional unit of ecology defined over a particular geographical location where biotic and abiotic components of the environment interact among themselves. The geographical surroundings of an ecosystem can be extended from a droplet of water to the entire world.

Ecosystem Resilience

Ecosystem resilience describes the inherent ability of any ecosystem to sustain its normal functioning by absorbing all the disturbances made to it. This ability, therefore, works as a buffer against all environmental damages made by anthropogenic activities. A resilient ecosystem can continuously renew its resources, protect and maintain its biodiversity and regenerate its functional capacity all the way.

Fossil Fuel

Fossil fuels are decomposed forms of plants and animals, are rich in carbon content and thereby are used as a primary source of energy. Coal, oil and natural gas are examples of fossil fuel. These fuels are non-renewable and the main culprit behind the current concern of the world for warming.

Non-use Value

Non-use values arise either out of current direct or of current indirect non-use of ecosystem services. For example, it may arise either out of the satisfaction an individual derives from its mere existence (existence value) or from his desire to preserve it for future generations (bequest value).

Option Value

Option value refers to some kind of future use value. This value is assigned to a resource in the expectation that its contribution to the society and environment that is yet to be explored can be obtained in the future. Provisioning, regulating and cultural services, which are not consumed presently but kept for future consumption, are all examples of option value.

Provisioning Services

Provisioning services refer to the benefits offered by various ecosystems in the form of various goods and services to the consumers.

Regulating Services

Regulating services are the biophysicochemical processes that help to sustain life-support systems.

Supporting Services

Supporting services are the services that do not directly provide for any services directly for the consumption of plants or animals. Rather, they help in proper functioning of the entire ecosystem. Examples of supporting services are soil formation, nutrient and water cycling, production of atmospheric oxygen, etc.

Use Value

Use values arise out of the present (direct or indirect use value) or future (option value) anthropogenic use of environmental resources. Direct use values are the outcome of direct use of ecosystem services by the people, which include consumptive use such as harvesting of crops or timbers and non-consumptive use such as enjoying ecotourism. Direct use values are best exemplified by provisioning and cultural services. Indirect use values, on the other hand, are derived from ecosystem services which provide benefits outside the ecosystem itself. Regulating and supporting services are examples of indirect use values. Finally, option values refer to the option of using the service by oneself in the future.

WTP (Willingness to Pay)

The maximum price which a consumer is ready to pay for any good or service is considered as his willingness to pay (WTP) for the good or service. The concept, developed in the context of welfare economics, is widely used in environmental valuation studies.

Index

S. Hazra, A. Bhukta (eds.), *The Blue Economy*,
https://doi.org/10.1007/978-3-030-96519-8

P

The manufacturer's authorised representative in the EU is Springer Nature Customer Service Centre GmbH, Europaplatz 3, 69115 Heidelberg, Germany. If you have any concerns regarding our products, please contact ProductSafety@springernature.com

Printed and bound by CPI Group (UK) Ltd, Croydon, CR0 4YY
28/09/2026
02224077-0008